Praise for *Wild New World*

"Never has there been so complete, so fascinating, and so accessible a telling of the long history of people with American wildlife."
—Obi Kaufmann, author of The California Lands trilogy

"[Dan] Flores relates this huge body of information about the birth of America with both style and clarity. . . . [*Wild New World*] enlightens readers about where we came from and where we might be headed in the future." —Leslie Doran, *Durango Herald*

"The future of conservation, and our own survival, depends on busting some of the most stubborn myths that have embedded themselves in Western belief systems—ideologies that have, for centuries, steered us down a course of overexploitation of our planet's resources. . . . In Flores' deft hands the facts, fortified by the latest findings in ecology, genetics, and archaeology, fly off the pages in vivid and fascinating detail." —Isabella Tree, *Bookpost*

"To see this book nominated for the National Book Award or nominated for a Pulitzer would not surprise me."
—Steven Rinella, author of *American Buffalo*

T0015212

WILD
NEW
WORLD

THE EPIC STORY OF ANIMALS AND
PEOPLE IN AMERICA

DAN FLORES

W. W. NORTON & COMPANY
Celebrating a Century of Independent Publishing

Title page image: *Caribou migration in the Arctic National Wildlife Refuge, Summer Solstice, 2019. Photograph by Dan Flores.*

For information about permission to reproduce selections from this book, write to Permissions, W. W. Norton & Company, Inc., 500 Fifth Avenue, New York, NY 10110

For information about special discounts for bulk purchases, please contact W. W. Norton Special Sales at specialsales@wwnorton.com or 800-233-4830

Manufacturing by Lakeside Book Company
Book design by Chris Welch
Production manager: Devon Zahn

Library of Congress Cataloging-in-Publication Data

Names: Flores, Dan L. (Dan Louie), 1948– author.
Title: Wild new world : the epic story of animals and people in America / Dan Flores.
Description: First edition. | New York : W. W. Norton & Company, Inc., [2022] | Includes bibliographical references and index.
Identifiers: LCCN 2022027350 | ISBN 9781324006169 (cloth) | ISBN 9781324006176 (epub)
Subjects: LCSH: Human-animal relationships—History—North America. | Ethnozoology—North America.
Classification: LCC QL85 .F59 2022 | DDC 591.97—dc23/eng/20220721
LC record available at https://lccn.loc.gov/2022027350

ISBN 978-1-324-06591-3 pbk.

W. W. Norton & Company, Inc., 500 Fifth Avenue, New York, N.Y. 10110
www.wwnorton.com

W. W. Norton & Company Ltd., 15 Carlisle Street, London W1D 3BS

1 2 3 4 5 6 7 8 9 0

As always, for Sara

That which happens to men also happens to animals;
and one thing happens to them both: as one dies so dies
the other, for they share the same breath; and man has
no preeminence above an animal: for all is vanity.

—ECCLESIASTES 3:19

There is grandeur in this view of life . . . whilst this
planet has gone cycling on according to the fixed
law of gravity, from so simple a beginning, endless
forms most beautiful and most wonderful have been,
and are being, evolved.

—CHARLES DARWIN
On the Origin of Species

CONTENTS

WILD
NEW
WORLD

ALL IS VANITY

It was the lead-up to a presidential election, an unsettling late summer for many Americans since the hugely popular incumbent, Theodore Roosevelt, was declining to run for president a second time. The national gossip centered around whether portly William Howard Taft, who Roosevelt handpicked as his successor at Chicago's Republican convention that June of 1908, could possibly follow the charismatic advocate of the strenuous life. Somehow Taft seemed an unlikely replacement for the president "who does not shrink from danger, from hardship." So the news story out of the remote Southwest late that August initially seemed little more than a momentary distraction from politics, even if for some it might have been a reminder of the kind of heroic leadership the country was losing.

What the nation read in its newspapers was that on the night of August 27, a sixty-five-year-old telephone operator named Sally Rooke had gotten a call that an immense thunderstorm hovering over the Colorado-New Mexico border had spawned a flash flood in the Dry Cimarron River. With a debris-choked wall of water ripping straight for the town of Folsom, New Mexico, Rooke had spent a crucial half hour frantically calling every local number on her switchboard, saving scores of people. Then the flood had torn through Folsom and swept her away, along with half the town. Sally Rooke's heroism became a national story. Telephone operators around the country contributed thousands of dimes for a memorial. But eventually the story faded from the papers. Folsom had imagined itself competing with Colorado Springs a hundred miles up the Rockies. But the town never recovered. Today Colorado Springs has half a million people. Folsom has eighty.

In the days immediately following the Dry Cimarron flood, an African American cowboy named George McJunkin was riding through grassy parkland a few hundred yards below the rimrock of a miles-long mesa that extended eastward from the Rocky Mountains, checking for ranch fencelines damaged by the flood. Suddenly McJunkin's horse braced, its hooves furrowing into foot-deep mud at the edge of a ragged scar flood-waters had cut into the slope below the mesa. McJunkin leaned out of his saddle to peer into a fresh chasm sliced into the brown shale. What he saw changed the story of America forever.

On a similar rainy August day in 2018, some thirty-five of us are step-ping through the lush grass of that same slope as it angles up toward the rimrock of Johnson Mesa. We're following David Eck, a New Mexico State Land Office archaeologist with a long ponytail halfway down his back, who is leading us toward the very spot where George McJunkin's horse had pulled up 110 years ago. A century of floods and cattle grazing has changed the look of the place, which the flood erosion of 1908 had exposed as an ancient box canyon. The topography is now a grassy, shal-low drain called Wild Horse Arroyo, and as we crowd around its edges it seems somehow too commonplace to be the scene of one of the conti-nent's most significant historical finds. Nonetheless, this, in the flesh, is the legendary Folsom Archaeological Site. It's the place where the world found irrefutable proof that we humans had been intimately involved with American wild animals that long ago went extinct.

What McJunkin had done, about where we now stood talking, was to spot in the flood-gashed arroyo bones of an immense size he had never seen before. The exposed skeletal materials turned out to be from a herd of *Bison antiquus*, an extinct form of giant bison. But the bones themselves weren't the pièce de résistance. At the time, the sciences of ethnology and archaeology in the United States were firm that American Indians had arrived in North America only a couple thousand years prior to the coming of Europeans. In the Old World, artifact hunters at France's La Madeleine rock shelter in 1864 had found a piece of ivory with the representation of a mammoth on it. That seemed certain evidence that in Europe humans

George McJunkin on his horse.   Courtesy Denver Museum of Nature & Science.

once coexisted with extinct Pleistocene animals. As early as the 1870s, popular magazines in Europe were carrying illustrations of humans doing heroic battle against monsters like cave bears and mammoths.

That such sites hadn't turned up in the United States fed into a bias going back to a famous debate in the late 1700s between the French naturalist Comte de Buffon and Thomas Jefferson about whether America really was marginal to the global biological and human story. History has taken Jefferson's side in their argument, but Buffon turned out to be right about one aspect. In contrast to Jefferson, who thought extinction impossible, the French naturalist convinced the world that the strange creatures quarrymen were unearthing in Europe and America were species that had once roamed Earth but were now long vanished. Buffon likewise annoyed Jefferson by casting aspersions on American Indians' assumed lack of a significant history compared to that of Europe. In 1908 the flood that swept away Sally Rooke, and George McJunkin's discovery in the wake of it, were about to change forever the narrative that America was marginal to humanity's deep story.

Since Jefferson's time, scientists and ordinary citizens had been looking for something to refute Old World snobbery about America's past. In the wake of the La Madeleine discovery in France, the Smithsonian had mailed a circular asking military officers, missionaries, and Indian agents to be on the lookout for ancient fossil sites that might also show human antiquity. The stakes for demonstrating a deep human past in America were huge, and over time amateurs and scientists of various stripes advocated for at least two dozen sites, from Florida to New Jersey, from Idaho to California, as possible evidence humans had been in America in the Pleistocene. They thought they'd found it in 1882, when footprints embedded in stone at the Nevada State Prison seemed to show mammoth tracks overlapping human prints. That led to an excited *New York Times* headline: "Footprints of Monster Men!" Famous paleontologist Edward Drinker Cope thought the account convincing. But when University of California geologist Joseph Le Conte went to investigate, he concluded that the "monster man" tracks were in fact those of a giant ground sloth.

A classic late-summer thunderstorm was brewing up over Johnson Mesa as David Eck was explaining to us that until his death in 1922, George McJunkin proselytized about his find to anyone who would listen. Four years later, in 1926, the Black cowboy's plea to have a scientist look at his bone pit finally reached Jesse Figgins, director of the Colorado Museum of Natural History in Denver. Something of an amateur at this game, Figgins was mostly interested in fossil bison that might make exhibits in his museum. His team began an excavation of the site in May of 1926 and quickly began finding the skeletal remains of bison of an enormous size. That was exciting enough. But in their second season of work, on August 29, 1927, Figgins's crew troweled up Big History paydirt.

As David Eck was gesturing to the dimensions of this near-century-year-old dig, in the pocket of my light Patagonia jacket my fingers closed over an object that I could fit into my palm. In shape it was oblate. Think a flattened football, but with an end bitten off. Beneath my fingers I could feel an irregular surface, made so by labor-intensive flaking to create a pointed blade that dwindled to a remarkably thin base. The delicacy of that base was a result of matching "flutes" skillfully popped from the flint on both sides. Actually, the object in my hand was not the real thing. It was a cast-bronze piece an artist friend once made for me. Today, residents of Folsom aver that McJunkin himself had found a real point of this kind. In that first summer of digging, Figgins's paleontologists had unearthed two of them in the loose dirt of the site. Eventually the Denver team would find eight of these stunning fluted points scattered among the bones. But it wasn't just the points that made Folsom what American Museum of Natural History scientist Henry Fairfield Osborn labeled "the greatest event in American discoveries."

In July of 1926 *Scientific American* had published an article by a famous Smithsonian anthropologist claiming that "there is no valid evidence that the Indian has long been in the New World." That view was consigned to the junkyard of scientific theory in August of 1927 when the second-season crew at Folsom flicked the dirt from the ribs of an extinct bison and were greeted by the sight of one of these fluted points solidly embedded to two-thirds its length in the bone. The bar for proof that humans were part of the American Pleistocene had always been an extinct animal preserving evidence that as a living creature it had been killed by human technology. Now, outside the tiny burg of Folsom, New Mexico, that bar was hurdled in a way no Smithsonian scientist could deny. Leaving their discovery exactly as they found it, Figgins's team at once sent telegrams urging archaeologists to come and see. As it happened, a star witness was available. The most famous archaeologist in America was Alfred Kidder, who that summer was hosting the first of his Pecos Conferences on the southwestern past. Kidder came, looked, and declared to the world that America, too, had an antiquity.

How much of an antiquity was still in question because radiocarbon dating was yet three decades in the future. For his part, Figgins claimed the site was 400,000 years old. Eventually archaeology and paleontology would agree that on an October day a band of three dozen humans had driven into a box canyon, killed, and butchered thirty-two giant bison of the species *Bison antiquus* in the spot where I was now standing, and they had done this 12,450 years ago.

No one knows now what these ancient bison hunters called themselves or their weapons. Their beautiful fluted points were likely attached to darts thrown by atlatls, or spear-throwers. But not knowing much about these early Americans didn't prevent the scientists from naming both the points and the people Folsom, after the nearby town.

Bronze replica of a fluted Folsom point.   Photograph by Dan Flores.

I worked the bronze replica in my pocket through my fingers like a card dealer in Vegas. It was exquisitely, almost deliciously crafted. It was also a replica of the kind of artful technology we humans have used to hunt and too often destroy much of the bestiary of the continent. But the Folsom era—there was indeed a "Folsom America"—was not the first time we humans had pursued distinctive American animals to their extinction, and it was very far from being the last. That long-term relationship between our species and wild animals in North America, which includes human-driven losses of other species in our deep history and staggeringly selfish and myopic destruction in our much more recent past, is an uncomfortable reveal about our species and our time on the planet. But our future demands we look it in the eye, look ourselves in the eye, and face this particular American story.

BECAUSE THEY LAY so far around the planet from humanity's origins in Africa, the Americas were the last continents we humans found in our grand explorations of planet Earth. North America was an unimaginable location that we initially struggled to find at all. Then, like some valuable heirloom we kept misplacing, America disappeared and receded from human knowledge again and again. The result was a lost continent that would function as a Wild New World more than once, and for more than one people. Folsom, it turned out, wasn't the American book of Genesis after all.

Six years after the Folsom discovery there was another dramatic revelation, and in an unlikely place. Out on the featureless sweeps of the country's middle, an ordinary gravel excavation near a tiny farming town named Clovis exposed the bones of long-extinct American elephants. Science and the reading public knew that America had harbored various kinds of giant elephants in the deep past. But unlike nineteenth-century mastodon finds in the East, this time the skeletons were intermixed with the projectile points and tools of an unknown culture that was apparently even more ancient than the Folsom people. Across the next decade, as these larger (and also fluted) flint points began to show up all over America, from Alaska to Florida, from New England and the Midwest to the Southwest, scientists puzzled over who these elephant hunters might be. At a conference in Santa Fe in 1941 they decided to name them after the locale where their inhabitation of the continent first came to light.

So, we added "Clovis" to America's lengthening human-animal story. When radiocarbon dating finally assigned them a time frame, the elephant hunters turned out to have occupied the continent Europeans once called "the New World" many thousands of years before any European city was born. The animals these hunters killed and butchered with brightly striped flint tools across every region of America died more than thirteen thousand years before the United States existed. Hitherto unsuspected American beginnings like Folsom and Clovis were turning what everyone thought a brief history—European colonization leading to the creation of the United States—into an epic older than anyone's written history or any human memory.

Americans hearing about the continent's new human antiquity couldn't help feeling uneasy. The elephants and giant bison were gone. So were the camels and horses and giant cats, whose remains fossil hunters had been excavating since the 1850s from places like the asphalt pools in downtown Los Angeles. Human weaponry and tools were now turning up in clear association with many of these extinct animals. That hardly seemed a coincidence in the 1920s and 1930s, decades when a remarkable number of species from the modern United States were also vanishing, one after another. Americans of the time barely recognized mammoths or camels as part of the continent's bestiary, but there were people still alive in the 1920s who had seen with their own eyes millions of bison and billions of passenger pigeons. As with camels and elephants, though, by the 1920s those scenes were no longer a part of America. Just a few years before, the last living individual of the most numerous bird species on Earth had died in her cage at the Cincinnati Zoo. Americans celebrated in the early 1930s when an organization called the American Bison Society announced that the buffalo, the most iconic of all our mammals, would not go extinct after all. But it was a pyrrhic victory. Bison lived, rescued on the brink of their demise, but not to return to the wild. Instead they were enclosed in preserves so curious citizens could marvel and wonder what had happened to an animal that once numbered in the millions.

And it wasn't just passenger pigeons and buffalo. A legacy of animal cleansing was visible everywhere you looked in the United States of the 1920s. The year Figgins's crew initiated the dig at Folsom was the same year gray wolves howled for the last time in America's first national park, Yellowstone. By then we had poisoned, trapped, and shot wolves into near extirpation across most of the country. That slaughter was more than symbolic. It meant that a dominant, keystone predatory animal whose mere presence had powerfully shaped North American ecologies across the past five million years of history was now almost gone from the continent south of Canada. At the same time, ornithologists announced that, like passenger pigeons, our giant ivory-billed woodpeckers appeared to be extinct. So did our sole native parrot, the Carolina parakeet. In New England, newspapers year by year tracked the demise of the eastern version of the prairie chicken, known as the heath hen, until the final bird, a last lonely male, was gone.

None of us born in the past century has gotten to see any form of a biologically complete America. We all exist in a world handed down by the prior occupants. Like coming generations, who will have to live with a planet our generations have overheated, we, too, suffer from the selfishness of those who lived before us. In our case our ancestors left us a simplified and devastated Earth. Sitting down to his journal one morning in

1856, Henry David Thoreau enumerated and lamented all the wild creatures missing from the New England of his time because of the myopia and self-interest of the colonial demigods who got there first. What he longed for enough that it ached, he wrote, was "an entire heaven and an entire earth."

IN THE YEARS that produced both the Folsom and Clovis discoveries along with widespread modern extinctions, Americans were among the most optimistic people on the planet. We basked in the sense of having turned the tide in the Great War, our institutions seemed to offer the best future for humanity, and our popular culture was on its way to global domination. Maybe there were a few dark clouds looming. But losing American animals was a minor thing in the big picture, especially since the country seemed to believe that no one and no institutions were really to blame. Birds and animal species were just casualties of progress and civilization, we told ourselves, collateral damage in the act of creating the best country in the world.

Explanations like this prevailed for decades, but a century later our excuses are crumbling in the face of powerful evidence of our culpability. The modern genetic revolution is a scientific breakthrough for the ages, rivaling the discovery of evolution itself, and is now at a sprint, fixed on writing a history that stretches far more deeply into the past than human memory. Genetics and genomics are not only transforming our knowledge of how America acquired its remarkable diversity of life; these sciences are increasingly suggesting what caused us to lose so many of the species that have disappeared.

The answers from both history and science are the subject of many of the following pages, and I won't say more here beyond preparing you for this. Woolly mammoths, Columbian mammoths, flightless sea ducks, great auks, heath hens, bison, pronghorns, beavers, sea otters, fur seals, bald eagles, hummingbirds, whooping cranes, snowy egrets, trumpeter swans, peregrine falcons, California condors, jaguars, cougars, alligators, rattlesnakes, gray wolves, red wolves, Mexican wolves, eastern wolves, grizzly bears, wild horses, passenger pigeons, Carolina parakeets, prairie dogs, black-footed ferrets, coyotes, spotted owls, and

ivory-billed woodpeckers all have both deep histories and modern stories that resemble one another more than you think.

✍

WE'RE UNCOMFORTABLE thinking of ourselves as having been carnivores,
as killers. For that matter we're not used to thinking of ourselves as a
species, preferring terms like "the human race." And we almost never
act as if we're another of Earth's animals. But of course we are all those
things, and all are a part of our Big History. The sweeping story I tell here,
then, is an extension of human history back to our origins as animals,
along with the evolutionary adaptations that helped make us who we are.
Because it draws a bigger circle around time and subjects, Big History has
advantages over conventional history. It can acknowledge that the destiny of a continent like North America lies not just with us but also with
our fellow creatures and the larger evolutionary stream in which we all
swim. That story unfolded beneath the grand forces that shaped American biology across the past sixty-six million years, from the Paleocene to
now. And those forces fashioned some of our planet's most remarkable
life-forms, from hummingbirds to Columbian mammoths. Late in that
larger trajectory was our own evolution, the appearance of an intensely
social omnivore who became master of the planet by reinventing itself as
an even more social carnivore.

Some will already know the shorthand version of this. As predators
we spread out of Africa, migrating all over the world in search of animal
prey. Our close ancestors pushed beyond their African homeland almost
a million years ago, and our species joined that quest to explore the larger
planet some sixty thousand years ago. In the past twenty-five thousand
years a few of us happened on the Americas, the last best place of human
exploration, and sometime after sixteen thousand years ago more of us
arrived to create America's first shore-to-shore human society, profoundly
altering the ancient bestiary. We did it all once more five centuries ago,
when Old Worlders transplanted themselves and their economies, cultures, and ideas to the continent's existing world of peoples and animals,
and then, like some new contagion spreading inland from the coasts, proceeded to effect a widespread demolition of almost all that was there.

This is our cross to bear and one of the inescapable truths of Big History. According to a 2018 National Academy of Sciences study, as our predatory genus *Homo* spread across the planet, Earth lost roughly three hundred mammal species, sacrificing more than two and a half billion years of unique evolutionary history. The migrations that ultimately led us to America during the Pleistocene erased a shocking two billion years of evolved mammalian genetic diversity. Across the past five hundred years human-caused extinctions have cost Earth another half-million years of cumulative genetics. And that's the mammals alone, never mind the birds and reptiles. As the Academy of Sciences authors put the matter, "This means that prehistoric and historic extinctions were close to worst-case scenarios."

To lay out the wildlife story of those last five centuries in another way, since 1500 we Americans have managed to commit the largest single destruction of wild animals discoverable in modern history. In the early twentieth century there were former market hunters who believed they ought to get national medals of commendation for their role in that.

So, fair warning, some of the stories in this book won't be easy to hear or process. But take heart. Our full awakening amid the rich life of the planet also bequeathed to humanity an undertow of wonder and fascination for other creatures in Earth's evolutionary stream. The late Harvard scientist E. O. Wilson has called this core human value *biophilia*, which he defined as a genetic memory of our emergence, a piercing love affair with the other life-forms that surround us on the planet. Our sense of this magic and romance is visible in the enthralled renderings of animals we painted on the walls of Europe's limestone caverns, and on miles of cliffs in the Serranía La Lindosa forests in South America. Biophilia was certainly there when Americans deliberately said No! to destruction and loss and passed the most significant law in wild-animal history. The Endangered Species Act of 1973 and its recovery programs, dedicated to the ultimate democratic proposition that every species has a right to exist, are this country at its best. They are also a course correction that those battling to rescue Earth from an overheated climate should feel some optimism about. We have it in ourselves to change.

A love of diverse life as old as our origins was always there to call on, and when we have done so we've bathed human history in a brighter light. This complex American story produced many inspiring, empathetic heroes, from Native peoples who preserved a wild continent largely intact for ten thousand years to naturalists compelled to know every detail of continental animal life, from visionaries who gave the world its first national parks to writers, activists, and politicians who confronted the destruction and loss of wild animals before it was too late. Always there have been scientists, describing and classifying exotic new species until accumulated experience and knowledge finally yielded up the grand insight about the diversity of life and our own origins. Others made breakthroughs that caused us to question our persecutions of America's anciently evolved predator guild, the canids, bears, and big cats. Most recently science is rewriting biological histories, while helping us understand that the self-awareness and cultural richness we celebrate as human place us *within* animal life, not outside it. All is vanity to think otherwise.

These and more are part of this story. But this is still a narrative leavened with pathos, because while we adore the diversity of our remaining birds and animals and marvel at them as our species always has, we've always been most concerned about our own self-interest. Those particular angels of our nature account for where we are in the twenty-first century, with a continuing sixth, and massive, extinction of wild lifeforms underway in America and across Earth, and with at least some of us dreading that outcome.

IN THE EARLY 2000S the chemist and Nobel laureate Paul Crutzen popularized a new term in global languages, the "Anthropocene epoch," designating the moment in history when human impacts on Earth had become so dominant as to transform our planet in ways formerly reserved for extraterrestrial impacts or geological upheavals. Scientists since have debated whether the designation is proper and, if it is (and most of the public that has heard of the idea thinks it is), when it might have begun.

At a time when climate change is on all our minds, Crutzen and his supporters have pointed to the industrial revolution and the onset of our fossil fuel economy as the obvious beginning point for an age when humans became Earth's most powerful force. Others have insisted on a starting date even more recent, in the 1940s with the dawn of the atomic age, which for the first time put a technology in the hands of the human species capable of destroying most life on Earth with the press of a button.

But standing here in Wild Horse Arroyo in 2018 as a massive storm cell gathers itself to shoot lightning daggers at the mesa overhead, I can't help entertaining an alternative theory for the Anthropocene. The replica Folsom point in the pocket of my jacket and the animal bones pulled out of the ground here imply to me a much deeper time frame for our species' transformation of the planet. Indeed, the Dutch curators of the History Database of the Global Environment have suggested that the Neolithic Revolution—widespread farming, animal domestication, and the emergence of cities a few thousand years ago—might offer a more apt beginning for the Anthropocene.

As for me, I'm more in line with the archaeologists like our ponytailed guide. A panel of his colleagues from around the world argued a few years ago that if we're really serious about dating a global Anthropocene, if we truly want to discover when our species' planet-altering impact began, we have to look much further back in time than even the Neolithic—at what happened to the world's animals, those victims of a love affair gone bad when we humans fully settled Earth.

CHAPTER 1

A PROLOGUE IN
DEEP TIME

What wakes me is a sound I've never heard. Or to be truer to the moment, two sounds. One is a soft, gentle sort of *chuffing*. Coming awake to the slanting light out the tent door, I register this one first, an auditory accompaniment to the angled light, which lands in my foggy brain as a language I can't quite translate. The other sound is almost as delicate but more percussive. It appears to reach from the faraway to the nearby, a *tinkling* with a thousand source-points.

It is June 22, the day after Summer Solstice, 2019, and with my wife, Sara, and eight of our friends I am waking in what seems to be primeval America. We are in Alaska above the Arctic Circle, at nearly seventy degrees north latitude, and that far up the curve of the planet on Summer Solstice we're experiencing a new cosmic reality, a powerful sense of being just off the apex of a gigantic sphere that's spinning beneath a light source the universe has forgotten to switch off. No matter how early or late you go to bed, or wake, the same fingers-of-dawn sunlight and shadowing await you. From our put-in high in the Brooks Range we're now nine days down the Hulahula River, named by—or for—a Hawaiian Islander on a ship pursuing wealth in furs here two centuries ago. By now we're out in the vast coastal plain of America's crown jewel, the Arctic National Wildlife Refuge, and while the oblique light angles never relent, these faraway/nearby sounds are new.

The source-points are hooves of thousands of caribou, migrating out of mountains filled with wolf packs down to the relative safety of the Arctic plain, where they are coming to drop their calves. Herd after herd is passing by within a hundred feet of our camp this morning, conversing in a murmur that audibly rises above the flow of the nearby river. Pulling the tent fabric aside, my first guess is that the *tinkling* is the clicking of caribou hooves passing through the stiff, eight-inch-high willows blanketing the riverside tundra. But the more I listen the more it sounds like a popping in their leg joints as they walk. The chuffing? That seems to be quiet caribou conversation.

For days we've floated downriver among these herds. And now everywhere we look they are the only upright things in an otherwise horizontal world, tan-and-white bodies rocking along in a kind of slow-motion parade. They seem aimed toward an indistinct assortment of white blocks, miles out, we're now grasping are icebergs in the Arctic Ocean. On the opposite riverbank, only yards away, herd males with antlers and faces black as squid-ink clack over the rocks. There are pregnant cows, also with antlers, posses of adolescents, and infrequently a black wolf or grizzly out in the opalescent distance. Some herds with brand-new calves—the little guys jetting and pogoing across the tundra—have abandoned the relentless march northward from the mountains. Ten days earlier, in Arctic Village, Neest'all Gwich'in tribal elder Sarah James told us that the Porcupine caribou herd is three hundred thousand strong. Funny, eloquent, with long silver-streaked hair, seventy-five-year-old Sarah travels the world fighting to keep oil drilling out of the refuge, a looming threat in 2019. She's got the bona fides for it. That tribal name means People of the Caribou. "The caribou are in our hearts and we are in theirs," she told us.

Standing by my tent surrounded by perhaps ten thousand caribou moving slowly through a vast, grassy landscape, I'm dumbfounded. I'm also struck by the realization that for the first time in my life I'm experiencing a version of an entire America, the continent human eyes have been marveling over for more than twenty thousand years. In the second decade of this century what's required for a primal experience of the natural planet is being in one of Africa's grand game parks, in parts of the Amazon, or in some wildly remote piece of North America. That spawns another thought. Are we capable anymore of remembering what we've sacrificed, that once these caribou and wolves would have

Migrating caribou in the Arctic National Wildlife Refuge, 2019. Photograph by Dan Flores.

been joined by grazing woolly mammoths and vast, zebra-like herds of wild horses?

In the United States of the twenty-first century, this wild coastal plain in sight of the Beaufort Sea is one of the last continental places to preserve a tantalizing echo of wild America as it waited across the millennia for its destiny to play out.

THE WILD NEW WORLD that greeted the first human arrivals was a more ancient place by far than humanity itself. Immense time and relentless evolution, plus at least one cosmic wild card that sent history off in a dramatic new direction, had painted a wondrous canvas of biological life in North America. Yet historical unpredictability lay in the seams of the America we humans found. The one certain truth about our Earth is that everything could have turned out very differently.

It was that shock to the timeline that sets this story in motion. Four times in the planet's early history it had suffered and ultimately survived global extinction events. Earth's fifth extinction event was a special one. Given the evidence for widespread planetary bombardment elsewhere in our solar system, we have to assume cosmic collisions on Earth aren't exactly rare. A few years ago I walked the rim of a remarkable half-mile-wide impact crater in northern Arizona that offers surviving evidence of that. But on planets where life has evolved, even fairly common impacts are dangerous because they threaten ecological disruptions. The one in question, though, wasn't common. And this was especially unlucky for North America because its landmasses were in the bull's-eye of one of Earth's biggest contingency events.

Sixty-six million years ago North America was not even a single piece of ground yet. The continent was in two major pieces, with the Western Interior Seaway inundating much of today's Southeast and from the Great Plains to the Great Basin. This shallow ocean entirely separated the eastern landmass, Appalachia, from the large western chunk, Laramidia. Dominated by dinosaurs and reptilian life, those halves were already hinting at some of the differences that would characterize eastern and western America. With the Appalachian uplift already in its past, the eastern piece was geologically calm, sedate, buttoned-up. The western landmass, closer to the active volcanic hot spots of the Pacific Ocean, churned with geological activity and reinvention. Although the Rocky Mountains had not yet soared skyward, tectonic-plate collisions were already lifting the Sierra Nevada and Cascade ranges. The simple fact that mountains existed both east and west would prove important in what was about to happen.

The approaching contingency was 7.5 miles in diameter, a rocky chunk ejected from the outer asteroid belt between Mars and Jupiter. It had almost certainly crossed Earth's orbit many times in the planet's history, whizzing by within sight of the life below. Millions of years in the future, scientists would recognize this death star as a carbonaceous chondrite, one of the most common of the various meteorites. What distinguished this one was its gargantuan size and the speed of its approach. The cosmic projectile was hurtling at twenty-seven thousand miles per hour and as it approached Earth the bottom end of the Western Interior Seaway assumed form directly below it. Today that's the northwestern tip of Mexico's Yucatán Peninsula, near the Mayan pueblo of Chicxulub, a village that now graces this Solar System–level event with its interesting, lyrical name ("CHICK-soo-lube").

Since the late 1990s we have assumed the Chicxulub asteroid's impact angle was low, that it was traveling from south to north with its primary

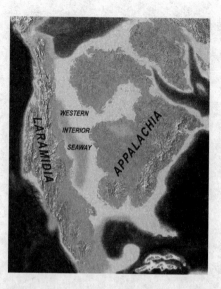

Appalachia, Laramidia, and the Western Interior Seaway. Wikimedia Commons.

destructive force aimed directly at North America. Extensive seismic and geophysical mapping of the site, downloaded into 3D-modeling programs, has now led to a different conclusion. The asteroid seems instead to have approached its strike zone from the northeast and at a steeply inclined angle of forty-five to sixty degrees. No glancing hit, this. The enormous rock smacked into an immovable backstop of water, mud, rock, and life at an angle guaranteed to pack the atmosphere with impact debris. It also hit during the rebirth of spring in the Northern Hemisphere, which made everything worse.

There has not been a spectacle even close to this on Earth for the past sixty-six million years, and that's a fortunate thing. According to a Los Alamos National Lab modeling of what happened, the unimaginable ferocity of the collision blew open a surface crater 18 miles deep and 124 miles wide within mere seconds. The hit instantly vaporized the asteroid itself, and because its strike was in ocean water it generated tsunamis a mile tall that ricocheted off Earth's shorelines—water sloshing to and fro across a planetary bowl—for weeks afterward. Those produced ripples on the ancient seafloor the height of five-story buildings. The next act was an ejecta of molten rock shot into Earth's atmosphere to such heights that

Arizona impact crater. Photograph by Dan Flores.

it rained down scalding glass that burned the planet's vegetation to ash. That column soared so high its top exited Earth's orbit, and over the next few million years deposited bits of our planet across the moon, Mars, and Jupiter and its moons. Some astrobiologists believe that while Chicxulub came close to extinguishing life on Earth, its strike on a planet teeming with living things may have seeded life in the form of microbes throughout much of the solar system.

In the aftermath, Earth famously experienced not just a nuclear winter but a toxic one. Dust and ash saturated the atmosphere. Trillions of tons of sulfur and carbon dioxide, along with billions of tons of methane and carbon monoxide, were hoisted into the biosphere by this giant petard. Everything that hadn't been blasted in a shock wave or burned by molten rain or frozen under dense cloud cover was now drizzled with sulfuric acid. Most terrestrial species were extinguished. Fewer than 25 percent of Earth's life-

forms survived. With such a steeply angled hit, Chicxulub's effects were distributed symmetrically across the planet. But the continental pieces that made up North America were basically fried. Chicxulub came desperately close to exterminating life on the continent. Only three spots—above the Arctic Circle and in mountainous recesses beyond the Appalachians and Sierras/Cascades—were sufficiently shielded to preserve continental plant life. The impact wiped clean the hard drive of evolutionary life all over the planet, but particularly in America. With dinosaurs now at an end, new life would be based on the relative handful of forms that survived.

Chicxulub was our premier contingency. Without it, no us, because critical among those few survivors were small, scurrying, rat-sized creatures, warm-blooded and capable of rapid generational turnover where natural selection could work on them quickly. They were mammals, of course, and with no warning they had now inherited a planet with thousands of vacant ecological niches to fill just as fast as evolution could shape them for the tasks.

※

ONE AMERICAN PLACE we know witnessed the miracle of rebirth in the Paleocene (as the epoch following Chicxulub is called) is a canyonated country just east of the Colorado Rockies called Corral Bluffs. The skulls paleontologists have found there tell a story of by-the-fingernails mammalian resilience. Ground-dwelling rat-sized mammals survived the extinction event far better than tree-dwelling raccoon-sized ones did. But within about one hundred thousand years after Chicxulub, mammals were finding Earth's new possibilities to their liking. By then creatures as large as raccoons were reappearing in Colorado. Some three hundred thousand years post-asteroid more warm-blooded terrestrials appear in the record. Some of them, like creatures scientists have named *Loxolophus* and *Carsioptychus*, were the size of small pigs. By fewer than a million years after Chicxulub there were mammals in America called *Eoconodons* as big as wolves or large dogs.

Of the grand forces shaping American nature, proximity may have been most important. The two landmasses that made up future North America were not completely isolated from the rest of the world. In a pattern that would continue down the arc of millions of years and would eventually bring humans to America, connections to other continents formed, vanished, formed again. North America East periodically connected via Greenland with the northern reaches of Europe, while other less ephemeral land bridges linked North America West to Asia. Those connections formed in cold-climate pulses, then melted away under rising seas in warm ones. And each time the connections persisted even for a few centuries, the possibilities for more diversity in North American life expanded, for elsewhere around the world mammalian and other life was re-creating an abundance of forms just as it was in future Colorado.

Five million years after the Chicxulub impact, another grand event—the Laramide uplift, this one precipitated by colliding continental plates in the West—elevated the Rocky Mountains and drained the interior seaway down the present Mississippi Valley. North America East and North America West then merged as a single continent. But the two former halves seemed to preserve their own particular geologies and life connections, the East more like Europe in its vegetation and life-forms, the West in truth a part of Asia.

In the wake of Chicxulub, North American life struggled through millions of years to regain its prior diversity. In the northern latitudes of Europe, Asia, and Africa a similar process was unfolding. But the closest connections to North America all lay through an Arctic filter that restricted the sorts of creatures that could colonize a recovering continent. The native species that had best made it through the impact were the aquatics—turtles, alligators, amphibians, reptiles, and ancestors of ancient fishes like gars and paddlefishes. Already nearly 150 million years old, American alligators somehow survived Chicxulub and eventually shared the southern coastline with giant, ocean-going American crocodiles. The ancestor of pit vipers also survived, although an American subfamily found nowhere else on Earth—the rattlesnakes—did not begin to differentiate into its three dozen separate species until twelve to fourteen million years ago. As with alligators and rattlesnakes, rewilding post-Chicxulub America from indigenous forms meant that with sufficient time the continent would again become a laboratory of explosive evolution.

Life is important to itself and to the ecologies it occupies without any reference to us. But many of the life-forms that emerged in North America from sixty-six million years ago down to our own recent arrival would become critical players in human history. However, for a striking number of them, recent human history has obscured their origins as distinctive American contributions to Earth's life. Within the first ten million years after Chicxulub, continental evolution had already begun shaping many such animals. Among the oldest were the earliest versions of what became the Equidae, the family of horses and burros. Emerging first as rabbit-sized forest creatures in the Paleocene epoch (66 to 56 million years ago), American horses rapidly evolved into much larger forms during the Miocene epoch (23.8 to 5.3 million years ago) in response to large-scale climate and vegetation alterations across the Northern Hemisphere. In time as many as a dozen different types shared the tundras and grasslands of America. Paleontologists have excavated horse remains from Alaska to Florida (where horses make up 60 percent of Miocene fossils) and in almost every state south of the Great Lakes. Today's genus *Equus*, which includes the horses we all recognize, was in place by four million years ago. At the time humans were first exploring America's wild new world, in some geographies those horses made up 25 percent of the biomass of all living creatures.

Horse history also corrects any supposition that the occasional connective bridges that linked America with the rest of the world were one-way streets. Whenever land bridges formed between America and Eurasia, bands of North American horses trekked far into Asia, Africa, and Europe. Eventually, once a connective bridge formed in the south, they would also emigrate to South America.

Ten million years after horses emerged, another family of recognizable animals developed from American mammal radiation during the Eocene (54.8 to 33.7 million years ago). Camels spent the first forty million years of their evolution exclusively in North America, evolving into many forms down the epochs until six million years ago, when they, too, crossed the Bering Land Bridge into Asia, eventually reaching Africa. *Camelops* (the one- and two-humped camels, along with the llamas and alpacas of South America) persisted in its original continent until roughly ten thousand years ago. The late species known as yesterday's camel ranged from Tennessee to Alaska and across almost every modern western state. With forty million years of presence and only ten thousand years of absence, camels are more American than many of the animals we now consider essential members of our bestiary.

Another characteristic post-Chicxulub native, although its size, form, and speed cause many to believe it some close cousin of the antelopes and gazelles of Africa, is the pronghorn. Despite convergent evolution independently fashioning gazelles and pronghorns, the American family did not produce true antelopes. The Antilocapridae evolved in North America twenty-five million years ago, just before the onset of the Miocene, but its earlier kinship is murky. Paleontologists don't agree on an older provenance of animals that are part antelope, part goat. Probably pronghorns are offspring of an older superfamily, the Cervoidea, which would make them distantly related to deer. But there is a theory that their closest living relatives are in the Giraffidae, since giraffe legs bear a resemblance to pronghorn legs. Fossil records in North America indicate there were two major subfamilies of these speedy grassland runners. One of those, the Antilocapridae, sported four and sometimes six horns. A dwarf, four-horned version not much larger than a jackrabbit still sprinted across the Great Plains ten thousand years ago.

Another indigenous American spawn whose history would eventually join with that of global humanity was the pinnipeds, the animals that became seals and sea lions. Pinnipeds evolved as coastal-dwelling, ocean-hunting carnivores along the Pacific shores of western North America in the very early Miocene. Numerous species among the sea lions and seals would spread from their North American origins around much of the world during the Pliocene (5.3 to 1.8 million years ago) and Pleistocene (1.8 million to 10,000 years ago). But the most important of the American pinnipeds historically was the species known as the northern fur seal. By a million years ago, fur seal rookeries extended along the whole of America's Pacific Coast and in an arc across the North Pacific to the Sea of Japan. Barking, swimming, hunting, breeding, pinnipeds joined members of the genus *Enhydra*, the sleek, beautiful sea otters, which emerged in the Miocene five to seven million years ago to become the characteristic shoreline otters of the Pacific by two million years ago.

Beavers and squirrels were yet another result of North American evolutionary radiation. Protobeavers began to appear in America early in the post-Chicxulub world, during the late Eocene (56 to 33.9 million years ago). They then made the passage across the Bering Land Bridge to Eurasia, only to return to their evolutionary homeland later, a pattern followed by several other American natives. Beaver radiation over the epochs produced versions that tunneled and versions that build dams. During the Pleistocene a type known as *Castoroides* that was roughly the size of a Volkswagen Beetle inhabited all America, Florida to Alaska.

Squirrels might seem too small and commonplace to have been historical actors, but they are on the stage frequently in the American story. The earliest fossil evidence for squirrels comes out of western North America at the end of the Eocene, thirty-six million years ago, followed by rapid divergence across the next five million years along with their appearance in distant Europe, although we're not sure just how they got there. But across several periods down to around seven million years ago there were forests on the land bridge between Alaska and Siberia, facilitating tree- and flying-squirrel passage. After that land bridge tundra would still have allowed ground-squirrel migrations between the continents.

All these American natives made up a rich world of horses and camels,

beavers and squirrels, seals and sea otters during the Miocene. Their predators, whose stories I tell below, were almost all migrants from Asia until
the Pliocene (5.3 to 2.6 million years ago). At that point another indigenous
family, the predatory canids, emerged to make America an even more
familiar place. The legendary Canidae family, the wolves and dogs of the
world, evolved in southwestern North America from ancestral forms at the
very start of the Pliocene. Early canid versions that became gray wolves,
jackals, and wild dogs eventually crossed to Asia, Europe, and Africa. The
canids that never left—or in the case of Pleistocene coyotes got no farther
than Siberia—became North America's distinctive coyotes, red wolves,
and eastern wolves, a clade of animals that make up what biologists call
"allotaxa": related species that readily hybridize.

Members of the Canidae shared the continent with another indigenous American wolf, the strapping (150-pound) and legendary dire wolf.
It evolved earlier (5.7 million years ago) and recent genetic analysis indicates it belonged to a different genus than *Canis*. Late in the Pleistocene,
between thirty thousand and fifteen thousand years ago, dire wolves and
the clade of coyotes and American wolves would be joined by migrating
waves of gray wolves returning home from Eurasia after three million
years of evolution there. All these canids were pack hunters and scavengers, and their interactions with one another around a found carcass, or
when one group made a kill that attracted others, must have been one of
the grand defining spectacles of Pleistocene America.

～

A DRAMATIC ASSORTMENT of post-Chicxulub creatures from elsewhere
on Earth eventually began joining those that evolved in America to create
the exotic bestiary humans found here. The conditions that made their
arrival possible form the next act of America's story.

Astrobiologists formulating scenarios for life on distant planets are
coming to believe that oceans may be an unsuspected key to the diversity
of life. Oceanic swirling transfers biotic energy around planets. In Earth's
case, its own aqueous swirling has been augmented by the fact that long
ago in Earth's history the planet's outer shell cracked. Once life evolved it
began drifting around on the giant crustal shards, whose island-like isolation amplified diversity. A fundamental principle of our planet's evolution

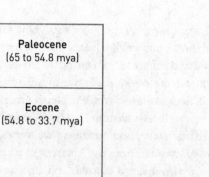

Chart of geologic epochs since the Chicxulub impact.

is that the larger shards, perhaps because of the intense competition for niches there, tend to produce the best colonizers. North America's receptivity to migrants has been shaped by planetary plate tectonics in another way. As the shard known as the Pacific Plate has dived into America's western coastline it has produced a singular deformation, crumpling up major mountain ranges that have oriented at topographical right angles to the pressure. North-south ranges mean the Northern Hemisphere's jet stream can funnel frigid Arctic air—the dreaded Polar Express—deep into America's southern latitudes. The consequence is that Asian species had an advantage over those from both Europe and South America in filling North American niches after Chicxulub.

Oceanic swirling continued as a grand force. As the shards of crustal Earth broke into smaller pieces their locations could alter the swirling of oceans and biotic energy. During the Miocene, one such plate orientation on the far side of the Earth transformed the possibilities for migrations and biological life as a whole in North America.

Forty million years after Chicxulub, down in the Southern Hemisphere the shard that became Australia rode free of Antarctica. That separation allowed icy water from the Antarctic to pour northward into the world's oceans, accompanied by cold, arid air that cooled and dried the planet and began to pile up ice atop the opposite pole. Miocene Earth had begun as a hothouse, roughly eight degrees Fahrenheit warmer than now. But as the epoch progressed, cooling and drying in Asia and North America began to promote extensive, mesic savanna grasslands. In much of eastern America deciduous forests of oak and beech began to replace conifers. This biotic makeover, precipitated by drifting crust and new oceanic currents, fashioned new ecosystems that drew new mammals and birds from both hemispheres.

Land connections between Asia and America had come and gone before, but as ice began to gobble up water the old pattern intensified. During cold episodes the ocean levels dropped, exposing a wide northern bridge, Beringia, from Siberia into Alaska. Seventeen million years ago that bridge brought North America a prodigious early emigrant, our continent's first elephants. These early arrivals were the great mastodons, browse-eaters who gravitated to heavy forests in the western mountains, the upper Midwest, and the whole eastern half of the continent. Among other important Miocene arrivals from Asia were the deer and elk that would become such crucial players in North American ecology and history.

The cervoids had emerged in Asia in the Oligocene epoch (33.7 to 23.8 million years ago). Various versions, including the ancestors of whitetail and mule deer, began arriving in North America about the same time as the mastodons. The largest, elk, were similar to red deer in Europe, but America's elk sprang from an Asian population in Kazakhstan, Mongolia, and Siberia (the line separating European red deer and American elk appears to lie through the Gobi Desert). As for the biggest deer, the moose, from Oligocene beginnings they ranged across Europe to the Caucasus Mountains, and from there through Manchuria and Siberia to

Alaska. A deer that prefers woods with a seasonal snow cover, moose took up residence from Alaska to New England and as far south as Colorado during the Miocene.

The caribou migrating by our camps in the Arctic Wildlife Refuge in the summer of 2019 occupy an intriguing niche in the story of cervoids and North America. While ancestral cervoids came to America out of Asia, caribou evolved into their modern form in Beringia itself, on the land bridge, during the late Pliocene, some two million years ago. A herding deer in which both males and females sport antlers, caribou colonized in both directions, westward into Mongolia and China, and eastward across Canada and Atlantic land bridges to Greenland, Ireland, Scotland, and Scandinavia. During the Pleistocene and after, isolated caribou herds in America became genetically and behaviorally distinct. The Alaskan herds we were among are closest to the original Beringia caribou, but there were woodland caribou that in time would get as far south as Nevada, New Mexico, Tennessee, and Alabama! Caribou, it turns out, make clicking sounds in their legs as they walk as a dominance display to establish hierarchy in the herds.

Another large family, the Bovidae, including sheep and goats as well as musk ox and bison, had its beginnings in Africa and Europe in the early Miocene. The late Miocene drought (climate historians call it the Messinian Crisis) produced an extensive global spread of savanna grasslands, the catalyst to an evolutionary radiation of Bovidae forms. In North America five million years ago Miocene drying produced an enormous interior grassland—the Great Plains—east of the Rocky Mountains. An American Serengeti of immense distances, with opalescent mountain uplifts on its western horizon and vast, rippling tracts of yellow grasslands, this country soon began to fill with grazers, many of which were different species of Bovidae dispersing from Asia. This migration continued for many millions of years and brought to America some of its most dramatic historic-era mammals.

Among these drought-propelled migrants were the North American wild sheep. Separating from goats some four million years in the past, wild sheep arrived in America in the Pleistocene, about a million and a half years ago. Paleontologists believe the first sheep in the New World were an offshoot of the argali sheep of the Asiatic steppes. Once here they evolved into three major types that inhabited America's arid western mountains, Great Plains badlands, and southwestern cliffs. Another late arrival was the musk ox, which evolved in Siberia and moved across the land bridge as far east as Hudson's Bay during the Pleistocene. Still another was an antelope-goat that in America became known as the Rocky Mountain goat. Once in America these pure white goats spread from the high mountains of southeastern Alaska through the Pacific Northwest to California and eastward into the mountains of Montana, adding a very late-arrival competitor in the high-country haunts of wild sheep.

Ironically in light of the iconic and symbolic nature of mammoths and bison in American history, these two were among the last animal immigrants here out of Asia. Recent DNA analysis of North American mammoth remains indicates that all our early mammoths sprang from populations of cold-adapted steppe mammoths that were native to northeastern Siberia and were descendants of northern Asia's grassland-adapted elephants. As grazers, in America mammoths and bison gravitated to the northern tundra as well as the immense savannas farther south. Both fossil and genetic evidence shows that mammoths found their way to America in the Pleistocene about 1.5 million years ago. The oldest genomic evidence science has so far analyzed from any species, extracted from tooth cores between four hundred thousand and roughly a million years old, indicate the Columbian mammoths that ranged south of the ice sheets were in fact hybrids of more northerly woolly mammoths and a previously unknown mammoth species. The explanation for these regional forms seems to be natural selection for versions adapted to the dramatically different ecologies north and south of the Pleistocene ice sheets, perhaps aided by gene flow from bulls that were particularly successful breeders. Given their late arrival, mammoths were never as distinctively American as camels or horses. But their size and roles as keystone vegetation engineers made them crucial players in our Big History.

Bison came to America later still. The oldest bison-like fossils are associated with southern Asia in the late Pliocene, 2 million years ago. A now-extinct species with a double hump and sweeping horns, the steppe bison, which appears in paleontological literature as *Bison priscus*, was likely the first Old World bison to enter North America. *Bison priscus* and its descendants spread eastward into Europe, too, but it was the American versions that would claim the spotlight. Latecomers they may have been, bison conveyed the idea of "America" to the world like no other animal. Along with genetic analysis, recent fossil work has produced the somewhat shocking conclusion that the earliest bison arrived here only 300,000 years ago and likely as recently as 130,000 years ago. Once they crossed into an America where horses and mammoths were their only grazing competition, steppe bison apparently underwent an ecological release, quickly adapting to widely divergent settings both north and south of the glacial sheets. An impressive longhorn version, B. *latifrons*, was extinct by 20,000 years ago, to be replaced by a southerly Pleistocene form known as B. *antiquus*, the very beasts taken down with fluted points at the Folsom site.

All those grazers attracted other Asian families that would become iconic Americans, our bears and cats. America's first bears reached the continent roughly eight million years ago. They separated into white bears in the circumpolar Arctic and black bears in tree country, where the latter could escape harassment from bigger bears by climbing. At least some inland white bears evolved into grizzly bears two hundred thousand years ago. Except for a population that clung to the coasts, feeding on beached whales and the fish that spawned in interior rivers—a type sometimes called brown bears—grizzlies would join that Pleistocene terror, the short-faced bear, to spread as far as open country (and ungulate populations) extended in the American interior. Eight-hundred-pound short-faced bears occupied their own genus, *Arctodus*. All the other American bears were related enough to be members of the genus *Ursus*.

Felis cats, the ancestors of cougars and lynx and bobcats, were in America long before the bears, their progenitor forms arriving from Asia about thirty-five million years ago, at the end of the Eocene. Among the spreading herds of ungulates in the late Miocene, one American Felidae replicated African savanna cat evolution, becoming the pronghorn's nightmare, the "American cheetah." It was most closely related to today's cougars. There were also the *Panthera* cats, evolved with a special morphology in their larynxes that enabled roaring. Pantherine cats emerged in northern Asia ten million years ago and got to America late in the Pleistocene, where they produced a magnificent lion along with a distinctive roaring American cat, the jaguar. The largest cat to evolve in American ecologies (it's still the third largest cat on Earth), jaguars sprang from a pantherine progenitor about a million and a half years ago, then spread from the Pacific Northwest through both American continents.

Impressive as jaguars were, and are, the most memorable and famous of all America's giant cats were the saber-tooths, which first became carnivorous terrors here 2.5 million years ago. *Smilodon gracilis* of eastern America was the runt of the group, reaching only about 225 pounds. But at 615 pounds, significantly larger than an African lion, *S. fatilis*, endemic to western America, was a creature out of horror fantasy, its massive canine teeth designed to slash into its prey and rip them open. These various cats, along with the bears and the several canids that had evolved in America, plus Old World hunting hyenas that followed the herds across Beringia, became hunters of pronghorns as well as of mammoth calves. They also pursued horses (especially colts), camels, and those other new migrants from Asia, elk, deer, and sheep.

⁓

FOUR AND A HALF million years ago, when ebbing seas allowed Panama's isthmus to form from several separate islands and link North and South America, Earth's newest connection provided another dramatic highway for the exchange of life. It allowed horses, camels, and canids to travel into the southern continent, and it brought ground sloths (including several species of giant ones), opossums, and armadillos northward. The isthmus also provided a flyway for some of the tropics' brightest, most colorful birds to join the bird life of North America.

Birds that became Americans followed all the patterns mammals did. The modern bird families evolved from ground-dwelling ancestors emerging after the Chicxulub strike, and with the occupants of so many niches destroyed, North America became the destination for a spread of bird life from elsewhere, particularly when ocean levels were low and land bridges existed.

Given their outsize roles in the historical story, some of America's classic bird life merits an especially close understanding. Until a century ago passenger pigeons were the most numerous birds on Earth; we ought to know them well. Eagles, condors, wild turkeys, hummingbirds, and a pair of natural superstars, our remarkable parrot and giant woodpecker, also deserve our full attention. The Northern Hemisphere penguin, the great auk, ranged across the North Atlantic, but it hunted and nested down America's Atlantic Coast to Florida. And there are the prairie chickens, including the eastern type known as the heath hen, that loom large in American history. The stories of many of these birds appear in the chapters to come.

Modern genetics is pushing America's bird past into new frontiers. From genomic work on passenger pigeon specimens starting in 2012 we now think America's most characteristic bird derived from a pigeon ancestor that arrived from Asia during the mid-Miocene, fifteen million years ago. Once here the ancestral pigeons diverged into two types, one (which became the bandtail) homing in on the more Asian-like West, the other adapting so well to the enormous biomass of the deciduous forests of eastern America that, like marine birds, it began to nest in giant communal flocks. No one is sure just how long ago passenger pigeon populations began to skyrocket. Genetic modeling indicates that they existed in immense flocks with a very healthy diversity for at least the past fifty thousand years. Pigeon flocks moving at sixty miles per hour and casting shadows hundreds of miles long across the American landscape were an ancient phenomenon.

America's biggest raptors, the eagles, trace their ancestral origins to the Eocene, when eagles separated from fish-hunting kites. Booted eagles with feathers to their talons eventually emerged as distinct types. The fossils of the classic American eagle, the bald eagle, appear here as far back as a million years ago. A fifteen-pound raptor that can live for almost forty years, the bald eagle has sailed the skies scanning for prey on an eight-foot wingspan since the Pleistocene. Like the golden eagle, its dryland relative, white-headed bald eagles were the avian *T. rexes* of post-dinosaur America. For a million years they scattered their thousand-pound nests from one side of North America to the other.

California condors are another purely American type, not related to the vultures of Eurasia and Africa. They and related forms evolved here during the mid-Miocene to fill a niche provided by growing numbers of ungulate carcasses. America's condor ancestors were related to Andean condors, but the California condor's genus, *Gymnogyps*, is distinctive and at least forty thousand years old. Early species were even larger than the existing bird, but fossils of today's condors are common from the rich American bestiary of the late Pleistocene, which provided them with coast-to-coast scavenging abundance, a circumstance in which they clearly thrived.

Closely associated with both Native America and colonial American history, wild turkeys present a complex story. The assumption has long been that America's wild turkeys are actually a kind of giant Eurasian pheasant. If so they have been hyphenated Eurasian-Americans for a very long time, likely developing their "turkey-ness" here. Early turkey ancestors migrated into North America around the beginning of the Miocene, but the oldest fossil evidence for them, curiously, is not from Alaska but from Florida. Fossils in the direct line to more modern wild turkeys date to the Pliocene, about five million years ago. By the Pleistocene there were three principal types in America, one associated with woodlands in the East, one a drylands turkey, and a third—*Meleagris californica*—a now-extinct wild turkey that paleontologists know well because its remains are common at La Brea Tar Pits.

Dramatic as so many were, no American birds would ultimately stun human eyes and ears like hummingbirds. Found in the Americas and nowhere else in the world, hummers originally evolved in Europe, arriving in North America via Beringia about a million years ago and evolving into more than three hundred different species. Meanwhile they went extinct in their original evolutionary home. Along with their tiny size, color vision, and rapier bills adapted to foraging for nectar among America's wildflowers, hummingbirds evolved a flight acceleration of as much as nine g-forces, the highest known for any vertebrate on Earth—all with minuscule wings that also took them on epic migrations up and down the continent.

Since most birds are fliers, land bridges weren't always essential for bird colonization. But they helped. The tropical parrots that spawned America's Carolina parakeets, or woodpeckers evolving in the tropics, could easily colonize northward into suitable American habitats. As for North America's dramatic parrots, the Carolina parakeet, a lucky sequencing of its genome from a preserved leg fragment in 2019 revealed its origins as a relative of South America's sun parakeet. Green-and-yellow Carolina parakeets split from that species three million years ago. Swooping through the forests in dense flocks like festive flying blankets, they colonized North America once the Isthmus of Panama pointed the way. By the time humans got here, America's parrots had separated into two subspecies, one associated with the rivers of the lower West, the other extending from the bayous of the Gulf South northward along the Atlantic Coast, providing brilliant life to forests all the way to the Hudson River.

The newest genomic work on America's giant woodpecker—the ivory-bill—indicates its ancestral antecedents came from the American tropics, as well. During the Pleistocene one population crossed to Cuba, another settled into the Sierra Madre in Mexico to become the imperial wood-pecker, and a third flew north to become America's legendary ivory-bills.

America's new land bridge in Central America was yet another contin-gency that changed the history of wild America. The closing of the isth-mus diverted ocean water northward, which sucked more moisture and rain up the coasts toward the Arctic. That contributed to another grand effect endemic to the planet. It fashioned the Earth of the ice ages.

Our planet does not rotate perfectly around its axis. Its gyroscopic twirling produces what's known as precession, which means our sphere wobbles slightly, like a spinning top losing momentum. Across time that wobbling spin tilts the poles away from exact true, which subtly alters the angles at which sunlight strikes the planet. With increased moisture now flowing up America's coasts to the Arctic, those tilts began to produce cli-matic swings at intervals from ten thousand years up to forty thousand. We know these as part of the Milankovitch cycles. Named after an early twentieth-century Serbian geographer who studied axial tilt, along with how gravitational pull from the giant planets, Jupiter and Saturn, alter Earth's orbit, these cycles and the enhanced moisture produced a new sit-uation for the Northern Hemisphere. An age of expanding, then retreat-ing, polar ice was at hand, a Grand Phenomenon that would by turns freeze North America and bury half of it in ice, then warm it dramatically, then freeze it again. Ice Ages and the warm pluvials between them stirred natural selection and repeatedly opened and closed the doorways for new life-forms passing from the rest of the world into America.

Life, for example, like that emerging in faraway Africa.

≈

TO CONSIDER the natural history of humans in our wild-animal form we have to drop early into our story. At some unremarkable time in Earth's Miocene epoch, between eight and six million years ago, a branch of pri-mates in Africa began to diverge from the limb that had produced chimps and apes of the deep forests. With their basic forest adaptations intact, the new hominids (the term for the various genera in our line) began to probe into the grassy, scattered forestlands that Miocene aridity was producing. They possessed a body structure, grasping hands, and stereoscopic vision adapted for moving through and standing upright in trees. Like birds and butterflies, they saw in color to recognize ripe fruits. As tropical apes, they liked seventy-two-degree-Fahrenheit temperatures. They preserved a troop-like life of intense and complex social interactions, which fostered a body/brain chemistry of dopamines and serotonins as catalysts for diverse, motivating emotions. Look at a sunset and appreciate its color. Grip a steering wheel with thumb and fingers. Check the thermostat setting. Every one of these is enabled by these holdovers from our long-ago African origins.

Science fiction is rife with scenarios wherein intelligent aliens observing Earth barely notice these primates of six million years ago. But the changes the planet was undergoing in the Miocene created new, more open ecological conditions all over the globe, and if the aliens paid the sort of attention smart aliens should, they might have noticed something interesting. By at minimum 3.5 million years ago, likely for half a million years before that, these primates were walking upright. An eighty-foot-long trail of two sets of fossilized footprints from then, found near Olduvai Gorge, attests to that. These bipedal primates were now on the road to becoming travelers, a cosmopolitan species for whom all of Earth beckoned.

At the height of the Pliocene African hominids were becoming something their descendants recognize. They were *australopithecines*, no longer quadrupeds balancing on their knuckles, and now living in open country full of predatory danger. Walking upright makes you look larger. It also enables seeing far and exposes less of the body to the sun. But most important, walking upright frees the arms and grasping hands to carry things, not only food things but any manner of objects. Australopithecine dental wear suggests their large intestines were still designed for a fibrous, chimp-like diet. But out in the drier grasslands hominid fare began including seeds and nuts, enhancing the small intestine's processing of lipid-like fatty acids. That turned out to be important. As for the things to carry, simple "found" tools for crushing nuts and seeds begin to show up about 3.4 million years ago. They were a step toward the tools we would later use to slit animal hides and crack bones.

Not everyone is thrilled to hear this, but this progression made us carnivores and predators. Every step fostered feedback loops that began to change us. In several different species the australopithecines survived across some two million years, during which time they began to *make* tools by chipping sharp flakes off rock cores (Oldowan tools, named after Olduvai Gorge). Isotope studies of their bones show a strong C4 signal, indicating a diet made up in part of grazing animals. They also appear to have acquired tapeworms, undoubtedly by eating flesh that cats and canids had also bitten into.

As the Pleistocene dawned two and a half million years ago, a taller, even more upright creature emerged in the hominid line and appears to have outplayed the australopithecines at this new game. The new beings were part of a genus we now call *Homo*, which includes several species (the latest count is twenty) of "hominins," different but related members of our own biological animal. Now, as *Homo*, our ancestors were about to undergo profound changes that enhanced and refined our success in the ecological niche we'd chosen. Those changes came directly from the deaths of other animals that surrounded us in the world. It really was quite remarkable. A primate undergoing rapid reinvention by feeding itself on animal fat and protein was becoming the newest, planet-changing contingency in Earth's history of life.

In the twenty-first century we do not look at one another and say to ourselves: "So, another confrontational scavenger to swap stories with." And yet there is good evidence that across australopithecine and early *Homo* stages we engaged in a lifestyle built around scavenging found or stolen animal carcasses. Chimps and bonobos and a great many species we don't consider carnivorous take advantage of fat and protein from meat whenever it's available to them. Chimps *make* meat available by hunting smaller primates like colobus monkeys and bush babies. But in an Africa that was much like the present continent, with great herds of ungulates and cat and canid predators, our ancestors' first adaptation was to become hyena-like scavengers. We weren't just another scavenger among the vultures, waiting our turn. Early hominins perfected the nuances of *confrontational* scavenging. Your stomach is fluttering for good reason. Primates are classic prey for big cats. The only chance hominins had of confronting big predators and stealing their kills meant organizing into groups, like hyena packs, large enough to be intimidating. Then doing the primate thing of brandishing sticks and hurling stones.

If imagining driving lions and leopards off their kills doesn't raise the hairs on your neck, contemplate this. The human genome includes a gene called APOE that, in four variants, is found among all human populations. About 13 percent of Americans from northern European backgrounds possess the ancestral hominin version of APOE that produces high cholesterol levels, coronary heart disease, and rapid cognitive decline, which is the primary reason we know so much about this gene. But in our situation two million years ago, those were side effects. The ancestral version of APOE actually evolved as an advantage, triggering the hominin immune system to fight off particular kinds of peptides found in foods swarming with bacteria. And why would we have needed that? Because this gene enabled us to survive scavenging meat and animal parts that were in a state of decay. The human disgust response, evolved to protect us from poisoning ourselves, appears to date from roughly the same time.

Not so long after *Homo* emerged, what paleoanthropologists call "persistent hominin carnivory" began to replace scavenging. That must have been momentous, because attacking living animals can be far more dangerous than scavenging. It was a feat our ancestors managed despite lacking the speed, power, and jaws and carnassial teeth of other predators. Predatory success implies that we performed this transformation not as solitary hunters like leopards but as pack hunters, now resembling lion prides or canid packs rather than hyenas. That kind of cooperation worked inside a feedback loop involving brainpower. In less than half a million years, *Homo* brains grew 70 percent by volume and were burning up a quarter of hominin intake energy. It was animal fat and protein that grew those outsize brains. And it was carnivory, along with the cooperative social lives this new savanna economy demanded, that selected for that brainy intelligence.

%

IN THE BACKYARD of my house a few miles south of Santa Fe, New Mexico, is a three-hundred-foot-deep canyon my dog and I cross to the mesa where we run. I have traversed this canyon and run on this mesa for years, and today with Kodi already on the trail home I indulge a slightly unusual act on my cool-down walk. It's a simple thing but it's a grand genetic inheritance from deep time. As I walk I scan the trail for rocks of a special type, preferably round, ideally the size and weight of a golf ball. These are not rocks to collect. They are rocks to throw, and six or eight times as I return from today's run I find the best grip I can on the rocks I've gathered and uncork throws at distant boulders.

For me the act of throwing is a muscle memory from childhood later refined by high school and college sports. It's a physicality I've always possessed and have never given up. As I perform the complex ballet of a throw—the foot opposite my throwing arm lifts and points at the target, my hips rotate as my weight shifts forward from my back foot, and with a whip-like motion from shoulder and arm I arc a projectile across sun-lit space—I think what an ancient act this is. We inherited our ability to throw from earlier primates, but in a change from their body design, evo-lution fashioned late hominins with shoulder sockets that attach our arms on the outside of our bodies rather than on the top or front. That design, plus the premium on balance and coordination our upright stance gave us, allowed hominins to throw objects with great force and accuracy.

Around 1.8 million years ago we started crafting weapons to throw, initially the grapefruit-sized, sharpened rock tools archaeologists call Acheulean handaxes. Either handaxes were awesome executors of homi-nin needs or no one could figure out how to improve them, because for the next million and a half years they hardly changed. The most ancient Acheulean handaxes are from locales like the Rift Valley of Kenya and Olduvai Gorge in Tanzania, where they seemed to serve as the origi-nal Swiss Army knife. Our ancestors used a version to grind seeds, and another to slice into animal hides, butcher flesh, break open bones. They served as weapons of defense, warfare, and murder, and supersized ones perhaps even for ritual. Archaeologists have found scores of them around ancient water holes, where they must have been thrown in volleys to fell gazelles and impalas.

Those moments, lost to us across a yawning abyss of time, were pro-found. Our relationship with the animals around us had changed. We had joined the predator guild and for the first time began to produce fear in other animals. Since we're unable to re-create the inner consciousness of a *Homo* species from a million years ago, there is no way of telling the psychological impact of this reversal. But it's a good guess that the initial idea of an afterlife—where death is not the final act—may have entered hominin minds with an emotion of remorse at confronting a beautiful animal that one instant was living and striving and in another lay sound-less and unmoving. Comparing body designs, along with butchering as the act that followed a kill, surely drove home how similar all other ani-mals were to ourselves.

The hominins inherited from primate ancestors a slight sexual dimor-phism, and that had consequences in the predatory economy. Certainly, capable females must have been good at the hunt. Planning hunt strategies no doubt was gender neutral, and carrying out those strategies included both females and children in hominin bands. But actual pursuit of prey was more often the task of the classic primate male troop, which over time may have hardened into a gender role. Male hominins hunted. Some females did, but others performed tasks that could be done in conjunction with childcare. Among modern hunter-gatherers, meat and other ani-mal products are a privileged resource, more important than plant foods. Sharing animal flesh with other members of the group would have been central to creating social status. No doubt that translated to sexual suc-cess for those who excelled, regardless of gender, which would be a driver that continuously refined the predatory economy.

Lives as predators forever changed us. Our primate ancestors had lived in dread of predators, a fear we yet feel. But now they had ascended the trophic pyramid. This wild, fierce life made hominins bigger, more phys-ical, better athletes. Yet another advantage of standing upright was that it made hominins much faster runners. Their twenty-mile-per-hour sprint-ing abilities weren't so impressive when a gazelle could run sixty miles per hour, but few creatures had such endurance. Their hearts changed from the thick chimp organs designed for brief spurts of energy to the thinner, supple, more elastic hearts of long-distance runners. They grad-ually lost their body hair, developed sweat glands. The ancient versions of us became taller, thinner, more gracile. And more wolflike. Both we and wolves retain from lives of running down prey a phenomenon we know today as "runner's high," a buzz from the body's release of natural endo-cannabinoids. Evidence is also growing that by moderating the effects of the APOE gene, the cursorial lifestyle led to an increase in the hominin life span beyond the reproductive stage, a phenomenon anthropologists refer to as the grandmother effect.

Around eight hundred thousand years ago our ancestors took a step that perhaps contributed to an initial sense that we were different from other animals. We began to control fire and put it to purpose. With fire, hominin bands could deflect predators. They could manipulate ecologies. They could look for prey animals in places no one had ever gone, explore colder regions by, in effect, taking the setting where hominins evolved along on the journey. And with fire they could transform the flesh of the animals by cooking, a step anthropologists from Claude Lévi-Strauss to Carleton Coon believe was crucial in the predatory economy.

So. Bringing fire, the flickering agent of destruction and creation, under control. Exploring a world more vast than any hominin could imagine. And killing other creatures, cooking and eating their flesh. How could these protohumans not develop some sense they were superior to all they beheld, a master life-form in a world brimming with creatures now delivered into their grasp?

〰

AS NORTH AMERICA took on a form we might recognize, in Africa various species of Homo were now on the move. Keystone carnivores, capable of affecting prey populations, we—for now our ancestors were becoming us—began to search for unexploited animal populations. We learned to construct shelters, and at some point the clever among us figured out how to make awls and needles to sew animal skins into fitted clothing. In a development eerily similar to the way NASA is today calculating how humans can survive on Mars, our technologies meant we could, in effect, take Africa with us wherever we went. These were technologies that allowed tropical apes to colonize new worlds.

Roughly eight hundred thousand years ago, hominins not very different from fellow passengers we see on the subway began to push out of the original homeland and into the bigger world. Earlier bands of hominins (Homo erectus) had already attempted an African exodus, getting to the Black Sea nearly two million years ago and eventually to Spain and Indonesia. But the movement into Eurasia of bands of hominins known as Homo antecessor and eventually another hominin derived from them, H. heidelbergensis, finally sent members of our genus deep into the largest landmass on Earth. That prepared history, at some fecund moment around half a million years ago, for two new hominins to begin their thousands of years of mastery of the different, colder, but animal-rich country north of Africa. They were the Denisovans and the Neanderthals, beings so closely related to us modern-day humans that most of us still carry their genes, coding that influences our height, our body hair (or lack of it), even our susceptibility or resistance to particular diseases.

Fossil sites and genomic analysis for Denisovan populations farther east in Europe and Asia are still in their early stages. But we've come to know a great deal about Neanderthals, since the first site with their fascinating remains surfaced in Germany in 1856. There are now scores of their sites from the Middle East to northern Europe, thousands of their bones in museum collections, and at least nineteen sequenced Neanderthal genomes. At every stage of discovery they become more intriguing and ever more enigmatic.

The first observation about them may come as a shock, but it rests on the inescapable fact that for many of us in the twenty-first century, as much as 2.5 percent of our genes (and not the same 2.5 percent from one of us to the next) derives from them. Genetic analysis is indicating so many episodes of hybridization between these late hominins and ourselves—at least four episodes of gene exchange between Denisovans and modern humans and another four between us and Neanderthals, the events starting as early as three hundred thousand years ago and continuing down to forty thousand years ago—that a 2021 study in *Science* posed the question of "whether archaic and modern human should be regarded as distinct lineages." If it is more accurate to think of Neanderthals–Denisovans–modern humans as a continuum of biological diversity instead of truly separate species, if we were all becoming versions of one another, then the story of these early humans becomes even more gravid.

The fossil evidence we have indicates they were on the whole slightly different, bigger than us (23andMe says I'm taller because of their genes), built as heavier long-distance walkers rather than runners, and designed by natural selection as predators. Nitrogen isotopes taken from Neanderthal bones show them to have been remarkably predacious, as carnivorous as wolves or hyenas. There is abundant evidence they preyed on red deer, caribou, aurochs, and wild horses, but some of their isotope markers indicate they were also eating mammoths, rhinos, even bears. With animals the size of caribou or horses a band of ten Neanderthals needed six of them a week, roughly twice what a wolf pack required. That was in excess of five thousand calories a day, yet they apparently butchered only for the choicest parts of their kills. Like their ancestors they still used handaxes, but the better weapon for close-in battles with large animals seemed to be wooden spears tipped with sharp flint. A site in Germany has yielded an eight-foot spear broken into pieces beneath a mammoth carcass, and other spears have come from a site with the remains of fifty horses. They also hunted and consumed birds and clearly were fascinated with raptors. Eagle talons show up frequently in their sites.

There's zero chance any of this was easy. These were people from thirty thousand generations of hominin predation, but it's abundantly clear their economy took a toll. Male Neanderthals (but not females) often had dramatically overdeveloped arms and shoulders on their dominant sides from a lifetime of thrusting spears, knapping flint tools, and butchering. Females didn't escape their own kind of wear, with worn and abraded front teeth from clenching down on the heavy hides they worked. As you could guess, Neanderthal hunters could end up remarkably battered from close-in encounters with big animals. When one individual from a site in Iraq died, he had a shattered upper arm, an amputated lower arm, a leg injury gone arthritic, and a crushed face with a blind eye. Other Neanderthal hunters had broken collarbones and broken legs, including thighbones. Broken jawbones were common. Another issue was that big Pleistocene hyenas were inclined to stalk Neanderthals as prey.

While a Neanderthal or a Denisovan might not attract undue attention in a modern city, they did possess traits not common among us. They lacked bony chins and their eyes and noses were large, the eyes so to take in more light in dim northern settings, their nostrils capable of inhaling twice the volume of air most of us can. To absorb the necessary vitamin D from cloudier high latitudes, their skin tones lightened, an adaptation for all hominins who went north. Neanderthal brains seem to have been slightly bigger than ours, but with a distinction that maybe made a difference. Their skulls, and the brains inside, were elongated. But while our skulls began that way, too, over time our braincases became round, allowing for additional myelin in our cerebellums and perhaps faster-growing neural connections.

Both Neanderthals and Denisovans were around for a very long time, although their total populations at any one time were probably only in the tens of thousands. The big question (other than "how many Neanderthal genes do I have?") is, what happened to them? To some degree, with a total of more than 20 percent of their genes surviving in modern humans, the answer is that they're still here. But no fully expressed Neanderthals or Denisovans exist, so the question remains. We know they made art of a kind, which implies an ability to create symbols to represent the real world. They were also highly advanced in planning hunts and migrations, and clearly employed theory of mind in understanding animals. Whether they thought there was some existence after death has been harder to determine. Their treatment of deceased companions almost exactly mirrored how they treated the animals they killed. In other words, they disarticulated them. While their purpose in doing so *might* have had its origins in grief and a desire to hold on to some part of the deceased, that they treated themselves just as they did other animals could in fact have represented the survival of an ancient hominin belief system.

The best answer to the big question involves language (scorned English majors—and I am one—our revenge is at hand). Language is one of the most complex neural and motor functions we perform. It requires coordination from both neural circuits and muscles at speeds in the milliseconds. These archaic humans probably did have language. Their inner-ear structure was different from ours, but seems to have been good enough for hearing the nuances of spoken complexity. Their vocal cords weren't unlike ours, and they had the FOXP2 genes critical to communication (although we inherited our version rather than theirs, implying adaptive advantage). But with a smaller frontal cortex and cerebellum, both involved in communication, their abilities could have been limited compared to ours. So language capable of transmitting culture richly looks as if it may have been our main advantage, especially since we occupied exactly the same niche they did. A recent study at the Max Planck Institute for Psycholinguistics speculates that the motor skills modern humans developed with advanced spoken language refined our brains in new ways that never took place inside Neanderthal skulls.

⁓

THE SEQUENCING of our genome early this century suggested that modern humans appeared as a new hominin at least 300,000 years ago, and fossil remains with modern human morphology found in present-day

Morocco have dated to 315,000 years ago. An Africa where multiple species of *Homo* had evolved and coexisted had now birthed a slender, clever hominin destined to populate the whole planet, including the Americas, that last grand prize at the end of primate wandering.

Like Neanderthals and all the hominins preceding them, *Homo sapiens* emerged as a keystone predator. Our out-of-Africa diaspora, likely motivated by our search for places where animals were numerous and innocent and guild competition light or nonexistent, proceeded in fits and starts. Except for the occasional lucky band—and there appear to have been a few—most modern human colonizers were bottled up by climate and geographic barriers for much of our early history, proceeding no farther north than the Middle East. But about sixty thousand years ago a significant wave of modern humans successfully pushed eastward into Indonesia, then northward into a Europe still vegetated by Ice Age tundra and populated by the herds and predators of the Pleistocene world. Then it was on into Asia, that vast landmass that ultimately, so many times for so many different animals, had provided an ancient road trip to America. Siberia and northern Asia witnessed modern human arrivals by at least forty-five thousand years ago, producing populations that would eventually probe farther eastward and lead to North America.

Of course Europe, Indonesia, and Asia were not empty of our Neanderthal and Denisovan cousins, and it was in the best river valleys for hunting mammoths, horses, red deer, and wild auroch cattle that we encountered them. Though archaic hominins were good as predators, we were better. Whatever fate befell our old African relatives, the diverse and plentiful animals of Eurasia clearly inspired us. The earliest human art is the most accessible way to discover what we thought important. And it was not portraiture of ourselves, or star-studded heavens, or the new landscapes we were encountering that compelled. Instead virtually all humanity's ancient artistic representations focus on the animal life around us.

The oldest figurative human imagery we've discovered so far, a colored pictograph, comes from Indonesia. Like much ancient human art it appears on the wall of a cave, places where we camped and spent leisure time. This earliest modern human artwork is a scene of eight hunters stalking prey that appears to be buffalo and wild pigs. Uranium dating has fixed its age at 43,900 years ago, which is intriguing enough. But most fascinating is the form the hunters take. They are what is known as "therianthropes," an anthropological term for figures where humans are merged with other animals. Therianthropes are common in Paleolithic art and stories. A mammoth ivory sculpture from Germany, the so-called Lion Man with a human body and the head of a lion, is a famous therianthrope from forty thousand years ago. This is imagery of a time when we still thought ourselves part of the wild bestiary, such close kin with other species that we could move readily from our bodies into theirs.

Imagery painted long ago, straight out of the human mind, brings us to a critical point in this story. In biological and even cultural terms there had not been any "great leap forward" from being animals to becoming something different. Modern humans elaborately buried their dead, but belief in an afterlife wasn't necessarily a step forward. They were still carnivores, engaged in the same predatory economy our hominin ancestors practiced. They were then who we are now, with brains just as big, just as intelligent and aware. But as modern science is finally instructing us, all the things we would believe unique about ourselves—that we are smart and self-aware, have language, accumulate information we pass on to others—are actually common qualities of all higher life-forms. They do not set us apart. Of course learning how to ascertain the reality of the world isn't easy. The difficulty of that has saddled us with many wild and dubious beliefs and practices across human history. But our ancient expertise as predators and now an enhanced ability with language accompanied our spread across the planet. The more of us there were, the greater the chance someone would invent or think of things that might be really important. Passing those ideas along via our linguistic skills and our growing numbers was about to make us the masters of Earth.

But we were still animals, hunters out of Africa, and what was most important to us were other animals. This wasn't so long ago in our history that the biology of it has faded, a fact made clear in a study done by neuroscientists in 2011. Among a large sample of patients, they found we twenty-first-century humans still retain a pronounced selectivity for imagery of animals in the amygdala of the human brain. Amygdalae are almond-shaped masses of gray nuclei inside each of our cerebral lobes. They are centers for emotional behavior and motivation, and what the science demonstrated was that our right-hemisphere amygdala evolved (and yet engages) a neural specialization "for processing visual information about animals."

Scene from Chauvet Cave. Courtesy Getty Images.

Which may explain why the great Paleolithic art done by our ancestors took on the subjects it did, and why three hundred centuries later we find this art from our past so irresistible and moving.

No part of Earth has preserved Paleolithic paintings as spectacular and gracefully executed as western Europe. One of the oldest identifiable human traditions there is the Aurignacian culture, dating from thirty-four thousand to about twenty-six thousand years ago. It was they, along with the Solutrean and Magdalenian cultures from western Europe of twelve thousand years ago, that left us the most dramatic of humanity's ancient visions of the observed world. These artists were of course *us*, albeit without the several hundred more generations of handed-down knowledge from which we benefit. With minds like ours, they sought out labyrinthine cave bear dens in places like the limestone gorges of the Ardèche Valley in France and rendered in paint the most important things about the first two million years of hominin life.

France's Chauvet Cave—the "Cave of Forgotten Dreams"—is our most profound look at their worldview. Chauvet is special not just because its paintings are thirty-four thousand years old but because one of its artists was a Paleolithic Picasso who reproduced on its walls the Sacred Bestiary, the animals we had come to consciousness with and then followed out of Africa. Chauvet's paintings showcase humanity's intimate and meticulous love of natural history. These rhinos and horses aren't just gorgeously drafted. There is perfect shading of pelage, astutely rendered expressions. The animals are muscular and appear in fluid motion, with even the rhythm of their footfall sequence correct. The art excels in three-dimensional distancing. The famous Lion Panel is so lifelike you expect to see sides heave and clouds of frosty breath.

We painted these animals with hands that shaped the tools that made us predators, reproduced them in primate living color informed by two million years of fascination. This was our oldest expression of biophilia, our worship of life in the form of elephants, aurochs, giant deer, bears, lions, and hyenas. Of course there is a therianthrope, a bison merging into a woman, a woman who is really a bison. At Chauvet, Lascaux, Altamira all this was testimony to the religion of our ancestors. It's why they—we—were exploring the planet, bound for America.

Earth must have trembled.

CHAPTER 2

CLOVISIA
THE BEAUTIFUL

We hardly know our actual beginnings in America. Even when the stories are set in places we recognize, the characters of our deep-time history can be alien to the point of fantasy. But while it may sound unlikely, in the 2020s there is no place quite like downtown Los Angeles for acquiring some sense of how the human story began on the continent. Rancho La Brea Tar Pits, just off Wilshire Boulevard, in the heart of a sprawling Pacific Coast city, today is the most accessible place in the country for picturing in the mind's eye the wild new world migrating humans found when they first saw America. True enough, there's a sense of time-travel shock having your Lyft drop you in the middle of swirling, honking LA traffic, only to stand face-to-face minutes later with Columbian mammoths fatally mired in tar, trumpeting their despair. Even if the mammoths are robots and their forlorn cries don't drown out the traffic, they and La Brea and the Page Museum still work a kind of magic. Twenty thousand years drops away if you let it, because La Brea preserves tangible remnants of a world at the far ends of the Earth for humans whose migrations had begun in Africa.

The Page Museum is a working laboratory of paleontology where visitors can watch scientists labor over the site's latest discoveries. Many of those are the remains of scavenger/predators once lured by the cries of snagged mammoths, or the scent of decomposing horses, camels, or

Mammoths in Rancho La Brea Tar Pits.   Photograph by Dan Flores.

ground sloths trapped by surface tar near what was once a water source in a dry landscape. The skulls and tusks of the elephants extracted from La Brea are impressive, but anyone who tours the museum has to admit the most stunning display is the wall, backlit in yellow, of hundreds of dire wolf skulls. The strapping canids, indigenous to America but memorably revived as fictional "Westeros" fauna in *Game of Thrones*, left the most remains here of any species—1,800 individuals. The fossils of hundreds of coyotes, a brawnier version than our modern animal, make up the third most common species here. But in second place are the ultimate ambush predators of the Pleistocene, the western subspecies of saber-tooths, heavily built cats with a fearsome, snake-like jaw gape and enormous fangs. The replica skull of a saber-tooth from La Brea sits a few feet away as I write this, its rapier-sharp canines, capable of tearing open a sloth or mammoth calf, gleaming in rich afternoon light. Each fang measures a full eight inches from gumline to tip.

The vast assemblages of hypercarnivore bones at La Brea join the skeletal remains of megaherbivores—mammoths and mastodons, giant bison, pronghorns, llamas, California turkeys, many more. The predator list is lengthier than just wolves, coyotes, and saber-tooths, as well. The cats whose remains have come out of the tar include American cheetahs, steppe lions, and giant jaguars. Immense, hyperactive short-faced bears twice the

Saber-toothed cat skull and fangs. Photograph by Dan Flores.

weight of a grizzly died in the asphalt. So did the enormous Merriam's ter-atorn, a Pleistocene bird of prey with a ten-and-a-half-foot wingspan. The remains span indigenous creatures spawned by continental evolution and migrants from Asia, some ancient to America, some recent arrivals. The mammals and birds may seem alien, or vaguely African, but in fact this bestiary was purely, classically American, the America of the Pleistocene.

The Rancho La Brea victims that left their bones and skulls encased in tar were once representatives of one of the grand ecologies of planet Earth. This was a different America than most of us conjure when we imagine the continent Europeans found five hundred years ago. But this La Brea world wasn't like the pre-Chicxulub Age of the Dinosaurs, absent of humans, either. Late in the Pleistocene our human forebears joined American ecologies as their newest predator. These first Americans lived their lives among La Brea creatures and created the first coast-to-coast human societies in American history. Their presence also left the conti-nent and this rich aggregate of impressive animals forever changed.

*I*t is long ago and a group of three dozen travelers, members of six families with their twenty dogs, are moving up a shallow, grassy incision in the earth. They are following a drain, threaded by a foot-wide stream, that courses in graceful arcs in the direction of the sun's westward progress in the sky. Out of sight above the sloping walls that enclose them is an immense, horizontal grassland with scattered trees, an almost featureless country but well known to these travelers. It is a place that elicits sensations of pleasure, anticipation, and some apprehension. Their navigator is an older woman in the group, but it is the men who are out front in this V-shaped advance up the drainage. Small herds of giant bison cease cropping grass and stare at the humans moving by, and bands of horses with zebra-striped legs pose motionless, silhouetted against an ice-blue sky. Camels and antelopes neither snort nor look. Their focus is on a nearby pride of steppe lions stretched out on a rim of the drainage.

The travelers move on with their dogs quieted and silently exchange glances when they hear unmistakable sounds, a cacophony of barks, shrieks, and roars. Birds, most of them black and of several sizes, flap and circle into view, swoop and disappear, then float above the grass again, airborne and protesting. The sounds are familiar ones. Wolves, cats, and bears are fighting over someone's kill. The birds are hoping for leftovers.

When they've proceeded some distance beside the rippling little stream, the sounds of the melee receding behind them, the band's navigator motions for a stop. What they listen for now is what set them on their journey days earlier, from a camp where striped rock, perfect for tools, erupted from the edges of a large valley. As they traveled their knapper had reduced blocks of reddish rock into beautiful hunting points, flaking slight channels into the bases of each point. These points with their channels are a huge confidence-builder for these human predators. What they are about to attempt is dangerous, but lavish in rewards if it works. Success will mean abundance, leisure, the opportunity to instruct their children, maybe beget more of them. These striped, razor-edged points are fixed solidly into their spears and put all prey within their reach.

The sound of trumpeting prompts the navigator to put her hand in the air. Around another winding of the shallow channel she knows that a lake fills a separate drainage coming in from the north. Their feet slipping on the grassy incline of the trough wall, the hunters crawl to the rim and peer in the direction of the trumpeting. At once they spot the angled gray hulks of mammoths at the lake. There is the herd bull, immense and solitary, cropping grass on the lakeshore with the afternoon sunlight glinting off long, curving ivory tusks that cross *each other beyond his trunk. And there are the cows, working their ears, alert and dangerous.*

The elephants are an extended family, a group much like the hunters' own, with adolescents and calves that with good fortune and the favor of the gods they might catch in deep mud on the shoreline of the lake, where they'll be easy prey. The cows they will have to fight.

〜〜〜

I RECONSTRUCTED the preceding scene out of evidence from the first place where we became aware of the Clovis people, Blackwater Draw on today's Texas–New Mexico border. Clovis was the first human culture to spread across all the Americas, which they did in a rapidly advancing wave thirteen thousand years ago, occupying every modern American state from Alaska to Florida for more than three centuries. Until a mature United States spread coast to coast, Clovis stood as the sole human culture that once draped across our entire country.

If you know about Clovis at all you probably think of it as a western phenomenon. To be sure, archaeologists have found 500 of their large, distinctive spear points—fluted like Folsom points but often five or six inches long—in California, 200 in the Great Basin, 120 in the Pacific Northwest, 800 on the Northern Plains, and more than 2,000 in the Southwest. But Clovis inhabitation of the eastern half of America was actually even more impressive. Some 2,600 fluted points have come out of the Midwest, another 1,400 from three or four huge Clovis sites in the Northeast, and a whopping 5,500 from the Southeast. For three centuries a very long time ago, America was Clovisia the Beautiful.

Yet we are still struggling to understand these early inhabitants who imposed themselves from Pacific to Atlantic to Gulf. Unlike many who came after them, they left no oral or written histories of their monarchs or defining events. We have no sense of the gods or philosophies they believed in or what language (or family of languages) they spoke, which means we're unable to call on their words to invoke them. We know a great deal about their tools, and we're developing a sense of the people themselves from their bones and more recently from their genetics. And thanks to a recent investigation in South America of panels of rock art stretching across miles, some of which apparently date to their time, we now may have a visual glimpse of how they saw themselves and the world of plants and animals in the continents they had taken.

A brand-new piece of information we do have about the Clovisians, though, is the certain knowledge they were not the first here. Much the way a sprinkling of modern humans filtered among Neanderthals long before *Homo sapiens'* big migration out of Africa, just as Vikings left evidence of settlements on Atlantic shores hundreds of years before Old Worlders began pouring into Massachusetts, Virginia, and New Mexico, humans had arrived from Asia as long as twenty-three thousand years ago. We know they were here because of a scattering of sites that reach eastward to Pennsylvania and Florida. But we're primarily convinced of their presence by a recent, astonishing discovery in what is now southern New Mexico.

"Before Clovis" is an exciting and unfolding story. But in Big History, pre-Clovis may be destined to play the Pleistocene version of Vikings versus Puritans. When America's climate finally warmed enough sixteen thousand years ago for an overland corridor to open between the coastal Cordilleran and inland Laurentide ice sheets, the people who became the Clovisians, bottled up in Beringia for more than fifteen thousand years, began to pour into North America and spread across the continent in one of the fastest geographic movements of people in the human record. If there were very many other human groups in the way of that advance, they certainly didn't slow it very much. In fact the rapidity of the Clovis migration suggests they encountered few human cultures along the way.

Starting 13,050 years ago and lasting until 12,750 years ago, the Clovisians placed their stamp on the country and its animals in a way that changed everything from then until now. Their Clovisia the Beautiful, or whatever they called their ghostly American civilization, looks to be the moment when the human hand seized the tiller of continental evolution. For this part of planet Earth, the search for the Anthropocene starts with Clovis.

Their name comes from the place where we first became aware of their existence, an ancient arroyo on the outskirts of the small town of Clovis on the windswept Southern High Plains. The original Clovis site has a continuing whiff of Indiana Jones romance about it, and driving across the Great Plains in the early autumn of 2019 with the plan of investigating this National Historic Site with its director affords me the opportunity to reflect on the still-unfolding movie script of these American ancestors. Why have they remained so obscure to the American public while they seem so critical in our past? Why, among those who know them, are they so wildly controversial? More salient to the human story, how could the three hundred years the Clovis people held America have transformed America so profoundly?

〽

GETTING IN CLOSE to wild creatures holds a fascination that resonates because it taps ancient imperatives still within us. The relationship between prey and their predators involves learning curves, and each side is very good at the algorithm. But prey do have to learn. In their pre-predatory stage, our primate ancestors had moved through the world among other creatures that did not fear them. Once hominins adopted carnivory, the prey we were interested in learned to allow close approaches at their peril. But unlike big cats or dire wolves that were obvious predators, upright primates didn't automatically fit the predator template for most animals. Numerous examples from around the world during the past five centuries testify that upon initially encountering humans, many wild creatures did not associate us with a threat. In their first encounters with humans scores of species reacted with trust and tameness. There is a term of art for this: biological first contact. When we appeared in new geographies for the first time, wild animals had to learn to be afraid of us. Many died standing and looking, never absorbing the lesson.

A recognition that animals without prior exposure to humans didn't perceive danger must have been a powerful motive for our migrations around the planet. Our hunting ancestors emerged among the bestiary of Africa. Those species became wary of us early on. But for animals elsewhere, tameness could be fatal. We shouldn't be surprised that hominins who'd been hunter-gatherers for two million years would seek out creatures that were naive about us. Searching out landscapes and prey unexploited by prior humans is a long-standing ecological pattern of ours. Researchers studying the phenomenon argue that even today, people all over the world react most positively to landscape art that portrays undisturbed wild animals and vegetation that shows no signs of previous utilization by humans. Scenes like those must have been a perpetual enticement for ancient humans, who endlessly sought out country with no footprints, no fire circles, and most of all, tame wild animals. It was a quest that led us around the globe.

In modern history the most striking examples of animals exhibiting fatal tameness are on islands. Seafaring fur hunters in the seventeenth and eighteenth centuries quickly understood how first-contact tameness in seals, otters, and birds like the great auk made it important to beat others to new island hunting grounds. Charles Darwin's experiences on the Falkland Islands when he was aboard the *Beagle* led him to write of a coyote-sized canid there so naive of humans that the gauchos killed them by offering a treat with one hand and stabbing them with the other. The metallic pigeon, a resident bird of Lord Howe Island, off Australia, went extinct early in the nineteenth century when sailors found them so incomprehensible of human danger they could be plucked like apples from the limbs of trees. The pigeons never seemed to grasp self-preservation. If they have no experience of being targeted, even creatures long familiar with humans can exhibit mortal tameness. One of the disturbing accounts from Lewis and Clark's journey is Clark's admission that once, on the Missouri River, he impulsively bayoneted a gray wolf that walked trustingly past within arm's reach.

Biological first contact, or something approximating it, with an unexploited landscape full of relatively tame and unsophisticated creatures would have been a powerful inducement for Pleistocene Age human hunters. But whether that was the America the Clovis people found thirteen thousand years ago has become a less certain, more complicated story of late. Those complications have to do with those pre-Clovis arrivals in America and whatever life they fashioned here.

%

UNRESOLVED SCIENCE certainly isn't new to the Clovis story. Although it was paleontologists looking for extinct creatures who found and worked several of the Clovis and Folsom discoveries in the mid-twentieth century, the presence of human artifacts quickly drew archaeologists interested in the human past. From the 1940s through the 1970s American archaeology became dominated by a paradigm called "Clovis first." With the radiocarbon revolution of the late 1940s allowing researchers to develop accurate site dates, the Clovis presence on the continent appeared to dovetail perfectly with a suite of big-picture developments. The last great (Wisconsin) ice age was at its maximum 22,500 years ago and locked enough ocean water up in ice that the Bering Land Bridge existed for several thousand more years. But as the ice age waned and warmed, by 16,000 years ago enough ice had melted to open a north-south corridor between the great coastal and interior ice sheets. That corridor would have allowed humans crossing Beringia to spill southward from Alaska to Alberta and all the country south of the ice. Scientists also noted another coincidence that converged with Clovis dates, a great extinction pulse—the so-called Pleistocene extinctions—that began taking out species after species at the same time Clovis people were spreading across America.

There was one question that challenged "Clovis first" from the beginning. If modern humans were in Siberia forty-five thousand years ago, why hadn't people gotten to America long before Clovis? To be sure, the Last Glacial Maximum had blocked the land route with ice for thousands of years either side of that date. But what about other possible routes? In graduate school in the late 1970s I was lucky enough to take a seminar with geographer George Carter, author of a widely read book, *Earlier Than You Think*. Carter was a pioneer of a movement to imagine a pre-Clovis human presence in America, and in fairly rapid succession a series of possible ancient human sites did begin to pop up across the Western Hemisphere. The candidate sites were found all over the country. Three were along the Tanana River Valley near today's Fairbanks. In the East were the Cactus Hill and Meadowcroft Rockshelter sites in Virginia and Pennsylvania, and in Texas the Friedkin site in the Hill Country. These and others produced apparent human cultural artifacts that seemed to date to between 15,500 years ago and as far back as 25,000 years ago.

During the sixty-year search that finally produced Folsom, archaeologists learned to indulge a healthy skepticism about evidence of humans in ancient America, and none of these sites was able to generate the kind of universal acceptance Folsom did in the 1920s. In Meadowcroft's case, for example, while its advocates claim its lithic scatter and 123 found tools might be twenty-one thousand years old, its dates have a plus/minus factor of as much as five thousand years.

Because of sites like Meadowcroft, along with the famous Monte Verde site far to the south, in Chile, which its discoverers say is fifteen thousand years old, one result has been to encourage speculation about alternative ways to get humans to the Americas. Back in the 1970s Professor Carter insisted to our seminar that because of the way ocean currents in the North Pacific flow, anything that floats and is tossed into the ocean around Japan has a good chance of ending up on a North American beach via the "Kelp Highway." Japanese garbage regularly washes ashore in Alaska and the Pacific Northwest even today.

Carter was prophetic. Like the coastal migrations that took us to Indonesia and Australia, the oldest human migrations to America likely skipped down the coastlines of the Pacific, in floating craft of some kind, with forays inland. Although few if any viable coastal sites have come to the support of the idea, nowadays paleoarchaeology assumes coastal migrations. A recently discovered site called Cooper's Ferry near the Salmon River in Idaho has dated to fifteen to sixteen thousand years ago and may typify how it worked. Asian migrants navigating the Pacific coastline could have gotten to the Salmon by finding the Columbia River and following it inland. And which Old World tools do the artifacts at this new Salmon River site most resemble? Ancestral Japanese lithics, the archaeologists believe. Good guess, Professor Carter.

What finally produced the "Folsom moment" for these ancient migrations to America was not artifacts, though, but human footprints. To be precise, sixty-one footprints, left primarily by children or adolescents in the soft mud of a lakeshore some twenty-three thousand years before the area became New Mexico's White Sands National Park. That blockbuster find by a park employee in 2019 ultimately drew a team of researchers from the U.S. Geological Survey to date the seeds of a species of grass crushed by the footprints. Their dating not only indicates a time frame at the height of the Glacial Maximum, it points to humans living in the region for nearly two thousand years.

The human footprints aren't the only tracks researchers are finding. There are also mammoth tracks, the prints of dire wolves, and those of giant ground sloths. In one fascinating interaction the tracks appear to show a young woman, carrying a child on her hip who she occasionally put down, walking a stretch of lakeshore and returning by the same path, which in the interval between her two passages both a mammoth and a ground sloth crossed. The mammoth paid no obvious attention but the sloth reacted, rearing on its hind legs in what may have been alarm.

So far, however, despite the growing knowledge that people were here,

White Sands footprints. "Tracks at Site 2," White Sands National Park. Photograph courtesy the National Archives.

despite footprints and a scattering of tools, despite the intense archaeo-logical search (much of it focused on America's ancient coastlines, some of it done in diving gear) and conference publications with titles like *Pre-Clovis in the Americas*, the search for a widespread, universal "First Culture" preceding Clovis has failed. Before Clovis there seem to have been a few people here and there, but so far as we can tell their numbers must have been small, with much of America empty of humans. Whatever the impact of these very first Americans on the ecology of the continent, it appears to have been negligible compared to what was to come. If pre-Clovis people were significant predators of Pleistocene animals, there's scant evidence. Aside from that alarmed sloth in New Mexico, the only dated pre-Clovis site for a kill of an animal larger than deer and elk is a site in Florida, where humans with pre-Clovis weapons apparently killed and butchered a single mastodon 14,500 years ago.

To the consternation of archaeologists who have hoped to replace Clovis with an even earlier continent-wide culture, in verifiable Big History we once again have to return to the Clovisians, whose presence in long-ago America is very apparent indeed.

SO, WHO WERE these Clovis people who left sites and tools and the bones of the animals they killed across America? One recent theory that briefly achieved traction in places like *National Geographic* came from the Smithsonian's Dennis Stanford, who believed that the direct ancestors of the Clovis people reached America eighteen thousand years ago . . . from Europe? To say that the scientific community scoffed at Stanford's "across Atlantic ice" claims barely does justice to the profound skepticism that followed them. While Paleolithic hunters in Europe and America did pursue similar megafauna, and flint points crafted by western Europe's Solutrean culture superficially resemble Clovis points, other researchers dismissed Stanford's claims that the two groups were the same people. Stanford's "Solutrean hypothesis" as explanation for Clovis culture did assemble a useful comparison between the collapses of Pleistocene animals in Europe and America. But linguistic and genetic conclusions refuted Stanford's larger argument.

The mystery of Clovis origins obviously drew attention once scientists were able to analyze genomic evidence from archaeological sites. That work quickly confirmed a trail of genetic kinship stretching from Siberia (rather than Europe) into the Americas. Eske Willerslev, a Cambridge geneticist who has sequenced ancient human genomes from Lake Baikal in Siberia, believes the people who ultimately swept into America sixteen thousand years ago first spent several thousand years on the Bering Land Bridge itself—a time he calls the "Beringian Standstill"—apparently awaiting more favorable conditions to move southward.

That long pause in Beringia may have produced humanity's first domestication of another animal. Engaged in their own return to America, gray wolves twenty-five thousand years ago were abundant in Beringia. Since human hunters ate only the fattest parts of the animals they killed, they had leftover lean portions they were willing to share. As for the wolves, some had a mutation that made them hypersocial enough that puppies with that gene were able to bond with humans. There probably also were wolves known today as "gifted word-learning animals," capable of picking up human language. "Dogor," the eighteen-thousand-year-old puppy whose remains melted out of the permafrost in northeastern Siberia in 2018, seems to have been such a wolf/dog. By the time the two species got to America, humans and tame wolves had formed a partnership for the rest of history.

The new genomic research has, naturally enough, laid out a complex human-migration narrative. Sometime around 17,500 years ago the common ancestors of two large early Native American groups separated from the Ancient Beringians. Around 14,600 years ago they began their passage through the ice-free corridor that funneled them southward from Alaska nearly to Montana. At that point the two groups appear to have separated. Between 13,000 and 10,000 years ago the group known as Southern Native Americans spread rapidly and widely all over North and South America. The second group, the Northern Native Americans, initially remained farther north in what is now Canada and the United States. Their genetics are best represented by a male toddler from a 12,800-year-old burial in Montana known as the Anzick site. But we know for certain that this population did not remain exclusively in North America. Genetic kin of that Montana child have turned up among peoples as far south as Argentina, Brazil, and Chile, proof that Northern Native Americans also migrated across all the Americas.

What human population was this early Montanan from? His genes link him directly back to Siberia. And what culture was he from? The boy from the Anzick site, located not far from today's Bozeman, was buried with a large cache of artifacts that included eight Clovis points painted in red ochre. His is the only known Clovis burial site in America. After he played an epic role in reconstructing the history of two continents, the Anzick boy was reburied by local tribes in Montana's Shields River, near where he'd lain for nearly thirteen thousand years, in 2014.

≈

I AM ON MY WAY to the source-point origin for Clovisia the Beautiful, that spot far out on the Great Plains where we first became aware of these ancient Americans. Looking through the windshield of my automobile at the endless flats on the approach to Clovis makes me feel as if I'm traveling across an immense pool table. No modern traveler would ever expect to see herds of mammoths trumpeting from lakes here, or saber-toothed cats noisily defending their kills through quivering prairie mirages. With its powerlines and highways extending to the curve of the earth, its wind generators and cell towers and center-pivot irrigation machinery reflecting the bright sunlight, this country doesn't look so much like the Serengeti as it does a stylized, sunnier version of Iowa farm country.

So you just have to accept that this was once mammoth country. To conjure mammoths in American landscapes you can't merely imagine escaped zoo elephants rumbling across Massachusetts or Arizona. Mammoths differed from modern elephants in both body shape and tusk growth. Mammoth backlines sloped so sharply they carried most of their bulk well forward of body centerline. And while their heads were comparatively gigantic, they had far smaller ears than modern elephants. While one genus of elephants we know as gomphotheres grew straight tusks, mammoth tusks curved dramatically, commonly crossing each other. Woolly and Columbian mammoths were the two most common types, the former a high-latitude, cold-country elephant whose pelage ranged from chestnut to ginger to blond. Columbian mammoths meanwhile ranged the middle and southern latitudes, south of the continent's ice sheets all the way to Arizona and Florida.

Examinations of fossilized mammoth dung indicate that sedges from lakeside marshes made up about a third of mammoth diet, followed by grasses, with some evidence of browsing. It's a pattern familiar from African elephants and leads to the conclusion that riverine/riparian areas were favored mammoth habitat (although a 2010 site near Snowmass, Colorado, revealed the remains of four mammoths—and thirty-five mastodons—that had roamed as high as nine thousand feet in the Rocky Mountains). Their grazing preference commonly took mammoths to different settings than mastodons. The latter, with their strikingly different dentition, were forest-dwelling and browsing elephants, common not just to mountain forests in the West but throughout the Midwest and East, from Florida to the Great Lakes.

While we have no surviving mammoth or mastodon populations to study, we do know a great deal about Asian and African elephant natural history, and if these relatives offer clues, America's ancient elephants might have been highly intelligent creatures, especially acute in what biologists call situational intelligence. Like modern Asian elephants, their closest living relatives, mammoths likely were also highly individualistic. Their trunks were elephant analogues to our opposable thumbs, and with as many as 150,000 muscle subunits those trunks would have been both delicate and immensely strong. As ecological keystone creatures whose activities shaped landscapes, mammoths and mastodons foraged in ways that likely transformed American vegetation the way modern elephants do in Africa. On the Southern Plains I'm driving, mammoth diet preferences probably suppressed certain tree species, like the mesquite that has spread here dramatically with our elephants gone.

Their home ranges likely stretched across hundreds of miles, which they traveled with an unusually powerful geographic memory. A recent study of a woolly mammoth's lifetime movements through Alaska seventeen thousand years ago, reconstructed by analyzing strontium isotope ratios (which reference geography) across the length of its tusk, gives us a sense of mammoth travel patterns and even its life stages. The isotopes preserved in the tusk preserved a record of this bull from calf to juvenile, to adult and aging elder. The mammoth began life as part of a herd that ranged mostly through the lower Yukon River basin. As a juvenile of two years to sixteen years of age he ranged across a larger landscape, moving from interior Alaska northward to the Brooks Range and the Seward Peninsula. Then, from age sixteen until he was about twenty-six, much like today's African and Asian bull elephants, the mammoth traveled a much larger region of Alaska, regularly trekking to the Arctic plain north of the Brooks Range likely in search of breeding opportunities. Finally, in the last two years of its life, the mammoth bull's tusk revealed he confined himself to a narrow range in the Arctic plain and Brooks Range foothills. Ultimately he passed at twenty-eight, leaving tusks that would survive him and eventually tell twenty-first-century humans about his life.

This mammoth's story is good evidence that, like modern elephants, mammoths, mastodons, and gomphotheres created family herds that were headed up and led by older alpha females. If they resembled surviving elephants they were naturally compassionate and also fiercely protective of their calves (reaching out to us from Clovis times there are certain sites that indicate as much). The Alaskan mammoth's travels also hint at a common pattern, where adolescent bulls separated from their family groups to travel together. We have additional information from ancient America how that could go bad. One of the tragic American mammoth sites is near Hot Springs, South Dakota, where without guidance from an older female, fifty young males all slid into a sinkhole and perished (along with a short-faced bear probably trying to scavenge them) some twenty-six thousand years ago.

All elephants are what biologists refer to as K-species, meaning they do not come into sexual maturity until they are fifteen years old or older, a state brought on by periodic *musth*, the pachyderm version of sexual heat. From insemination to giving birth probably took two years, a generational turnover slow enough to make population recovery difficult in the face of a threat. By the time humans were entering America, mammoths, mastodons, and other archaic elephant species were already beginning to suffer from a background rate of extinctions. In North America beginning around seventy-five thousand years ago ancient elephant species had begun a slow decline.

A great many sites preserve mammoth and mastodon remains in the United States, from Florida to Alaska, Arizona to New England. Some, like the Big Bone Lick site near the Ohio River in Kentucky, or the mastodon remains dug up in Ulster and Orange Counties, New York, created consternation in colonial and early Republic times. We also know of seventy mammoths and mastodons that are associated with human kills, in some twenty Clovis sites across America. So elephants and big cats and many other remarkable creatures did once occupy the ground where we now commute and go to sleep in our suburbs. Only they all disappeared, quite suddenly and mysteriously, long, long ago. That disappearance is one of the most profound ecological and historical events in the continental story.

Because of how we Americans tend to think about history, namely that nothing important happened in North America until the rest of the world "found it" five hundred years ago, when we think about our indigenous bestiary at all, our assumption is that those that were here in 1492 represent our normal baseline ecology. But as Alfred Russel Wallace, Darwin's ally in the discovery of evolution, once wrote, in fact we present-day Americans live "in a zoologically impoverished world, from which all the hugest, and fiercest, and strangest forms have recently disappeared."

Wallace said that in 1876 and he was using "recently" in a figurative, Big History sense. All those hugest, fiercest, and strangest animals vanished from America between about thirteen thousand and nine thousand years ago when we lost thirty genera and forty species, virtually all of them our very largest creatures. In the past five centuries America has suffered additional massive bird and mammal extinctions, near extinctions, and reductions. But right down to our present moment in the twenty-first century, these ancient losses during the last years of the Pleistocene make up the most dramatic extinction event since humans arrived here.

But science has never grouped the Pleistocene extinctions with the five great planetary extinctions of Earth history. It is different from all those, which were global, extinguished life both on land and in the oceans, and showed no particular size bias in the creatures they marked for disappearance. The Pleistocene losses look unlike the global events that preceded them. They didn't happen in oceans, in Africa, or in southern Asia. They devastated life on Earth only in Eurasia, North America, South America, and Australia. And the extinctions targeted almost exclusively large mammals and carnivorous, scavenging birds. Most birds, reptiles, fishes, amphibians, small mammals, and sea life went unaffected.

Something very odd was unfolding in specific parts of the planet during the late Pleistocene. But there is a common thread. These were all places where human predators out of Africa, seeking out large animals to hunt, were arriving for the first time. The Pleistocene extinctions, in other words, look very much like the first act of the Anthropocene, the beginnings of what we now call the Sixth Extinction.

～

LATE IN THE EIGHTEENTH CENTURY, European scientists who were interested in the planet's animals confronted an unprecedented riddle that struck at their understanding of the world. Why were the bones of mysterious creatures no one had ever seen alive turning up across the globe when the Old World's model of life-forms—it was called the Great Chain of Being—claimed that no animal created by God could ever disappear? If all the species on Earth had been in place since God made them, what were these strange relict creatures eroding from the ground?

Although the first excavated bones of extinct elephants came out of a sand pit in Germany in 1695, America had actually hinted at this conundrum earlier on. According to Bernal Díaz's account of the Spanish invasion of the Aztec Empire, in 1519 headmen of the Tlaxcalans showed Cortés's officers their special manitou, a fossilized leg bone the height of a man, evidence (they said) that their world once held "men and women of giant size." Almost certainly that manitou was the femur of a mastodon. Two centuries later, in 1705, an English farmer from Claverack, south of Albany, New York, yielded up (for a quantity of rum) a mysterious prize from the Hudson River country, a brick-sized tooth that eventually landed in the Royal Society, in London. Near Charleston, South Carolina, slaves found the "Teeth of a large Animal" that their African upbringing reminded them of elephant molars. What appeared to be other elephant remains, prized ivory tusks, were also arriving at European scientific academies from Siberia, utterly confounding the ancient understanding of elephants as creatures of the tropics. "Mammoth," in fact, comes from a Siberian language and means "earth horn."

The next evidence of animals no one had ever seen was a cache of bones that included an immense femur, tusks, and molars as heavy as bowling balls, gathered by a French officer from Big Bone Lick near the Ohio River in 1739. These went to the Royal Cabinet of Louis XV and eventually to what is now known as the Muséum national d'histoire naturelle, in the Jardin des plantes on Paris's Left Bank. The Jardin happens to be near a small hotel where my wife and I like to stay in Paris. We've often dined in the open air of the museum café, watching wallaby mothers and babies on the green before losing ourselves for an afternoon in the Gallery of Paleontology and Comparative Anatomy. With its jaw-dropping exhibit of skeletal mammals in a parade five creatures abreast, from humans/primates to whales and mammoths—a half football field's distance of meticulously assembled bone and enamel—this is the repository that stores the fossilized treasures that changed humanity's sense of the world. Outside the hall, where tree-canopied promenades funnel you across the grounds, are statues of the naturalist thinkers, Comte de Buffon and Georges Cuvier, who puzzled through these bones and teeth to a grand epiphany they were the first to name: *extinction*.

It was remains of a bizarre American giant, which scientists named the Incognitum, that set Buffon and Cuvier on the track of this discovery. More bones and tusks from Big Bone Lick had given English anatomist William Hunter the idea in 1769 that there was an American Incognitum that was "some carnivorous animal, larger than an elephant." This was when the American advocate and U.S. ambassador to France in coming decades, Thomas Jefferson, had his famous engagement with Buffon about extinction. Unable to imagine that "the economy of nature" countenanced the loss of any species—"For if one link in nature's chain might be lost, another and another might be"—Jefferson told Buffon he expected the elephant-like creature still existed and would turn up beyond the Mississippi River. With some trepidation, the French author of the multivolume *Natural History* advanced a novel new idea: the Incognitum in fact probably no longer existed.

When a preternaturally self-confident twenty-five-year-old, Georges Cuvier, landed a position at the Paris museum in 1795 and began looking over its recent acquisitions, it took him less than a year to conclude that the elephant-like remains from America and Siberia, along with the bones and teeth of what appeared to be a gigantic sloth from South America (he named it *Megatherium*), represented two unknown kinds of elephants and a giant ground sloth. Like Buffon, Cuvier concluded in a Paris lecture in 1796 that these were all "lost species," no longer alive on Earth.

If America once had elephants that no longer existed, what had happened to them? Why had elephants survived in Africa and Asia but not in America or Siberia? When scientists like Benjamin Smith Barton, mentor of the initial generation of American academic naturalists at the University of Pennsylvania, accepted extinction as early as 1807—"THERE IS NO SUCH THING AS A CHAIN OF NATURE. . . . I speak of these animals as *extinct*"—the operative questions now became *why?* and *how?* By then Europeans were becoming aware of the reduced state of Old World wildlife, and in the 1820s a Scottish naturalist named John Fleming offered an explanation that would anticipate the debate over extinction for the next two centuries. "Man," Fleming argued, "is induced, by various motives, to carry on a destructive warfare against many animals." If science was now searching for a cause of animal extinctions, we should commence with humans, he averred, for with the "weapons of the huntsman" we ourselves "may have succeeded in effecting the total destruction of a few."

Fleming's instinctive take (he had little to go on) laid down what has become a smoldering set-piece argument in science. Cuvier's counter-explanation was his theory of periodic catastrophes, on the scale of the biblical flood (maybe elephants had been *washed* into Siberia). But in the wake of a breakthrough book, Sir Charles Lyell's 1832 *Principles of Geology*, the reading public soon understood that the Earth was far older than a reverse countdown to Genesis accounted for, that topography was created not in catastrophes but in the drip-drip of daily processes. Understood across a vast scale of time, then, extinctions were "a regular and constant order of Nature." In 1864 came the discovery at La Madeleine rock shelter in France revealing a piece of ivory engraved with the image of a galloping mammoth. Convinced that modern humans had indeed coexisted and had a relationship with animals that no longer roamed the Earth, even Charles Lyell now was willing to indulge the dark suspicion that humanity was involved in the disappearance of so many of Earth's grand animals. Lyell put it this way: "We wield the sword of extermination as we advance."

The late 1920s announcement of the Folsom discovery left no doubt that humans in the Americas had interacted with—that is, *killed*—extinct forms of bison. The discovery at Blackwater Draw of slain, butchered mammoths added more evidence that humans had hunted another American beast now extinct. At that point the American Museum of Natural History's Henry Fairfield Osborn joined the emerging view that because humans had come to America last in our global journeys, the animals here were tragically susceptible. In Africa, Europe, and Asia, Osborn wrote, humans and Pleistocene beasts had coevolved in a way that preserved "a state of ecological balance" across a long span of Old World history. But by Clovis and Folsom times humans weren't just highly skilled carnivores, they were also a late, and new, addition to American ecology. "It may have been the entrance of this destructive animal . . . that caused the extinction of so many great mammals," he wrote. Biological first contact was becoming the best guess for the destruction of America's Pleistocene fauna.

As Robert Ardrey laid out the case in a popular 1963 book, *African Genesis: A Personal Investigation into the Animal Origins and Nature of Man*, we needed to realize who we were, "a killer ape."

~

ALL THIS WAS MERELY prelude to a scientist who was able to imagine how it might have happened. Paul Martin, who passed away in 2010, was one of the country's late-twentieth-century intellectual giants. He also was lucky enough to have a brand-new tool to play with, radiocarbon dating, invented in 1946 by Willard Libby, who won the Nobel Prize for it. That new tool almost overnight allowed an understanding of something very

Map of dated Clovis sites. Reprinted with the permission of the AAAS from *Science Advances.*

crucial about the Pleistocene extinctions: When did the various animals disappear, exactly, and how did the arrival of humans in America line up with the dates?

Martin had a unique career. I'm not sure I've ever known another thinker as brilliant and innovative whose work so many attempted to undermine with so little success. I got to meet him at a point in his career when he seemed to bear a resemblance to a target at a shooting range. When I spent a couple of too-brief days with him in the 1990s, Martin already had three decades of working out his theories for the Pleistocene extinctions. At a time when politics and many university departments embraced the idea of ancient peoples as ecological examples for the modern world, there were those who saw Martin's argument that early humans out of Africa were responsible for extinctions everywhere they went as politically incorrect. The popular Native American writer, Vine DeLoria Jr., who I also knew, was vitriolic in his condemnation of Martin, which I could tell mortified and baffled the paleobiologist. The Solutrean Culture of southwestern Europe eighteen thousand years ago, named for a site in France that featured the human-killed remains of twenty thousand horses, had converged on a similar mastery of hunting strategies that wiped out Europe's remaining Pleistocene creatures. Clovis and Folsom were not "Indian" stories, Martin insisted. They were Big History human stories.

Martin and I arranged to get together on his visit to the University of Montana, where I taught. I had drained beers with enough archaeologists to absorb the skepticism of some of them about his theory that humans had laid waste to America's Pleistocene megafauna. But I had also read Martin and was open to listening to his arguments. I was also surprised on first meeting him. I'd not realized he walked, with considerable difficulty, with canes. (In his last book, *Twilight of the Mammoths*, he referred to the polio he'd contracted in 1950, as a young man of twenty-two, as "a chronic if minor handicap.") After two days of wide-ranging conversation, I began to think of Martin in the manner of a Stephen Hawking. When his body had slowed, his vast energy had lit a turbocharger that accelerated his mind.

The crux of the Pleistocene story, Martin told me, was that North America had functioned as a continental island remote from the evolution of humans, and when we finally arrived in numbers in the form of the Clovis people, the widely known slaughter humans had made on island biologies all over the world came to America. We were a brilliant new predator with sophisticated weapons, dogs, and fire (and baggage like rats). Everywhere we experienced biological first contact—New Zealand with its many species of giant, flightless moas serves as one well-known example—we rapidly exterminated local creatures. Likely, the predation we engaged in (today we call human predation "hunting") changed local ecologies so substantially that animals evolved before we were on the scene couldn't survive once we were. I realized Martin was giving me a command performance of his "Planet of Doom" theory, a modern version—now buttressed with science, history, and details—of Lyell's gloomy insight: "We wield the sword of extermination as we advance." As Martin put it in his 2006 *Twilight of the Mammoths*: "I argue that virtually all extinctions of wild animals in the last 50,000 years are anthropogenic." Here, simply, is where our present "Sixth Extinction" began.

Martin's version of how this early stage of American history unfolded is that from the moment the *Clovis people* (specifically) entered an unsuspecting North America 13,050 years ago, it took them only three centuries to sweep across both American continents. Like pythons spreading across modern Florida, our Clovis ancestors in North America experienced an unimpeded ecological release—or to use phrasing from human history, a "successful colonization." Naive species that had little experience with human predators, and the ones with especially slow population recovery, like the elephants, melted away before the onslaught. So did their predators, the saber-tooths and hunting hyenas, along with giant scavenging birds like teratorns. The reason condors are called *California* condors is that the loss of so many big animals left them able to hang on only near Pacific shores where they could scavenge beached sea life.

By the time the destruction was over, which Martin thought took only one thousand years if you included the Folsom giant bison hunt, only a handful of America's biggest animals remained. Most of those were either European or Asian, like caribou or bison, that had prior experience with humans, or native American ones like pronghorns that carried so little fat they offered little inducement for hunters. Otherwise, the Clovisians erased millions of years of evolution.

In 2001, independently of Martin, an Australian paleobiologist at the Smithsonian, John Alroy, developed a computer model to test this American extinction story. It began with the premise that a mere one hundred individuals bringing a highly refined hunting economy entered America from Asia fourteen thousand years ago. Alroy's computers modeled an absolutely classic ecological release. Within a thousand years that founding population had spread across the Americas and was now a million strong. By six hundred years after that, excepting a few scattered remnants overlooked by hunters but now too separated to exchange their genes, 75 percent of America's Pleistocene bestiary had been gutted. His model correctly predicted the extinction, or survival, of thirty-two of forty-one Clovis prey species.

Of this first stanza of the Anthropocene, Alroy concluded this: "Long before the dawn of written history, human impacts were responsible for a fantastically destructive wave of extinctions around the globe."

~

IN 1972, many millennia after Clovis, the predator specialist Hans Kruuk came upon a scene in the Serengeti involving a single pack of spotted hyenas that had attacked a herd of Thompson's gazelles during a storm. They

Paleontologists working the Colby site in Wyoming, a Clovis kill featuring multiple butchered mammoths. Courtesy Paleoindian Research Lab, University of Wyoming.

had killed eighty-two, injured twenty-seven more, then walked through the downed animals taking occasional bites out of the soft parts. Kruuk went on to argue this as a sometime response by predators to hapless prey with few, or no, defenses. Biology has ever since referred to the phenomenon as surplus killing, or the henhouse syndrome. I experienced it as a kid in Louisiana when a fox got among my grandfather's chickens and killed sixty of them in one night. It ate three, and of those three only the heads and entrails.

Southeast of present-day Tucson, along the Santa Cruz River, there are three famous Clovis sites. You suspect in long-ago Clovis lore this may have been a legendary event. Or, given that many similar stories follow it in the historical record of America, maybe what transpired here wasn't legendary at all, just the way things were done. What seems to have happened is that at the most westerly location (now called the Lehner site), a Clovis band surrounded a family group of fifteen mammoths. The herd apparently huddled together for defense against the assault. But thirteen of them, all adolescents and calves, died in the spot. Archaeologists found exactly thirteen Clovis points in their remains.

But it must not have been an easy thing. In different locations a few miles away (the Escapule and Naco sites), archaeologists would also find two adult mammoths who had apparently fled the slaughter. The large male had died with two Clovis points in his body. But the female must have put up a tremendous fight to protect her young before, mortally wounded, she fled. In her remains there were no fewer than eight embedded Clovis points.

The hunters who killed those mammoths neither tracked nor butchered the two adults.

THE PRIMARY CHALLENGE to the human overkill hypothesis in America has blamed a changing Pleistocene climate for the extinctions, but it begins with disbelief. There is no way (or words to that effect) an expanding human population could have killed all those animals. Or so the climate proponents insisted. There had to be some other explanation why a whole bestiary vanished. Amid a fairly small set of possibilities, a changing climate has appeared to many the likeliest alternative. That disbelief in humans as a cause, which has fueled the search for several optional explanations for the Pleistocene extinctions, seems to involve an instinctive rejection of humanity's ability to alter the planet. It's the same motive that has fueled the knee-jerk reaction against human-caused climate change in our time.

Further, the changing-climate explanation has confronted several problems of logic. Since radiocarbon dates have shown the extinctions occurring quite suddenly between 13,000 and 10,000 years ago, scientists hoping to promote climate as cause were obliged to look for some extreme, perhaps unprecedented, swing to new conditions in that time frame. The best candidate seemed to be the so-called Younger Dryas cold snap, which plunged Northern Hemisphere temperatures for 1,300 years right in the extinction sweet spot. This last hiccup of the Ice Ages spread tundra across northern Europe and northeastern North America, then around 12,000 years ago its sudden reversal saw forests colonize New England once again. Unfortunately, the Younger Dryas theory has faced an explanatory hurdle. Its advocates had to clarify how bestiaries that had just survived the much colder conditions of the Last

Glacial Maximum keeled over from a short cool episode a few thousand years later.

What about a change to warmer conditions as the Pleistocene ebbed into the Holocene—our modern climate regime? That argument can't account for why the thirty-five mastodons at the Snowmastodon site in Colorado were doing so well during a Holocene-like warm spell 120,000 years ago. Or explain why elephants did well in tropics elsewhere in the world, why camels that vanished here preferred deserts elsewhere, or why horses experienced almost instantaneous ecological release in the modern, warmer American West when Europeans reintroduced them to their evolutionary home 400 years ago. Finally, if climate was the culprit, why did it affect only *big* American animals? And why would climate disruptions severe enough to kill big animals somehow track human migrations around the world?

Two other recent efforts to dislodge humans from a primary role in the Pleistocene extinctions seem more like desperation Hail Marys. In 1997, inspired by the new understanding of widespread death from pathogens during the time of colonial settlement, Ross MacPhee of the American Museum of Natural History speculated that possibly unknown disease epidemics brought by the first human arrivals had wiped out the continent's big creatures. Without a candidate infection, though, and with little evidence that new panzootics destroy entire species without leaving survivors, the disease hypothesis won few converts. Then, in 2007, an article in the *Proceedings of the National Academy of Sciences* advanced what has since come to be called "the fireball hypothesis." In an attempted reprise of Chicxulub, its authors proposed that an extraterrestrial impact broke up over North America 12,900 years ago, setting large parts of the continent on fire, causing the Younger Dryas cold snap, and wiping out America's megafauna. This theory was unable to explain why a fireball would have destroyed large creatures but not smaller ones, though, or why there were similar selective extinctions in South America and Europe in the same time frame, far from any impact. It, too, has faded as an explanation.

Thus, as with Clovis as the first coast-to-coast culture in America, we are left with ourselves, us humans, as the trigger for the first grand ecological transformation here, 130 centuries ago. It's an old argument but it has not been easy to defeat. Science has been refining it over time, however. Although radiocarbon dates for the mammoth-kill sites do cluster around thirteen thousand years ago, we now think that rather than dying in a shock wave of killing, at least remnants of some species lingered on for thousands of years after humans arrived. It's not difficult to wrap the mind around that. Biological first contact must have been brief. Cultural transmission about humans no doubt made surviving animals far warier. Some would have fled to places remote from human activity. But separated from others of their kind, their genetic diversity and health would have waned.

As evidence for lingering remnant populations, mastodons seem to have still been in Michigan until 12,400 years ago. And some mammoths survived in Saskatchewan until 11,700 years ago. Camels and llamas were in Arizona at least down to 10,300 years ago. As for saber-toothed cats, American lions, and short-faced bears, they all seem to have disappeared between 12,000 and 11,000 years ago. Dire wolves apparently were still roaming present Missouri down to about 11,450 years ago, and there were at least some horses still in Alaska between (roughly) 10,500 and 7,600 years ago. Recent research on predator tooth wear at La Brea shows late saber-tooths and lions to have been well fed, which does run contrary to the argument that these megapredators were dying out because human predators were depriving them of prey. So plenty of mysteries remain, not least whether those pre-Clovis humans in America provided our animals some advance training in recognizing humans as a danger.

Our best strategy for understanding America's Pleistocene extinctions may be on an animal-by-animal basis. Clovis hunters almost certainly wiped out the elephants, and Folsom people the giant bison, but animals like dire wolves, giant beavers, and big cats may have simply been outcompeted by gray wolves and modern beavers and cougars. Smaller size and earlier sexual maturity fitted the replacements better for an America inhabited by human predators, the first examples on the continent for what biologists call "anthropogenic evolution." Horses and camels do remain enigmas. Sites of Clovis age in southern Alberta and Colorado show horse and camel kills, but nothing like the vast numbers of horses from Solutrean sites in Europe. Why did various camelids survive in South America anyway, providing later Native people domestic possibilities, but not farther north?

There is a coda to this story of passed-over remnants, though. When the seas began to rise again in the Holocene, on small chunks of Beringia that emerged as islands in the watery Arctic expanse, a few mammoths survived. As the waters closed around Wrangel Island, to be named thousands of years in the future for an explorer who never quite discovered it, the island captured a population of some three hundred woolly mammoths. They were isolated from human predators, although not, of course, from the changing climate. These Wrangel mammoths survived the extinction of the mainland herds by more than 5,000 years. Then, around 3,600 years ago, we lost the last woolly mammoth.

Why, though, did they disappear if climate was incapable of wiping them out and humans did not find them? Only recently have we discovered the reason. "Genomic meltdown" is a new term of art, but that apparently is what happened to these last New World elephants. Isolated in a tiny gene pool on a single island, over time so many genetic mutations built up in their genomes that the last mammoths failed at the prime directive. They became unable to reproduce. That's a lonely ending for the last herds and packs of America's Pleistocene animals, and it was likely a more common demise than just among the mammoths of Wrangel Island.

~

POWERFUL PUMPS to suck water from the Ogallala Aquifer, and center-pivot irrigation machinery to spread it across the surface, have enabled modern settlement in the High Plains of Texas and New Mexico where elephants once roamed. But those last few miles driving to the spot where the world first discovered Clovis culture in the early 1930s do not stir many feelings of the technological sublime. The Blackwater Draw site may be a UNESCO World Heritage Site but it has only state and local-college funding. Not only are there almost no highway signs to direct a visitor, the site itself is lost among the sameness of cotton, peanut, and sorghum fields and dairy farms all inhaling the fossil groundwater. Even inside the car I hear planes roaring overhead from the local air force base. As I turn off the highway and bounce over a badly rutted dirt road down to the visitors' center, it's hard not to notice that this internationally famous place looks down-at-the-heels, with its interpretive signs disintegrating under the High Plains sun.

Dr. Brendon Asher, whose position at Eastern New Mexico University includes the directorship of Blackwater, is sitting on a bench outside, waiting for me. Although it's a Friday and the site is open to the public for the weekend, ours are the only vehicles in sight. Bearded, slight, only a couple of years out of graduate school at the University of Kansas, Dr. Asher is wearing a baseball cap and sunglasses, a checkered shirt, and jeans and boots for our walk of the site. It turns out he's worked as an archaeologist on Paleolithic sites across much of the country, which strikes me as perfect for my task of trying to put flesh on the Clovisians.

We walk away from the tiny visitors' center into a cliff-rimmed depression of a hundred acres or more, all of it inset below the level of the plain. Once upon a time, Blackwater's director is telling me, this was a deflation basin that harbored a shallow, eighty-acre marshy lake that drained into the main arroyo, Blackwater Draw, just to the south. The headwaters of the Mississippi it's not, but Blackwater Draw actually is a headwaters source drainage for the Brazos River, whose waters reach all the way to the Gulf of Mexico.

"The site was actually first discovered in 1929," Brendon is telling me. "A local rancher, only nineteen years old then, Ridgley Whiteman was his name, was out here when the start of the Dust Bowl was blowing sand out of this basin, and he was finding what he called 'warheads' and elephant bones on the surface. His letter to the Smithsonian about what he'd found got a paleontologist out, who declared there was no evidence of any human association with the bones. And that was that."

"So, that must have been about when the highway department decided to mine here for gravel?" The thought that the Blackwater site was a gravel mine both before and after the discoveries here still blanches me.

"Right, that was in 1932, but fortunately that happened just when the archaeologist Edgar Howard, a master's student at the University of Pennsylvania, showed up. Howard had started a Southwest project in 1929. His idea was that if there were Paleolithic sites in America they were going to be in caves, like in Europe, so he'd been working in the Carlsbad and Guadalupe Mountains country. Ironically, Howard actually found a Clovis point in one of his cave sites, but at the time no one knew what it was. One of the things that eventually made Blackwater so important was that it showed a clear gradation between Folsom materials and Clovis ones."

As we walk a whitetail buck bounces away from us through the shrubbery, and Brendon pauses in his story to point out a spot in the cliff where his field school has exposed the remains of a Pleistocene camel. "If we could find a Clovis camel kill, that would be significant."

"But that lack of clear Clovis involvement with camel or horse kills is still one of the great mysteries of the American Pleistocene, wouldn't you say?"

"Yes, and if you can answer that one, all of us are going to be grateful. There are 'background' camel and horse materials here and in most of these sites, but it's hard to find a smoking gun." We've reached a metal structure erected over many of the early excavations here and Brendon is sorting through his keys. "A lot of people will disagree with me, but I don't like the overkill idea so much. Hunting was part of it, but I tend to think the extinctions were caused by a variety of factors, maybe environmental change mostly."

The hope of getting humans off the hook has fueled this debate for a century, I'm thinking.

We step inside to Blackwater's showcase of bone strata, with Clovis at the bottom and extensive Folsom giant bison sites above. More than eight thousand Folsom artifacts from a time when mammoths were already gone have come out of excavations here.

"So in front of us in 1932 the gravel operation opened two trenches that quickly got into bone material. That summer, news of the finds brought Howard through, and he decided there was a connection between the bones and the large points he'd seen. So that was how the news of Clovis broke."

I know Howard spent the next five summers, with varied teams of assistants that included Ivy League students, racing the gravel miners at Blackwater. By 1937 they had found three mammoths, one with a point embedded in a vertebra, along with several other beautiful, fluted points among the remains, significantly longer and more robust than Folsom versions.

"How many mammoths have you found here by now?" I ask Brendon.

"Twenty-eight. We've also found two Clovis cache sites here, where thirteen thousand years ago someone in a Clovis band cached blades nearly ten inches long, then never retrieved them. So who knows what big discoveries are still waiting out there."

Dr. Brendon Asher, director of the Blackwater Draw UNESCO World Heritage Site in Clovis, New Mexico. Photograph by Dan Flores.

DRIVING OUT OF this plains country where the first record of our Clovis ancestors emerged, I know that scientists like Brendon Asher are excited that we're pushing at the frontiers of Clovis knowledge. We've been studying these people for almost a century now, yet however important they are in America's history, the Clovisians themselves have always been maddeningly elusive. They are *us*, of course, but it's difficult to know even your recent relatives if all you have to go on are their tools and diet preferences. We know that with the fluted point, a purely American invention not found in Siberia, their thinkers had solved the ancient technology hurdle of affixing points solidly to wooden spears or darts. We also know that they were consumer connoisseurs of the best the world had to offer. Clovis artisans fashioned their tool kit from the hardest, sharpest, most vividly colored flints and cherts in North America, whose outcrops existed as a geographic atlas in their heads. They journeyed hundreds of miles to those sources, as if on quests to special magic. Some of their tool caches— the spectacular deposit found in an apple orchard near East Wenatchee, Washington, is one—feature multiple, gorgeous, unused points of eight to nine inches in length, with sacred red ochre still adhering to them.

One conclusion to draw from them, maybe, is that as consummate hunters, their hunting tools weren't just their technological sublime but their art. Beyond those stunning tools testifying to their coast-to-coast presence, what we have learned most recently about them is the genetic affirmation of their origins, their phenomenally rapid expansion across the Americas, a little about their burial practices, and that their genes are in the genomes of virtually all Native peoples in the Americas (save those last migrations that brought Inuit peoples to the Far North). Until very recently, one Clovis mystery was, why no art? Why nothing like the grand paintings of animals on the cave walls of Chauvet, Lascaux, and Altamira in Europe? There are pebbles incised with crosshatching, an elephant carved into a piece of ivory. Otherwise we had no hints what they thought of the animals they hunted, of America, of their lives in general.

That may be changing with a new (2019–20) investigation of the rock art of a region in the Colombian Amazon known as the Serranía La Lindosa. Local and Indigenous peoples have long known that tens of thousands of pictographs, rendered in red pigment, adorn as much as eight miles of rock shelters and faces on limestone buttes called *tepuis*. Rock art itself is notoriously difficult to date, but new archaeological excavations in three local rock shelters produced datable materials showing human arrival 12,600 years ago, the junction between Clovis and Folsom. The assumption is that the early humans did at least some of these thousands of images, for some portray animals that look like mastodons, ground sloths, and ice age camels and horses. Humans appear, too, many equipped with hunting spears, others holding hands, some dancing, all hatless. Those very social portrayals put flesh on the Clovisians or their descendants. Maybe. If the images actually do depict Clovis or their Folsom descendants.

One recent theory is that the Clovisians may have been a "Northern Hemisphere wild-type," a group of hyperaggressive "Siberian Vikings." According to modern science, a high-fat diet *is* a strong trigger for enhanced testosterone. But who they were, really, is us. My 23andMe profile shows 3 percent of my genes are Native American, a common figure for those of us whose European ancestors arrived in North America three hundred or more years ago. Clovis heredity is within us. And the Clovis story resonates because we imagine them as ancient versions of ourselves, explorers of hidden continents, the last of the masterful hunters of enormous animals. Those beautiful point caches suggest that's how they saw themselves, too. Who knows what they thought about history or what stories they told about themselves. But they were the culmination of forty thousand generations of hunters and they must have had a sense of that timeless tradition.

What did they think, what did they *do*, when so many of the animals they and other hominins coevolved with began to disappear, to dwindle to a last few scattered survivors until there were none? We can't know what they thought, and whatever actions they took, if any, didn't work. In some respects, what the Clovisians faced is mirrored by our own twenty-first-century circumstances. Like us, they lived as their ancestors had lived and no doubt had every expectation that the world would continue as it always had. And so long as there was a Siberia, or a Beringia, or an America out there, it did. But Earth proved finite, and so did its animals. Much as we are doing today, the Clovisians ran into a wall of limits. You could always pivot, which is what they did when they evolved into the Folsom buffalo hunters, and that got them a couple thousand more years before the giant bison gave out. But having chased our first great economy around the planet, humanity for the first time had to reassess. The world as we'd always known it was no more.

I'm driving west now, the setting sun in my eyes, squinting at vast spaces that no longer hold elephants, or camels, or ground sloths, or dire wolves. Or, for that matter, the bison or gray wolves or many other species that were here only 150 years ago. What lesson might we draw from Clovisia the Beautiful? That 2018 study in the *Proceedings of the National Academy of Sciences* concluded that on our way out of Africa and around the world we destroyed most of the largest and oldest animals on Earth. The human mind may be capable of prodigious feats, but judging from our reaction to our own extinction crisis, we struggle to grasp the enormity of our effects on the planet and maybe we always have.

What lesson do we humans thirteen thousand years down the timeline draw from Clovisia the Beautiful? The whole planet is still holding its breath to know.

CHAPTER 3

RAVEN'S AND COYOTE'S AMERICA

I am walking the edge of a sharp-rimmed cliff in outback Montana before sunrise, moving through a twilight of grays and blacks and outlines. Large, graceful birds, sandhill cranes, are fluting their strange Pleistocene cries in the pastel sky overhead, but I am focused on the lines of the topography in front of me, especially the way the mesa I'm walking narrows up ahead. Seeing that narrowing, my walking pace quickens.

This is a historic piece of ground. Starting some two thousand years in the past and continuing down to two hundred years ago, it was the scene of frenzied, albeit sporadic, human activity. Like most historic places there is something maddeningly mute about the spot now. It is why we often stand and gawk numbly in such places, unable to connect to the events we're supposed to marvel over. But this morning I'm not going to be stymied by my usual lack of imagination. I'm here with a purpose, my intent to experience at least some part of what a buffalo-jump drive was all about.

It was fully dark when I arrived here an hour earlier, parked my car at an interpretive sign, finished a cup of coffee, then slowly worked through the boulders to the top of this mesa. While I walked eastward to the luxuriant grassland of a high meadow, the sky had gradually lightened. Now, turning back toward the car and the cliff I climbed in the dark, I'm becoming caught up in what I tell myself are echoes of the place. Pointing myself down the contracting mesa toward the far rimrock, I start to jog.

I am running a track that men and other animals have run many times in the past, but in contrast to my lope beneath the fluting cranes, then there would have been the pounding thunder of sharp black hooves cutting through the grass and the alarmed grunting of animals, their huge forms wrapped in billowing clouds of dust that must have made for a ghostly stampede. Now I hear only my footfalls and my breathing, but in the real thing the air would have been rent by the exultant shouts of the drivers, urging on runners wearing the skins of both wolves and red-coated bison calves and leading the herd to its destiny, their costuming a ruse to fool buffalo cows into thinking that wolves were selecting out defenseless young ones. Listening to the rhythm of my feet, I wonder if the herd's noise wouldn't have been so overwhelming it would have morphed into silence, becoming dreamlike.

The whole affair would have commenced days earlier with a religious ceremony, and careful maneuvering of a bison herd in that high meadow into position for a stampede. Then if all went well—and it went well enough times in the past to accumulate a bone layer five feet deep at the base of the cliff I climbed—the runners who led the herd to the cliff edge would escape, if they could, by darting aside at the last moment, dodging the relentless brown river of animals hurtling into space in a dream of wild, frozen action.

Where I've begun my run is a half mile back from the cliff, and soon enough I cross to descending benches and realize I am at the point of no return in this bison jump. Get the animals here, and have them running, and the downhill pitch steepens so quickly there would be no way for the herd leaders to either stop or turn aside.

I am running harder now, pulled faster by the angling slope, and I register that out in the valley dawn color has arrived. Chrome-yellow light cast by the rising sun is lighting the white cliffs on the far side of the river, a scene of great beauty—one last soothing sight of Earth, perhaps, as the lip of the plunge is scarcely 120 feet away now. Beyond that is windmilling motion and the silence of 40 feet of free space, then the jarring stop among the boulders.

I slide to a stop a few feet from the cliff edge and stand panting for a few minutes, looking down on the slope below. By modern standards the scene would not have been pretty. In 1797 the British trader Peter Fidler described such a concluding set piece: "The young men kill the [crippled animals] with arrows, bayonets tied upon the end of a pole &c. The hatchet is frequently used & it is shocking to see the poor animals thus pent up without any way of escaping." However pod-like their behavior as classic herd animals, all these bison were individuals, of course, and that is the way they died.

Slanting sunlight, throwing morning shadows hundreds of feet long across the Madison Valley of Montana, lights my face. Over the mountains I see a jet glinting silver, a mobile diamond slicing through the blue, its motion fetching me back to my climb down to the car, back to my commitments. But before I start I stand for a moment thinking of the bison that died among the boulders below. Humans drove buffalo off cliffs in America for twelve thousand years, and despite knowing something about it, I find it a shock to be in this space where it happened, this close to how it worked. I have visited Head-Smashed-In Jump in Alberta, and absorbed archaeologist friends' accounts of Bonfire Shelter Jump in the gray limestone canyons of the Pecos River in Texas, hearing at the visitors' center in Canada that Indians carefully utilized every part of the animals, yet knowing that in Texas the cliff at Bonfire Shelter is scorched hundreds of feet high from the spontaneous combustion of an enormous, mangled heap of unutilized bison Native hunters drove off the rim above.

Those two sites beg a big question. Putting aside whatever fantasies of the past we have, what kind of relationship did humans and animals fashion over the one hundred centuries of Native America that followed the Pleistocene? And if it *was* different—more ecologically benign or balanced than what came before and what came after—then why?

*

CLOVISIA THE BEAUTIFUL ended with the demise of elephants and the majority of America's big animals. The Paleolithic economy then devolved into one final form, the Folsom and the similar Plano, Cody, and handful of other cultures pursuing the last of the giant bison, a hunt that persisted for about 2,000 years. But by 10,000 years ago, a new reality had settled across North America. People were here, but most of the original animals—now even including *Bison antiquus*—were not. How many generations did it take for the great Pleistocene fauna to fade from human memory? The haunting stories of endings must have lasted far down the timeline, because the trauma of having so many charismatic creatures disappear seems to have shifted human behavior. A phase of history in America lasting millennia now ensued that implies a more careful, thoughtful use of nature. The span wasn't entirely extinction-free, but thousands of years later many arriving Old Worlders would describe the Wild New World that greeted them as a paradise of animals. The image of a continent existing into the modern age as an Animal Eden out of prehistory has shaped colonial America's sense of itself ever since. But was that truly the reality of Native America?

When nineteenth-century ethnographers began to assemble a linguistic map of Native America and mulled the seemingly chaotic distribution of languages, the conclusion anyone would draw is that over the preceding ten thousand years there had been a tremendous movement of peoples around the continent. Athabaskan speakers lived in interior Alaska and also way down in the Southwest. There were pools of Algonquian speakers in New England, in the Ohio Valley, and in the foothills of the Rocky Mountains. All this was in contrast to Australia, for example, where Aboriginal populations have stayed in place for fifty thousand years. The American story implies significant experimentation with different locales and ways of life.

In most of history the spur to migration has been a search for something better. Some of those human migrations may have been related to the reshuffling of American nature that took place in the echoes of the Pleistocene extinctions. The biology of the continent was reinventing itself. The vegetation was changing. Without ground sloths to disperse their seeds, the range of Joshua trees began to contract. And without mammoths to curb them, honey mesquite began to spread. There were so many missing animals that a remarkable number of ecological niches were either vacant or newly filling. Reinvention was not so dramatic in the woodlands east of the Mississippi, where members of the deer family like moose and elk and whitetail deer continued as the primary ungulates, with new browse possibilities available to them with the demise of mastodons. When the last short-faced bears disappeared in Ohio, black bears emerged as the dominant ursines, and grizzlies and polar bears did the same farther west and north. With dire wolves now extinct, gray wolves and the trio of ancient American wolves emerged as the primary canid predators. On streams east and west, smaller modern beavers replaced the extinct Pleistocene versions and fashioned marshy, wet landscapes. A watery interior is not an image of the past we hold in our heads, but beavers created exactly that.

The ecological rebirth was most dramatic in the western half of America. The loss of mammoths, giant bison, horses, camels, ground sloths, dire wolves, short-faced bears, scavenging birds, and a range of cat predators opened niches at every level. In cases like wolves and bears there were ready replacement species. But with nearly 70 percent of America's ancient grazers gone, niches for replacements were wide open. So with almost no competition, a new, smaller bison supplanted horses, mammoths, and its huge bison ancestors. Within a few centuries this new "dwarf" bison grew into a biomass of animals that had almost no analogue anywhere else on Earth. Seals, sea otters, and sea lions excepted, along with the one pronghorn species that survived to browse the forbs camels once ate, most of the large animals left west of the Mississippi— bison, elk, mule deer, moose, bighorn sheep, musk ox, caribou—were Asian immigrants.

On some parts of the planet the warmer climate that marked the end of the Ice Ages allowed hard-pressed humans to try out some new things. For more than half a million years the migrations out of Africa into new parts of the Earth had let humans continue a focus on hunting. But Eurasia ten thousand years ago was becoming old and picked over. Earth as humanity had known it for so long was now changed. Millennia of human pursuit had used up once-teeming populations of big creatures. Horses, camels, aurochs, sheep, and goats survived because Old World humans figured out how to tame them into domestication before they entirely disappeared. Domesticating and herding animals was an enormous alteration in human economies, but it was only one step to the actual new order. The warming world grew more plants, and with wild animals ever scarcer, a whole new human adaptation in the form of herding and agriculture—with cleared ground, settlements, eventually cities and empires—was at hand.

In the Old World many human cultures in the Middle East, Southeast Asia, and Europe took that step either side of ten thousand years ago. But since humans extensively settled America only thirteen thousand years ago rather than fifty-five thousand, North America didn't force the issue the way the Old World did. The giant animals of the Pleistocene might be gone, but America's replacement fauna remained for a very long time a rich and diverse food base for a human population still growing into the continent. With plant life also flourishing in a warming climate, America's human cultures segued to a stage where animals were still of primary importance economically, culturally, religiously, but plants were taking on a much more significant role. "Archaics" is the term anthropologists and archaeologists have long used for humans living this way, by which they mean people existing as hunter-gatherers.

So while the Old World experimented with agriculture and domestication, in North America the hunting-gathering lifestyle continued over vast spans of time and diverse geographies. In the Great Lakes country, Northern New England, the High Plains, the Great Basin, and almost all of the West Coast, hunting-gathering continued as the primary human economy all the way into modern history. We ancient hunters of animals surrendered our oldest lifeway with extreme reluctance.

It wasn't that America's post-Pleistocene hunter-gatherers had inherited an Eden. The linguistic map argues for a great deal of movement and experimentation. And America wasn't just in the throes of biological re-creation. Around 8,500 years ago there was another potent change, in some parts of the continent serious enough to be an emergency that demanded an extreme response. The warming cycle that ended the last ice age didn't relent, and America's climate swung into a hot, dry phase that stayed in place for a mind-blowing 3,700 years. This was the depths of the last interglacial, the long slide out of the frozen Wisconsin Ice Age. Now the Earth's rotational wobble had the Northern Hemisphere slightly closer to the sun and for almost forty centuries some parts of America cooked. The Altithermal, as it is called, came close to turning large parts of the continent into a true desert, and a vacant one. Many species of animals left for wetter settings. So did many human groups. As animal populations shifted eastward and westward, the country off the southern end of the Rocky Mountains, which had drawn humans twenty-three thousand years ago and where Clovis and Folsom people had thrived, nearly emptied of humans.

In other parts of America, and in the West, too, once the Altithermal subsided and animals and humans returned, generations of some hunter-gatherers occupied the same landscapes for centuries, even returning to the same camps again and again. That kind of close familiarity gave them bodies of handed-down ecological insights about how to live in particular places. The feedback they read from place-based living enabled them to come up with a striking epiphany, one allowing them to live well without using up their world.

The breakthrough—what seems to have been the real key to success in Native America—sprang from generations of acknowledgment of America's new circumstances. No matter how far you migrated or which direction you traveled, there was no longer a Wild New World, empty of other people, out there. Clovis-like expansion across a virtually uninhabited continent was over. But adapting to local conditions and accumulating ecological knowledge about places led to a universal conclusion. Humans now had to learn to deliberately, carefully, manage their own numbers to avoid overshooting local resources when times turned bad. In a variable world, good times inevitably give way to bad times. That was an ancient lesson. Basing your numbers on the good times set you up for disaster.

Given our modern difficulties regulating human population growth, how did these ancient Americans manage to pull off controlling their populations so they could live well on locally regenerating resources? Birth-spacing was one common strategy. Breastfeeding the young for a span of years suppresses ovulation during a mother's fertile years, preventing a rapid succession of pregnancies. Child mortality was high among ancient humans anyway, ensuring that only a fraction of new additions to a population factored into its ecological fit. But most hunter-gatherers also freely practiced forms of abortion, to which they sometimes added infanticide for excess or unwanted newborns. The idea is draconian enough to shock us, although the evidence is that as an ecological strategy it worked, and worked well. But particularly for the women who carried babies to term, infanticide was a psychological burden. Ultimately many hunter-gatherers sought to escape it.

But the larger equation was relentless. The hunting-gathering economy was still the predator's economy, and predators of whatever kind are always few compared to prey. Hunting and gathering required space to roam, habitats for birds and mammals. Living the good life meant you could not overburden the world with people.

Populations in Native America in 1500. From *Fire, Native Peoples, and the Natural Landscape,* edited by Thomas R. Dale. Copyright 2002 Island Press. Reproduced by permission of Island Press, Washington, DC.

THERE WAS ONE possibility to increase human numbers, but it meant giving up much of humanity's ancient life and investing in an entirely new economy. Halfway through the ten thousand years of Native America, though, around the time the Altithermal ended, an agricultural revolution similar to the one that swept the Old World slowly began to spread into North America from the south. In the north this new approach focused almost exclusively on plants. The selection and training of wild plants first emerged where human populations were densest and animal populations lowest, namely crowded Mesoamerica: the Yucatán Peninsula, and the high valleys of Central Mexico. Unsung, unknown traders and travelers first carried ideas about domestication northward about four thousand years ago, and later the actual seed stocks of Mexican agriculture. Over the ensuing millennia crop fields and farming towns began to dot Native America from the South to southern New England, then along the river valleys of the Midwest, and even in scattered locations in the desert Southwest. Once the agricultural transformation took root, populations began to grow and sometimes centralized governing bodies, usually religious ones, organized towns into regional empires we know as Cahokia, Spiro Mounds, and the Chacoan Empire. All of these were late experiments in the last thousand years before Old Worlders stumbled onto America.

Despite its regional successes, the move to agriculture was unhurried and spotty across much of what is now the United States. Agriculture's late arrival meant that hunting-gathering remained the sole economy across the entirety of America for the first six thousand years after the Pleistocene ended, and in vast regions agriculture never replaced it at all. So wedded were Native people to the hunt that even as agricultural towns emerged, many of the farmers continued to hunt, at least seasonally. Some returned exclusively to hunting when circumstances allowed. All this means that living off wild animals and gathering plants is the longest sustained economy our species has ever practiced in America. No other way of life gets close. New technologies did emerge, along with wonderfully rich cultural and religious traditions interweaving the lives of people and animals. But by and large, especially among Archaic hunter-gatherers, an ur-conservatism prevailed in Native America. Life among other animals was ancient, exciting, mythic. What human existence could be better?

＊

FOR TEN MILLENNIA these first Americans experienced lives that don't much resemble classic Western clichés about humans in a state of nature. For one thing, some brilliant thinkers revolutionized the technology of the hunt and the ideas spread. Atlatls were predatory tools that doubled the length of the human arm in its throwing motion. Their launchers required more preparation than spear thrusting, but they propelled flint-tipped darts with twice the power and range of a thrown javelin. Later on, when the bow diffused southward out of the American Arctic, it produced a similar revolution. These technologies didn't just extend the reach of a human throw. They meant far fewer injuries from close-in grappling with big animals. Butchering and processing animals was still work, though, and as with plant gathering, women and children did much of it. Making clothing and skin lodges was a formidable job. A shirt and pants required

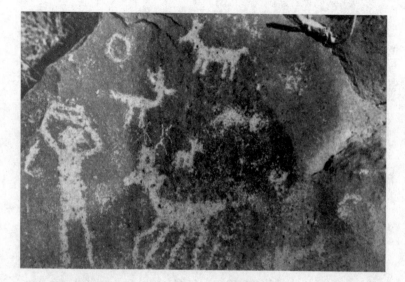

Hunter-gatherer rock art. Photograph by Dan Flores.

at least five deer hides that could take as many as eighty hours of effort to soften and tan. Making a skin lodge required thirty-five elk or caribou skins. It's no surprise that women in hunter-gatherer groups show the same pattern of tooth wear from clenching and working animal hides that Neanderthal females did.

Archaic America can seem like a runeless slate, difficult to fix in the mind. For one, it's a mistake to imagine everyone on the continent living in small bands of a hundred people. America was large. It was environmentally diverse. People then lived far differently, one group from the other, than we do today. Caribou hunters in Alaska and bison hunters on the Great Plains shared lifestyles, but hunter-gatherers who could exploit ocean coasts and inland resources, or could also tap the fishing possibilities of big rivers like the Columbia or Mississippi, fashioned different lives. The Haida and Tlingit were Pacific Northwest hunter-gatherers but they built and lived in fixed villages featuring wooden homes and elaborately carved poles bearing representations of sacred ravens, whales, elk, eagles. Other Archaic groups constructed a wide variety of built structures, like spoked rock circles known as medicine wheels that dotted the Northern Plains (a famous one sits atop the Bighorn Mountains in Wyoming). They marked solstices and other events that attest to a sensory-based human integration with the skies, planets, and cosmos.

The most wondrous built structures American hunter-gatherers erected were their earthen mounds. Starting around 5,500 years ago, cultural groups along several rivers in the South and Midwest initiated a new lifestyle: urban life in the midst of built mound works that undoubtedly were religious in nature and often were effigies of special animals. In Louisiana I grew up around mounds, badly damaged ones, from the final phase of this 4,500-year-long phenomenon, Mississippian Culture farm villages from a thousand years ago, with raised, multilevel platforms. Two aspects of those remnant mounds always stood out to me as remarkable mysteries. First was that America's earliest earthen mounds pre-dated the oldest pyramids in Mesoamerica by thousands of years. Equally unexpected, it was hunter-gatherers who had initiated this mound-building phenomenon and bequeathed the practice to farmers down the timeline!

Since it became a UNESCO World Heritage Site in 2014, the best-known of the Archaic mound towns is Poverty Point in the Mississippi floodplain of northeastern Louisiana. Built 3,700 years ago, Poverty Point in that far-off time was a city laid out on precise, C-shaped residential terraces that were raised one above the other. Below the terraces was a plaza, nearby was a court for sporting events, and centrally located above the hemispheric terraces was a towering mound known as Bird Mound, possibly built to represent a totemic bird of prey. From an aerial view the town appears to have originally been a circle that lost half its size to a flood. Archaeological work suggests Poverty Point was a major trade hub moving goods from as far away as the Great Lakes.

The mound-building tradition diffused far up America's great central river drainages and endured the transition to agriculture. Cultural groups along the Ohio River we call the Fort Ancient, Adena, and Hopewell peoples—hunter-gatherers who eventually began domesticating weedy local forbs before adopting Mexican crops—built animal-effigy mounds that have captured the human imagination for 1,500 years. One called Serpent Mound winds in seven undulations along an Ohio ridgetop for 1,330 feet. It's the largest representation of a snake in the world.

There's a third mystery here. How did hunter-gatherers with a concentrated-enough population to build a mound city create a town that lasted a surprising five centuries? Almost none of Native America's agricultural cities lasted that long. In most circumstances hunter-gatherers require space and small populations to prevent overshooting resources. Human populations that exceeded available animals were the Achilles heel of hunting-gathering. That deflection point was why Eurasians became farmers and herders, and why many Native people in America eventually did adopt agriculture. Jared Diamond once observed of farming that it was "the worst mistake in the history of the human race." But except where there were truly remarkable numbers of animals like caribou or bison, hunter-gatherers were rarely able to aggregate in the kind of numbers Poverty Point's mound town implies.

The likely answer to Poverty Point enduring for five centuries is that bison and caribou herds weren't the only possibilities for an especially rich hunter-gatherer life in Native America. There were other unusually fecund places such as ecologically diverse California with all its microhabitats, a world so rich even hunter-gatherers were able to reside in fixed villages, specialize in various crafts, and produce inheritable wealth and status. The same was true of the sea-and-salmon-fecund Pacific Northwest. And in Poverty Point's case, along the Mississippi River, the North American version of the Amazon. That seemed yet another locale where human cultures that relied on the bounty of nature could live for hundreds or thousands of years without using up the world. Or at least not much of it.

∿

ALL THE EVIDENCE indicates that America's Native people, whether hunters or farmers, lived immersed in art, stories, and observations designed around the grand theme of understanding themselves in a sometimes impenetrable world. A fundamental way to probe those kinds of understandings is through stories of gods.

The oldest named characters in North American history are deities. Heroes of origin narratives with many variations and nuances, these gods created the continent and its life and set in motion the stages of human life with all its victories and tragedies. Stories of these deities make up the continent's oldest literature. With few exceptions these ancient American gods were animals, although the stories describe some as anthropomorphic animals, perhaps descendants of those part-human, part-animal therianthropes from the Paleolithic. In western America, the deity who

Coyote, Beaver, and other deities in a rock-art scene from Native America. Photograph by
Dan Flores.

acquired the universal epithet Coyote ("Coyotl" in the original Nahuatl,
the language of the Aztecs), stood upright on its legs and brandished human
hands, but had the fur, sharp nose, erect ears, and tail of a coyote. The deities who made it into modern English as Coyote, Raven, Spider Man, Skeleton Man, Master Rabbit—along with humanlike gods, Glooskap, Old
Man, and Nanabozho—possessed a far more fundamental human quality
than opposable thumbs. They shared a basic human nature with their followers, and not just our most benevolent, inspirational nature. Our vices,
our lusts and our jealousies, our selfishness and our narcissism, resided in
America's ancient gods, there to witness and there for good reason.

We've often referred to these gods as "trickster figures." I think we've
been missing the point. While there are sometimes tricks in the ancient
stories about them, the trick is more often misdirection. It's why the trick
works that's important, and invariably the reason a trick works is because
of who we are, because of the essence of human nature. These are stories that survived thousands of years because they're such penetrating
exposés of all our foibles. The deities not only explained to listeners why
North America was the kind of world it was, they were at their best when
they taught lessons, often uncomfortable or funny ones, about human
behavior and motives.

Despite all their varying details, the stories of Spider Man of the Siouan speakers in the Midwest, and Master Rabbit of the Native farmers in the Southeast, are similar in their purpose. They set creation in motion, then hang around for adventures among us so they can teach us about the human animal. Algonquian-speaking people saw their deity/instructor as manlike, not an animal avatar, although in the case of Glooskap of northeastern America, he was formidable, at ten times the size of ordinary men. But Glooskap, Nanabozho of the Great Lakes country, and Napiwa (Old Man) of the Rocky Mountain Algonquians, were still lords of bird and beast. Skeleton Man of the Hopis in the Far Southwest took human form, too, but like the others he is instantly recognizable—"a god, a creator . . . a thief, a liar, and a lecher"—as a classic professor of human nature.

Coyote, who emerges from the stories as a kind of whirlwind biophysical force with an enormous appetite for pleasure and sensuality, is one of the most widely known gods out of ancient America. It was the actual canid, the jackal-like coyote who ranged across most of the western interior of Indian America, who inspired this celebrated figure. Regardless of cultural differences, all the people in its range regarded the deity version of these small wolves as an avatar, a stand-in for humans in the world. Among the Navajos of the Southwest, Athabaskan speakers who migrated south from the Arctic and picked up the Coyote figure ("Ma'ii") along the way, Coyote represented an ongoing chaotic force in the universe. In most Native traditions, Coyote's role in the supernatural realm is actually that of a *semi*-deity. He is the on-the-scene conductor of a master plan set in motion by an aloof "first cause." This more knowable, approachable kind of god was common in Native America. Another, known from the Tlingit and Haida of the Pacific Northwest northward to the Inuit and Alaskan Athabaskans (like Sarah James's Neest'all Gwich'in people from our Arctic river trip), is Raven.

If there are mysteries in the world you have wondered about, let Raven's adventures explain them. Raven knows things, like why there are tides, why crows (and ravens) are black, why people wear hats and hold feasts, why marmots have five toes, why mountains are white on top, why people should cremate the dead. Raven was yet another merged animal/human deity, who told the Tlingits, "I was born before this world was known." One of his tasks at the beginning of time was to bring light to the world. The Tlingit story of how that unfolds has Raven opening a

Raven.   Photograph by Dan Flores.

bag to loose the sun, whose intense light causes some animals to jump into the sea, others to hide in the forest. Uttering his monosyllabic *"gaah,"* Raven proceeds to shape each animal in a slightly different way and to name them all: Whale, Seal, Eagle, Bear, Caribou, Beaver, Salmon, Sea Otter, Land Otter, Wolf. The birds he paints in bright colors because he wants them to be pretty.

But lighting the world and creating and naming all its animals does not finish Raven's sacred tasks. America is still incomplete. The Inuits preserve this Raven story. Raven makes a pea pod that births the first human, and with clay produces a female companion. He then teaches them the use of fire and how to make food and clothing from animals. Walking a shoreline he sees objects scattered across the beach. Coming closer Raven sees they are human vaginas, scattered like shells on the sand. With a cocked eyebrow he gathers them up, continues on his way. Soon he comes to a village of humans whose women are despondent. "Why are you so sad?" Raven asks. "Because we have no genitals!" they cry. So Raven distributes the vaginas he's collected and now women can enjoy life more. And reproduce. *Gaah!*

It's good to remember to make prayers to the Raven.

There is one worrisome thread that runs through Raven and Coyote stories. In early times, the Inuit explained, Raven is concerned that humans are becoming too numerous. Human villages are growing too large and subsequently their residents are killing too many animals. The Inuit First Man agrees, and tells Raven: "If the people do not stop killing so many animals, they will kill everything you have made." In Coyote's case, as both the Yanas of California and the Navajos of the Southwest told the story, it is at this point the gods realize humans have to die, because if they do not, human overpopulation will result in the destruction of all the animals and of Earth itself.

WHEN I FIRST lived in Montana I had the opportunity to do a sweat-lodge ceremony led by two prominent Blackfeet men, named Woody Kipp and Gordon Bellecourt. The sweat took place on a November night in the Bitterroot Valley, on land owned by someone Woody kept referring to as "the old man." The lodge was gigantic, capable of swallowing the twenty of us, and stood only ten feet from where gurgling Kootenai Creek spilled out of the mountains. Away from the fire, where rocks were heating, bare cottonwood limbs and mountain ridges stood in silhouette under shrouded moonlight. Woody, lean and angular as a blade of grass, directed us into the lodge. His long, black hair hung in braids and his upper front teeth were on vacation, and he was at once the most likable person there. There among the hissing rounds of water splashed on glowing-hot lava rocks we sweaters spoke our prayers and hopes aloud, and a consistent theme from the group was that we were troubled. We had won the continent but lost our souls along the way. Surely there was some ancient way out of the environmental mess we had made of things.

At the end of the sweat we slipped quickly down the muddy bank and into the rock-bottomed river, a caress of opposites as steamy bodies crashed like calving lava globs into the glacial waters of the Bitterroots. I went full length into the astonishingly cold water, leading with face and chest, then gasped and clamored over the rocks and was out and up the bank in much less time than it takes to read this.

"Thank you, old man," Woody muttered in the direction of house lights on the ridge above us. The old man, it turned out, was a historian of religion named Joseph Eppes Brown, author of a famous book, *The*

Sacred Pipe. As a young scholar in the 1930s Brown had become fascinated by Native religions. He interviewed traditional Lakota elders, including the legendary Black Elk, and ultimately set down the ideas that made up part of hunter-gatherer knowledge about America's animals.

Brown's informants told him they had always perceived the essential nature of animal species as much through dreams and visions as through Native science. That combination of evidence led them to a clear ranking of animals in their taxonomy, with bears, bison, and eagles accorded highest ranking and respect. Bears ruled the underground, as bison did the surface and eagles the air. Certain animals illustrated particular traits useful to the human animal. Members of a wolf clan sought to invoke the wolf's cooperative skills in hunting and killing. If a young man on a vision quest heard a bull elk bugle for cows in the crisp air of autumn, he might then regard the elk as a totem animal whose potent sexual symbolism he could internalize.

These elders also recalled a connection, involving energy flow, among creatures. These were connections neither eighteenth-century Linnaean science nor twenty-first-century genetic science would ever think to link together. What the Lakotas called *Umi* or *Yum* was whirlwind power, the unrestrained residue of the energy of the four winds. They remembered whirlwind power as much-sought, in part because possessing it made one difficult to attack in battle. But only a small number of special animals—spiders, and also moths, dragonflies, and bears, elk, and bison—possessed the whirlwind secret.

As for bison, seasonal winds coming from the north or south seemed part of their mystery, bringing them or taking them away. A south wind might produce herds that blanketed the landscape from horizon to horizon. But they could entirely disappear, which led to a widespread belief in Native America that bison had their origins underground and sometimes returned there. The precise location of these regeneration places moved as people did. As tribes migrated onto the Great Plains in the 1700s, among the Kiowas the place where bison poured from the Earth was the Wichita Mountains. For the Comanches, bison regenerated in the canyons of a plateau, the Llano Estacado. The Lakotas believed this mysterious renewal happened in caves, like Ludlow Cave, in and near the Black Hills, which Native people surrounded with petroglyphs of buffalo tracks and human vaginas, enjoined symbols of fertility.

However incomplete a glimpse Brown got, his interviews give us some sense of a long span of time in America when Indian peoples possessed a philosophy with rich linkages between the human animal and the other animals surrounding them. As had been true of our first hunting ancestors in Africa, true of the Neanderthals, true of the Clovis people, these cultures' ceremonial lives centered on an ancient human desire to control nature. But they did so primarily as part of a religious philosophy, not a scientific one. Managing animals based on population modeling, carrying capacity, or selective/sustainable harvests as modern ecologists do would have been incomprehensible, and for a simple reason. Their cause-effect explanations for why things happened relied on different premises. Native people were never "the original ecologists," as many admirers have claimed, because the science of ecology didn't exist in human perception until the late 1800s.

The religions through which Native people understood animals in a larger context were superb at apprehending the kinship between animals and humans. The stories they told about their deities were equally superb at drawing insights about human nature, enabled by shrewd psychological observations both of human motives and of animal traits that mirrored ours. Those were augmented by a form of Native science—that is to say, a cultural knowledge of animal natural history—based on thousands of generations of handed-down observations about animal personalities, cultures, reactions, and migratory movements.

But crucial in Native America was knowledge about how to influence animals in a realm usually defined as the supernatural, an essential part of religion. Another friend from my years in Montana, the historian Rosalyn LaPier, has done the best insider account so far of the "invisible reality" that was central to this ten-thousand-year world. Rosalyn's own Blackfeet people, in her case the Algonquian-speaking Amskapi Pikuni, or Southern Piegans, were never secretive about what they believed, and their willingness to share imparts something profound about humanity's ancient belief systems. The Blackfeet never considered themselves living "in harmony and balance with nature." To the contrary, they attempted to control animals and to alter nature to their benefit. Rather than submitting to storms, they tried to change the weather. This "powerful worldview," as Rosalyn calls it, suggests that the Blackfeet desire to manipulate animals and nature is a deep-seated human impulse, and the strategies they used to effect it reach equally far back in time. For the Blackfeet, accessing the animals they hunted, encouraging or discouraging weather they hoped for, was all available through the assistance of supernatural allies. The degree of power one possessed to call on those allies determined how much you could make happen.

Blackfeet control over the buffalo shows this worldview in action. For thousands of years the Blackfeet lived in the country where the great, arid prairies abut the Northern Rocky Mountains. That Raven was one of their supernatural helpers ties them to the Northwest, but as prairie people they faced eastward, toward country where water often came in the form of beaver ponds. They perceived a triple world of powerful beings. Some, like Raven, lived in the Above World. Others, like Beaver, resided in the Water World. Humans and normal animals existed in the Below World, where events or dreams were partial reflections of what was happening in the "invisible reality" of the realms above and below.

Human beings could become vectors of power from these supernatural realms if a sacred being sought them out, or through a vision quest or other effort to find a sympathetic animal ally, or even by purchasing power from someone who already had it. Power from a supernatural being like Beaver commonly took the form of a "bundle," a package of sacred objects. The Beaver Bundle resonated power, as its possessor could call on Beaver to control the behavior of the buffalo in the visible Below World. "Charming the Buffalo" was a ceremony including songs that invoked Beaver, Raven, Sun, Moon, and Morning Star to bring buffalo to Blackfeet camps—along with a weather change, a warm wind to ease the tasks of skinning and butchering to further control nature for the benefit of humans.

One other intriguing look at Native religious traditions with respect to animals comes from the work of an anthropologist who lived with the Athabaskan-speaking Koyukon peoples of Alaska. The Koyukons preserve an ideology with powerful echoes of how life in the ten-millennia span of Native America must have worked. Keen observational naturalists with a highly refined knowledge of animal behavior, the Koyukons traced their link with animals back to what they called Distant Time, when animals were human and spoke human languages. Once again the deity animal was ever-watching Raven. Raven rarely missed anything, and was always alert to violations of taboos about how to treat animals and respect them. Many Raven stories were about the bad luck that befell people who transgressed against the animal world.

Stories from Distant Time defined the characteristics of particular animal species for the Koyukon people. Bears and porcupines, for example, were cousins in Distant Time, and that relatedness was evident in that they sometimes still shared a den. Animals remained cognizant of what humans were thinking and saying, and some of them—bears, lynx, wolves, wolverines—possessed dangerous spirits that could rob a human who offended them of skill, luck, even life. Shamans who visited animal towns in dreams could "call in" animals when they were scarce. For fear of insulting animals, Koyukon people did not point at them, and never boasted about hunting skill. There were animal-related rules about butchering, about how entrails were disposed of, meat was stored, about who could eat particular cuts of an animal. The latter was most commonly enforced against women because women had special power, especially when they were of childbearing age.

Animals were critical to a major life force—luck—that could make or break a person's life. Demonstrated luck in hunting or gathering was a very special power and easily lost. Luck could be transferred to someone else if the giver was of advanced years and less in need. But luck, the Koyukons believed, was an award from ever-watching Raven as a result of correct behavior toward animals. And the most correct behavior was treating them as kin.

⁂

HUNTER-GATHERER relations with animals were never exclusively philosophical. They also took place in the material world. One manifestation in Indian America proved truly significant. Beyond dogs and turkeys, across those vast spans of time Native people never attempted to domesticate the wild animals around them. Deer resisted domestication in both the Old World and America, so moose, elk, and whitetails were never real possibilities. America possessed no suitable wild cattle. Europeans domesticated aurochs but bison never wore the yoke there. America had lost its horses and camels at the end of the Pleistocene, while Europe's remnant wild horses survived just long enough to be saved by domestication 4,500 years ago. In North America bighorn sheep, caribou, and pronghorns were probably the best candidates. But while humans in Eurasia were domesticating a wide variety of animals for various tasks, the idea never took hold the same way in North America. Groups that embraced agriculture, like the Pueblo peoples of the Southwest, did widely domesticate wild turkeys. But in America big mammals remained wild, preserving their magic down the centuries.

On the one hand, lacking domesticated animals America missed out on the development and possibilities of the wheel. On the other, it also meant that Indian peoples never indulged the hatred of predators that became such a feature of Old World herding. So in America, wolves, bears, mountain lions, and coyotes never stopped playing their ecological roles as keystone animals, and for humans predators never ceased serving as totem animals and inspirations for clans and even tribes. Predators remained animals worthy of respect, even sacred respect. An important result was that Native peoples' lack of interference with predators preserved the continent's ancient ecologies intact down to the very moment Old Worlders arrived.

One of the long-term consequences of the Pleistocene extinctions was animal survivors that benefited from the loss of competition. The primary benefactor was the new, smaller bison, whose numbers had exploded. Half the size of their Pleistocene ancestors, reaching reproductive maturity far faster, buffalo adapted perfectly to the grasslands of the interior of the continent. Their population was no doubt highly variable, but based on how many livestock eventually replaced them, their numbers in their core range likely ranged between about twenty and thirty million. Great climate swings like the Altithermal redistributed them and shrank or grew their populations, but never pushed them toward extinction. Buffalo herds grew so enormous, and were such a perfect fit to post-Pleistocene conditions, that no amount of predation from either gray wolves or humans seemed to diminish them. Biologists now believe modern bison are a classic example of anthropogenic selection, their size and rapid reproduction (a natural increase of 18 percent) selected by human and wolf predation. That made the modern bison one of the most perfectly adapted American species ever fashioned by natural selection for a continent where humans were present.

For more than eight thousand years a lengthy sequence of different human cultural groups—they go by fanciful names like Mummy Cave, Oxbow, McKean, Pelican Lake, Besant, Avonlea, and Old Women's—

lived on bison, drove rivers of bison over cliffs, corralled and stalked and built religions around bison. Two thousand years ago, when Rome was transitioning from a republic to an empire, Besant and Avonlea hunters were undergoing a transition of their own on America's bison plains. The Besant people still relied on atlatl technology invented by Folsom hunters. But the Avonlea had the newest hunt technology, the bow, introduced to America by the ancestors of the Inuit. Even so, the bow hardly dented the enormous bison herds.

Like many prey animals, bison evolved to be highly social herd creatures. Numbers mean lots of eyes on predators and enhanced chances you're not the target. The herds varied in size and makeup across the seasons. At the macrolevel, three massive groupings spread across the western landscapes of the continent. In timbered parts of Alberta, the Yukon, and Alaska there was a distinctive type we now call the wood bison. Out on the grassy sweeps east of the Rockies a northern herd of plains bison ranged from Alberta to Nebraska. From there to the yellow expanses of Texas another mass worked across the Southern Plains in search of rains and greening grasses. These big aggregations of animals, groupings really made up of thousands of smaller herds, drifted southward in winter, then reversed direction to shift northward in summer.

To human observers, these movements and the incomprehensible masses of animals must have existed in the sensory plane much like stars and planets, or the sun's progression across the horizon from Winter Solstice to Summer Solstice. Like those cosmic spectacles, the organic flow of animal life was on such a scale only the supernatural seemed capable of explaining it.

The way to imagine these immense herds is by understanding their seasonal rounds, and the proper beginning is in the scorching heat of late summer when bison cows became receptive to sex. Over the next chaotic few weeks the rumbling bellows of two-thousand-pound bulls created a din heard nowhere else on the planet, audible for miles across the boundless plains. The oddly front-weighted males jousted, head-butted, and hooked at one another in dust-shrouded battles for females. Half the size of the bulls, cows didn't always honor the winners of these contests, often rejecting both strivers for a higher-ranking bull elsewhere. Over the few weeks of rut some bulls bred as many as forty cows. Others completely struck out.

Once the rut was over, bison would begin their general seasonal drift southward, the small groupings led by high-ranking cows (until they were eight years of age, younger cows were subordinate to older females). Whether southward or somewhere more local, the destinations for these migrations were forested river valleys, where bison spent months of snow and cold protected from the winds that swept open-country snow into drifts. As winter wound down in April pregnant cows dropped their young, and while eagles waddled around among them picking at afterbirth, the cows urged their bright red calves to stand, pop their tails over their backs, and run. Gray wolves, knowing well when a bison herd was vulnerable, were certain to be trotting by, yellow eyes fixed, red tongues lolling. Following the spring green-up bison moved into open country, where the herds sorted themselves into gender groups through spring and early summer. Bachelor bulls worked their way across the upland plains in all-boy posses while cow-calf herds stayed separate and distant until the pheromones of late summer began to drift through the hot air once again.

While rituals that charmed and lured bison may have been under the sway of supernatural animal deities, all those bison hunters over all those thousands of years understood from observation that the animals' movements were predictable. They also understood that bison preferred green grasses from freshly burned country. Humans had been using fire to alter the world to their advantage for a million years. In the eastern woodlands, regular human firing produced patchy ecotones whose rebounding forest created browse for whitetail deer. In the West fires produced wildlife park savannas for bison, pronghorns, elk, and wolves. Those fires actually pushed the areal extent of the savannas and their animals nearly to the Mississippi.

Archaeologists have mapped out a predatory human pattern that mimicked the prey wherever bison herds ranged. In the fall the hunters set fire to specific upland grasslands they wanted to hunt in the spring, knowing this would draw the herds. In winter those same hunters moved into the forested river valleys to set up their camps, aware that bison, elk, and deer would congregate there, allowing local hunts to take place throughout the cold months. These hunters were pedestrians whose only beasts of burden were dogs, and preserving meat by air-drying was a huge undertaking. Nonetheless, in suitable topography like Head-Smashed-In

(Alberta) and First Peoples' and Madison buffalo jumps (Montana), under the supervision of hunt managers they ran bison off cliffs, a strategy they learned by observing wolves. They also knew buffalo were entirely capable of exchanging cultural information. At these jumps they attempted to kill every last animal to prevent buffalo survivors from passing on cultural knowledge about the strategy.

The Great Bison Belt of the savannas east of the Rockies was the modern animal's evolutionary home. But bison were not just of the interior. Archaeologists reconstructing past climates have mapped out a whole sequence of bison "presence/absence" periods across ancient America. The Altithermal, that 3,700-year heat wave, was one of the absence periods in much of the Great Plains. That huge drought, and another only a thousand years ago that cycled off and on for six centuries, shriveled the western grasslands. Bison numbers likely plunged as the herds sought out better-watered refuges both east and west of the Great Plains, then trickled back in when weather improved. Then, between 1500 and 1600, as Old Worlders were settling America, a climate change to wet and cool conditions grew bison into vast aggregations again, sending teeming herds in the West eastward beyond the Mississippi River, convincing Europeans that in America they had found the Eden of the Animals.

～

AS IN THE OLD WORLD, eventually agriculture became an economic choice that was hard to resist. For those groups that took up farming, the catalyst seems to have been a human population growth that made living by hunting ever more difficult. Successful farming meant it was possible to set aside birth control and expedite population growth. Attractive as that was, it meant that parts of North America were no longer sparsely populated wildlands.

Ten thousand years ago the entire human population of planet Earth numbered only about 4 million. Across all the Americas humans then likely made up only a quarter of that number. North America probably had barely 500,000 people then, fewer than a single large city in our time. Agriculture changed that, but because big parts of the continent were unsuited to farming, and because farming was a new development, America wasn't entirely remade the way Europe or Asia was. By five hundred years ago the best guess is that America north of Mexico had grown its population to just under 4 million people. More than half of those, roughly 2.3 million, were farmers. One of the largest agricultural populations in America, an estimated 908,000 people, was in the desert Southwest. Indian farmers in the Northeast and Midwest regions combined had a population of a bit less than a million, while agriculturalists in the Southeast numbered slightly more than 400,000.

But five hundred years ago, on the eve of European arrival, there were still a million and a half hunter-gatherers in America. Many of them, more than 350,000, were in the Arctic and Subarctic, while others (roughly the same number, about 350,000) lived in the interior West. Another group of 350,000 or so people harvested the products of the sea, the coast, and the adjacent rainforests in the Pacific Northwest. California's more than 440,000 hunter-gatherers, living in a region with a mild climate and a remarkable number of microhabitats, outnumbered the hunter-gatherer population of every other region in America. With five hundred species of animals and plants in California, life was so easy no one was willing to do the work of farming.

Four million people spread across a landscape that in the twenty-first century supports four hundred million seems explanation enough for why humans and wild animals coexisted well for so long in Native America. But the simple numbers are misleading. As a consequence of the global economy, modern Americans and Canadians exist off the resources of the whole planet. In ancient times America did feature economic trade networks that funneled resources from region to region. And there were groups—bison hunters on the plains and farmers along the Rio Grande in New Mexico—that eventually engaged in mutualistic trade, exchanging protein from the hunters for plant-based carbs from the farmers. But most of the trade items that crossed long distances in Native America tended to be luxury goods, traded for fashion or status, not food to support larger populations. Copper ornaments from the Midwest, conch shells from both coasts, tanned bison robes from the plains, turquoise from southwestern mines, caged parrots from the jungles of Mexico, all were luxury trade items. Most Native groups depended on the resources of their regional ecosystems for food, so local places still defined the limits of human populations.

That might be an argument that for hunting-gathering and subsistence-farming economies, four million people was just about the carrying capacity of the American landscape. The effects accumulated, though. Across the final 1,500 years of Native America before Old Worlders arrived, a cumulative total of 150 to 200 million people lived out their lives north of Mexico. America was no howling wilderness. It was a long-inhabited, lived-in world. As a consequence, surely there were stresses on wildlife populations, at least occasionally.

In fact, there's no doubt of it. Humans are biological and no species gets a free ride in nature. In some American archaeological sites animal remains show a significant decline over time. The massive Emeryville mound site on the shore of San Francisco Bay portrays a steady decline in the bones of sturgeon, salmon, deer, elk, and pronghorns, demonstrating a drawdown of local wildlife as human populations grew in Native California. Elk remains in many continental archaeological sites are so scarce that some scientists suggest that elk numbers must have been suppressed, and the almost-certain cause was human hunting.

There was also at least one human-caused wildlife extinction in Native America. As humans spread around the world, flightless birds were always particularly vulnerable, and the Pacific Coast of California and Oregon, along with the Channel Islands, held one, a flightless sea duck in the genus *Chendytes*. In the past decade researchers dating the remains of these goose-sized ducks from six coastal sites concluded that humans began killing them 10,000 years ago, just as the Pleistocene gave way to Native America. Wiping them out was hardly the three-century blitzkriegs that took out mammoths or passenger pigeons, but by 2,400 years ago West Coast Natives had hunted Pacific flightless sea ducks to extinction.

Judging from the stories people preserved of their culture heroes, the most common environmental overreach was what the Inuit Raven story feared: overhunting brought on by growing human numbers. Localized overhunting happened for reasons that reach deeply into human psychology. To explain it science employs a phrase from biology. "Optimal foraging strategy" describes a use of nature that makes immediate sense. To satisfy desires of all kinds, living organisms tend first to reach for low-hanging fruit. Only when resources within easy reach are exhausted do we exert more effort, going farther afield or turning to other possibilities. On a macroscale, optimal foraging strategy explains why in the twenty-first century we're struggling to wean ourselves from fossil fuels. In America's past the principle is equally insightful about why wildlife around Native villages was soon hunted out. Anyone who has camped in the same spot and gathered firewood for several days grasps the basic idea.

In Indian country this phenomenon was well known from the 1600s through the 1800s. Wild animals grew rare in close proximity to Native villages. On Lewis and Clark's coast-to-coast journey across America in 1806, William Clark wrote that a consistent theme in their travels was that wildlife was always more abundant distant from Indian villages. A common feature of colonial America was the existence of "wildlife buffer zones" between villages or regular human camps, where unmolested populations of animals and birds built up in large numbers. In cases where adjoining Native groups happened to be unfriendly or in conflict, their hunters often avoided dangerous territory in between, which allowed wildlife populations to grow undisturbed. Peace agreements, truces, and new alliances often permitted the exploitation of these buffer zones stocked with animals, a pragmatic reason for seeking peace.

The biological impulse to engage optimal foraging strategy encouraged another cultural development that actually resembled an Indigenous endangered species policy. Indian religions commonly featured taboos against pursuing or killing certain creatures. As with the Blackfeet's historic refusal to trap beavers during the fur-trade period, explanations for taboos sometimes rested on an animal's critical position in religion or in nature. But more frequently than just chance, those taboos were on animals that were scarce, difficult to hunt, or nutritionally suspect. A taboo directed group energy away from animals less cost-effective in terms of effort, a classic calculation of optimal foraging. Once the Pleistocene extinctions had run their course, taboos must have played some role in the admirable paucity of human-caused extinctions across those ten millennia of Native America. Or at least we can guess as much.

Taboos provided protection for some American animals, and the passenger pigeon may be the best example. Native people in the East and South where passenger pigeons roosted and nested had long taken advantage of the birds' presence and numbers. The Senecas held that in ancient times the bird people had met in council and decided that the pigeons would furnish a tribute to humanity. Like other northeastern tribes, the Senecas strung nets to snag pigeon flocks that winged through the forests like arrows in flight. Midwestern peoples such as the Ojibwes, Shawnees, and Osages did the same. In the South, Cherokees and Creeks blinded the roosting birds with torches and swept long poles through the trees to topple them. So there was significant harvest, although as with bison, it's fairly certain that passenger pigeons were so well adapted and numerous that human numbers in Native America didn't threaten them. But there were restrictions Europeans never employed. A widespread taboo among Native people banned molesting the adult birds on their nests. That taboo guaranteed survival for birds with the responsibility to create new generations of pigeons.

COYOTE'S AND RAVEN'S America existed for seventy-five times longer than the United States has so far, so it shouldn't be a surprise that a history reaching beyond human memory would provoke a religious awe from its human inhabitants. Native people interviewed in the first centuries when European note-takers were on the scene appeared to conceive the continent as operating on a mysterious plane. Among their scores of different creation stories a common theme was that animals like salmon, buffalo, deer, beavers, and wolves joined the other great forces of the universe—the sun and moon, the sky overhead—as present from the beginning of time. Those stories also described the measures one should take when, because of some transgression, relationships with animals became strained and environmental disasters loomed. Bison might have their origins in the Earth and thus couldn't disappear permanently. But sometimes they seemed to abandon the world. Still, there were measures to resolve crises like this that revealed an ancient and crucial connection between animals and humans.

The village-dwelling peoples of the Midwest, the Siouan-speaking Mandans and Hidatsas, the Caddo-speaking Arikaras, and the Algonquian-speaking Cheyennes all joined the agricultural revolution and became farmers of corn, beans, and squash. But their need for protein and their older traditions as hunter-gatherers meant that like other Native farmers they never gave up hunting. Their stories offer us a window into

how the human-wildlife arrangement worked in Native America when
the relationship grew troubled. Each had a culture hero or heroes who
taught essential lessons for repairing those relationships. The great
heroes for various subdivisions of the Cheyennes were Sweet Medicine,
Erect Horns, Yellow-Haired Woman, and Coyote Man. Among the Mandans and the Hidatsas the most famous heroes were Lone Man, the first
human, and Hoita, or Speckled Eagle. For the Arikaras the primary hero
was Man-Who-Kills-Game-Easily.

The culture heroes taught that the key to the animal-human relationship was kinship. Animals *were* people. They had families and societies, opinions and cultural memories. Like people, they also possessed
something essential to them, a breath or spirit that survived death. The
nineteenth-century ethnographer James Mooney wrote of a common
belief by groups such as the Crees and the Lakotas that, treated properly,
animals could even reincarnate on the spot. He quoted a Native infor-
mant: "We came to a herd of buffalo. We killed one and took everything
except the four feet, head, and tail and when we came a little ways from it
there was the buffaloes come to life again and went off."

Centuries of living among America's wild animals convinced Native
peoples that supernatural entities, like Beaver among the Blackfeet, controlled human access to animals. When animals yielded it was because
humans respected them and honored taboos. Respect took several
forms—Coyote Man and Yellow-Haired Woman advised the Cheyennes
that they should never "express pity" for any animal. But primarily, respect
came from honoring that humans and animals were kin and acknowledging that we and they could move between each other's cultures because
we sprang from the same source. This became the key when, because of
some human hubris that violated the arrangement, the animals retali-
ated by withdrawing from humans' presence. Pleading with bison, elk,
and deer to return and rebalance the world became a focus of some of
the grand ceremonies Native peoples developed in Native America, and
it was the culture heroes, conferring with the animal supernaturals from
Beginning Time, who established the rituals that corrected the world and
reaffirmed proper reciprocity.

There's evidence that ceremonies to reestablish ties and regenerate
animals were ancient. Erect Horns, a Cheyenne hero, supposedly learned
a ritual for doing so during creation. Sweet Medicine also possessed such

knowledge, and whenever insufficient respect for him as special animal envoy caused the animals to withdraw, the Cheyennes apologized and repented and performed a ceremony called the Massaum that would call on the animals to reappear. The Mandans had a similar story about the first human, Lone Man, quarreling with Speckled Eagle, who controlled access to animals. As punishment Speckled Eagle withheld the animals inside a mountain called Dog Den Butte. Finally Speckled Eagle taught a properly penitent Lone Man the correct ceremony to arrange the animals' return.

In every case these ceremonies were about reaffirming kinship, dating back to their conjoined origins, between humans and other animals, about reassuring the supernatural entities that humans did see other creatures as real people and respected their willingness to die so humans could live. The way to renew this relationship was through ritualized erotic encounters, adoptions, and marriages between people and animals. Thanks in part to the artists George Catlin and Karl Bodmer, who traveled among these farming peoples in the 1830s and observed these ceremonies, we know they featured animal-costumed dancers re-creating repentance, and special lodges and altars representing the mountains or caves where the animal supernaturals hid their charges. The Cheyenne Massaum was a great animal dance that re-created Coyote Man's and Yellow-Haired Woman's release of the animals to spread across the world. The Mandans effected the return of the animals through the Snow Owl and Okipa ceremonies, while the Hidatsas did so with a ritual known as Red Stick.

Many Native peoples across America likely had ceremonies like these, grand mythic re-creations of the ties that bound the human animal to the diversity of life. But the next human arrivals to North America, driving domesticated Old World creatures before them as they unloaded from seagoing wooden ships, would dismiss all these ceremonies out of hand as Satan-worshipping superstition.

ON A SUN-DRENCHED November afternoon I sit in T-shirt and shorts a few feet from the edge of a canyon rimrock, looking through four hundred feet of transparent desert air on a thousand-year-old city. My

Cahokia in its heyday. Mural by L. K. Townsend. Courtesy Cahokia Mounds State Historic Site.

wife, Sara, is pulling a water bottle from her pack a few feet away. Various friends are scattered along rock-cairn-marked trails through the uplands behind us, where the faint indentations of ancient highways, four hundred miles of them, extend to horizons miles distant. The whole country—sagebrush uplands, the canyon floor, the enclosing rimrocks, and the ruins with odd names that lie in every direction below—is a uniform tannish brown, the color of dust. Or perhaps the color of abandonment.

During the time of the Crusades in Europe this spot, and another on the east bank of the Mississippi River, just across from today's St. Louis, held the two largest cities in North America. Both, interestingly, were religious centers. With its ceremonial effigy mounds of lizards and serpents and a Stonehenge-like circle of upright timbers planted to mark out solstices and equinoxes, the city in the eastern woods—today we call it Cahokia—probably held a fairly permanent population of thirty thousand people, larger than London at the time. I first saw Cahokia in the early 1990s with

a girlfriend who had Missouri roots and insisted we visit the place. I'd seen mounds but never anything on the scale of Monks Mound, towering up out of the American Bottoms like an earthen Chichén Itzá pyramid. After three hundred years of urban life an earthquake mostly destroyed Cahokia city, but not before its population had gone through twenty thousand trees and almost all the wildlife for scores of miles around.

As for the city whose ruins lie below us now, either side of ten centuries ago (AD 800 to 1140) it was the Vatican of the American desert. We call it Chaco and it is another of our UNESCO World Heritage Sites. Chaco was the closest Native America ever got to an empire like those of the Aztecs, Mayans, or Incas. But this was not an empire of warrior armies and conquered provinces. It was an empire of priests, who organized many thousands of scattered farming hamlets across fifty thousand square miles of today's Four Corners Southwest into an economic and religious network. No European principality of the age matched it. What the priests promised was direct intervention with the deities who controlled rain, crops, and animals, those grand imponderables whose presence made life good and whose absence ruined it.

The city of Chaco housed the priests, their families, and a resident population of thousands. It stored and distributed surplus crops. Then, at solstices and other special times of year, it hosted grand ceremonies to which the outlying residents made holy pilgrimages. At those times Chaco gathered a population of as many as forty thousand. Looking down now on its buildings and avenues, one suspects both the ceremonies and the nightlife must have been epic.

Chaco America almost seems foreign in the modern United States, as if lifted from the Middle East. The agricultural revolution arrived in this region 1,300 years before the city existed, and pollen studies indicate this development produced two immediate environmental effects. Human populations skyrocketed, and crops that needed to be boiled before you could eat them meant that daily cooking fires soon reduced a robust piñon-juniper woodland to desert. This became a world in need of priests who could intervene with the gods. Sitting and admiring the sprawling, hemispherical architecture of Chaco's largest structure, Pueblo Bonito, as its lines and shadows and religious kivas shimmer in the afternoon sun, I know this is a place that reveals much about humanity. Sara passes the water bottle over to me and, reading my mind, sums it up. "It wasn't until the 1880s that anyone built a larger building than that in America. In its time this city lasted longer than Washington, DC, has so far. But ultimately it just couldn't continue."

Chaco and its satellite hamlets survived for 340 years. The shorthand version of its collapse is that it all ended with a series of droughts across the Southwest, and that is true. But the many archaeologists who have interpreted Chaco know much more happened here. When the rains stopped coming, the farmers seemed to act abruptly, dropping their digging sticks in the fields, turning their backs on the grand religious gatherings at Chaco, and relocating across the Southwest. Some went north to what we now call Mesa Verde's Cliff Palace and long-since-abandoned towns like Sand Canyon and Castle Rock, in present Colorado. Most of the people who abandoned the Chacoan world congregated along the upper Rio Grande River, eventually founding towns still home to their descendants, the Pueblo peoples famous for their adobe apartment complexes, geometrically painted pottery, and turquoise jewelry.

Why did Chaco collapse in what sounds like a fit of pique? The evidence—and ultimately the response of the Pueblos afterward—points to a crisis we should recognize. Down there in Pueblo Bonito, a single room (out of 650) yielded the remains of fourteen people, whose funerary items indicated they represented Chaco's religious/political elites. In the room were flutes, ceremonial staffs, thousands of pieces of turquoise jewelry, conch-shell trumpets from America's West Coast, the remains of macaw parrots from the tropics. The oldest burial dated to AD 800 and the last from Chaco's abandonment, so those fourteen spanned the entire life of the city. And not just that. The genetics of nine of the fourteen showed them to be descended from the same matrilineal line, from a woman who evidently had been there at Chaco's founding. Most of them had an intriguing physical anomaly, a sixth toe on one foot.

Disparities in wealth and quality of life, along with the resentments they produce, are familiar to modern Americans. Isotope comparisons of the bones of the priestly class in Chaco's great houses with those of farmers from the villages indicates the elites consumed far more protein from the meat of deer and pronghorns. They were better fed, grew almost two inches taller, suffered less from disease, had three times the survival rate for children under five, and lived longer. They were also obviously conspicuous consumers of high-status goods. In the late 1800s an early archaeologist working in Chaco shipped more than seventy thousand high-status items just from Pueblo Bonito to the American Museum of Natural History.

The farming class suffered this gap between rich and poor as long as the elites delivered on their promise to make it rain. But when drought came and the priests were powerless to stop it, the lower classes attacked and killed many in the upper class. They also embraced a new belief, the Kachina religion. Diet studies in the collapse's aftermath imply that by then rabbits and rodents were almost the only huntable animals left. The need for protein perhaps explains why some of the new villages were founded close to bison plains. At the most easterly of them, called Pecos, the diet and health of ordinary farmers soon approached that of the former Chacoan elites.

Those weren't the only changes the Pueblos made post-Chaco. The new communities they established were far more egalitarian, almost to the point of being collections of conformists. And a population that

Ruins of the Chacoan Empire, Chaco National Park. Photograph by Dan Flores.

had been growing for a thousand years now shrank and stabilized. An initial Spanish expedition among them in the early 1580s thought there might be 130,000 people in all their towns, but the founder of colonial New Mexico, Juan de Oñate, estimated only 60,000 in 1598. Chaco was in their memories, and it was disaster enough for them to change how they lived.

TEN THOUSAND YEARS after the Pleistocene ended, the human wooing of the Americas had produced large farming populations and empires in numerous places in Mesoamerica and South America. In what would become the United States there were echoes of those: Chaco, the Hohokam villages of present Arizona, the Hopewell tradition and Cahokia in the Midwest, and a version known as Spiro Mounds across the South. Agriculture meant dramatic population growth, so on the eve of European arrival, considering all the Americas as a whole—from the Arctic to

Tierra del Fuego—we think there were then sixty million people living out their lives, 20 percent of the global human population.

But north of the Rio Grande River, in what is now the United States and Canada, hunting/gathering cultures still prevailed across vast stretches of North America, and here the human population had not yet reached five million. Even with human numbers seemingly so slight, five hundred generations of humans had physically transformed North America. To the Native peoples the continent was occupied, settled, its birds, reptiles, and mammals all intimately known, considered kin. Even with fewer than five million inhabitants, parts of America held large enough numbers of people that wild animals weren't always abundant, as the earliest of the Old Worlders would discover. But, except for that extinct Pacific Coast flightless duck, all the species that had survived the Pleistocene still existed. Beavers continued to engineer a watery landscape, shorebirds and ducks filled the skies, and bears, wolves, and other predators still played their crucial role in American ecologies. Even with thousands of years of human harvest, bison and passenger pigeons were among the most numerous species on Earth.

But the change that was coming was on a scale no one could possibly fathom. Since the ice age ebbed and northern seas flooded Beringia, America's animals and humans had lived in near total isolation from the rest of Earth. No one on either side of the Atlantic had any inkling the other existed, or that such biological isolation sat ready to deliver one of the most profound tragedies in all history. When the planet's human populations finally rejoined after parting thousands of generations earlier, the America of Clovis and Folsom, Poverty Point and Chaco, bison and passenger pigeons, confronted a devastating transformation.

Raven and Coyote, Beaver and Speckled Eagle would never be able to turn things back to the way they had been.

CHAPTER 4

TO KNOW AN ENTIRE HEAVEN AND AN ENTIRE EARTH

After ten thousand years, how would you intuit that the world as you've known it is about to change forever?

Maybe you dream it, which at some insightful moment in the 1500s is how a group of Native people, located far inland on today's Lake Superior, first understood that everything was on the verge of shifting. In this band of Ojibwes, a prophet had a dream that disturbed him, and after taking a long time analyzing and thinking, eventually he shared it. In his dream strange people with baffling customs had crossed the big waters in canoes of giant size and landed on the distant coast, he said. They had pale skins and bushy hair on their faces, and knives that were frightfully sharp.

What else? his listeners asked. The strangers also possessed long, black tubes, the dreamer related, which they pointed at animals and birds, and from which smoke and a noise so terrific emanated that he was startled even in his dream.

In the version that has come down to us, the dreamer did not say what had befallen those animals and birds when the smoke cleared and the din faded.

THERE'S A GOOD CHANCE that by the time the Ojibwes sensed this impending alteration in their world, the unprecedented newcomers were already probing at various parts of North America. Fishermen and whalers from ports hundreds of miles across the great eastern ocean, which for so many centuries had isolated America from the human history unfolding on Earth's largest landmass, had been sailing west to exploit Outer Banks whaling and fisheries. Already by the 1520s, but far to the south of the Ojibwes, there had been shipwrecks of the giant canoes on North American coasts. By the 1580s the strange people with hair on their faces and black tubes that spewed thunder and smoke were coming ashore up and down the coasts, wearing bizarre garments and speaking languages no one had ever heard before. The natives tried to make themselves understood: Do you come for firewood to warm yourselves? some of them asked.

There is a theory of how first contact between cultures unfamiliar with each other unfolds and it bears a resemblance to biological first contact. The theory argues that making sense of new beings proceeds in two stages. In the initial stage you can interpret someone standing before you only through the history or knowledge already in your head. They may be in strange (or no) garb, vocalizing unintelligibly, gesticulating or making expressions and body-language signals you cannot decipher, but your reading of them rests on how your culture has prepared you to see. As a result, misses at this stage can be epic. In the science-fiction classic *The Sparrow*, written during the five hundredth anniversary of Europeans arriving in America, first contact with a civilization among the stars happens when a scientific and religious order on Earth receives a marvelous and beautiful auditory signal from space. Interpreting it as a musical tribute to the deity, Earth dispatches a ship that includes a religious ambassador entrusted to make first contact. He discovers that the gorgeous music that lured Earthlings across vast distances is indeed poems sung into space by a planet's beings, who are joyous he has come and expect joy from him.

In first contact's second stage, both sides produce a more realistic assessment of each other. In *The Sparrow* Earth's ambassador realizes the songs that drew him across space are not in fact celebrations of God but tributes to lurid sexual encounters with the alien species attracted by the songs' aural splendor. The singers discover that a celibate Jesuit from

Earth doesn't react so joyously as expected. Similarly, five centuries ago American natives and Old Worlders widely misinterpreted one another and their motives at the outset. Over time they came to see each other with more discerning eyes.

In the next thousand years we Earthlings may have a chance to see first-contact theory in action if we encounter some new form of intelligence. For now, the confrontations that took place in America and on Pacific islands in the 1500s and 1600s are the most instructive examples we have. In the Caribbean, and on Atlantic, Gulf, and Pacific shores, natives isolated from the rest of humanity for fifteen thousand years now stood face-to-face with Europeans and Africans on missions of discovery, exploitation, colonization, and survival. The stakes for accurate mutual understanding could not have been higher.

But every encounter was filtered and mediated by the information already in peoples' heads. The most famous instance in colonial history was the conviction on the part of the Aztec king, Moctezuma Xocoyotzin, that the shocking appearance and technological wizardry of the Spaniard, Hernán Cortés, was a fulfillment of a prophecy that their culture hero, Quetzalcoatl, would one day return from the sky. There are other remembrances from North America. On first encountering Europeans, the New England Abenakis believed they were seeing the cannibal giants of their ancient stories. Witnessing strange newcomers apparently unharmed by bewildering new afflictions devastating their own villages, Native peoples in Virginia concluded that Europeans were the spirits of their dead come back to life. Cabeza de Vaca, a shipwrecked Spaniard who would lead a small group of survivors on a trans-Southwest overland crossing in the 1530s, said the local tribes allowed them safe passage because the Europeans were wizards who could cure the sick and bring the dead back to life. Cabeza de Vaca didn't argue against that interpretation.

The misses were far from a one-way street. Encountering American natives, Europeans also struggled. Once they figured out America was not Asia but a new, unsuspected continent, they had to decode their religious and cultural beliefs that neither the continent nor its inhabitants were supposed to exist. As did these peoples they called "Indians," Europeans of the age interpreted the world through religion, in their case a religion that had spread during the Roman Empire's colonization efforts throughout Europe a thousand years earlier. The Judeo-Christian tradition sprang from Middle Eastern agricultural and herding origins. Its first-cause deity resembled a human and he presided over Earth from somewhere in space. Springing from a highly localized beginning, this religion had nothing to say about the wider world Europeans now confronted. America, its peoples, its often bizarre new wild animals were entire mysteries.

Ransacking their religious stories for clues, Europeans could come up with only one, a reference in the first volume of their sacred text (the Old Testament, 2 Kings 17:6) to "ten lost tribes" that had left the Middle East, were never heard of in the known world again, and never returned. It was well into the 1800s before the idea that Indians must be descendants of Old World Hebrew speakers crashed on the rocks of reality, but reality didn't prevent the American-born religion of Mormonism from emerging out of the puzzlement of first contact.

With so little to go on, Europeans, seeing tattooed Americans wearing odd hairstyles, clothing fashioned from the skins of strange animals, and feathers from unfamiliar birds, drew on the memories of their religious traditions and history. This was how European pagans whose nature religions Christianity had long ago defeated once looked and acted. On the southern shores of the new continent, the shipwrecked sailor Cabeza de Vaca wondered if these Indians were in fact rational beings. The Gulf Coast peoples he was among regarded their dreams as more real than waking reality. The Spaniards discovered that if the natives dreamed you had committed some offense, the dream itself was sufficient evidence of guilt.

The conclusion came readily. These were people who had never heard of the deity Old Worlders called "Jesus." They apparently regarded animals as sacred and godlike and believed dreams could convict someone of a crime. They believed in sexual freedom for both men *and* women, and readily acceded women's right to divorce. Therefore, the newcomers concluded, these Indians must be disciples of the supernatural force the colonists believed a source of evil. Almost no Native peoples, with the exception of a northeastern group Europeans called Iroquois (Haudenosaunee), believed there was "an evil force" in the world. But if Indians embraced animal deities and knew nothing of Jesus then they must worship Satan. The European conviction that people holding nature sacred worshipped the dark forces of the world led to tragedy on numerous occasions. It also produced violent backlash by Native people, like the Pueblo Revolt in the Spanish Southwest, when persecuted Native shamans led an uprising that banished all Europeans for more than a decade.

First-contact theory predicts these kinds of wide misses, cultural misconstructions based on the idea that "truth" is found in one's own belief system. But with more experience with each other, evidence-based conclusions eventually replaced most predetermined reactions. Spending time among Europeans, Native people determined the newcomers were not a realization of their storied traditions and prophecies. They were merely a group of ordinary humans, previously unknown, who for some reason possessed interesting and effective tools that possibly could provide new ways of controlling the world. By the 1620s and 1630s, as Native people watched the newcomers struggle to learn Native languages and saw them appear helpless living off the land, savvy observers formed an even more unflattering impression. The highly intelligent Huron (or Wendat) headman named Kandiaronk, who debated the Jesuits who were attempting to convert his people to Christianity, apparently traveled to Europe and appeared as Native antagonist in a widely read 1703 book about the relative merits of Native and European worldviews. He concluded that while Europeans possessed many remarkable technologies, as a group they seemed generally dim-witted, greedy, and ignoble.

In this second stage, European conclusions about the Natives soon enough settled on a new assessment as well. Native religions might matter to members of the European clergy, who would stubbornly continue efforts at converting tribal people to Christianity for centuries. But in the larger game that was now afoot in North America, the worldview of the Natives wasn't nearly as important as their obvious interest in the Old World technological marvels Europeans were unloading on American shores. The more experiences Europeans had with Native peoples, the more they concluded that, like themselves, Indians were self-interested and therefore rational. First contact, stage two, now revolved around a new question: Exactly what were these Indigenous Americans willing to trade for a transformative technology? Looking around themselves at this wild new world, the newcomers thought they saw an answer to that question.

WHEN HE LEFT his hometown in southern Spain in 1527, Álvar Núñez Cabeza de Vaca never imagined himself the future author of a North American first-contact chronicle. A young man who curiously bore his mother's last name ("Cow's Head"), he had somewhere acquired a well-rounded education. He also had good contacts. The king himself had made him treasurer of a major expedition to the southern coast of North America, no small thing. It turned out a good head for figures was wasted on the Narváez expedition to colonize Florida, which ended up a running disaster of Shakespearean proportions. Six hundred settlers aboard five sailing ships set out from Mexico. Four years later three Spaniards and an African remained.

History's revered first impression of North America, handed down by French, English, and Dutch observers in the century to come, would be of a "virgin" continent, with exotic animal and bird life in staggering abundance. But Spaniards like Cabeza de Vaca were in America earlier, and for both Native people and wildlife those few decades between the 1520s and 1620s were more critical than anyone knew. The dwindling cadre of shipwrecked Spaniards who struggled in sodden horsehide rafts along the Gulf Coast from Tampa Bay to Galveston Island in the late 1520s, then pitched up on the Texas coast, experienced a different America than later Europeans would. A century before the Pilgrims and Puritans, this version of the "New World" seemed less like a virginal Eden-of-the-Animals than a place long lived in and a bit used.

Cabeza de Vaca and his companions experienced a combination adventure/ordeal that stretched out to eight years, from 1528 to 1536. Initially, as their ships wrecked and their expedition collapsed between Tampa and Apalachicola Bays, the Spaniards found themselves under attack by the natives as the strangers rummaged the coast for cornfields, fan palms, and whatever wildlife they could find. There wasn't much, and it largely consisted of animals the Spaniards recognized from Europe. There were deer, bears, and lions, along with rabbits, hares, and "civet-martens." An "animal with a pocket on its belly" dismayed Europeans, who had never seen a marsupial before, but then the American opossum would puzzle colonists for the next 150 years. The coastal lagoons did hold vast numbers of birds. They saw geese, ducks, herons, flycatchers, and quail in great numbers, along with many different birds of prey soaring above the waters. But the wild creatures they saw weren't numerous enough to feed them. They ended up killing and eating all their horses to keep themselves alive.

Eventually storm-wrecked on the sandy islands of the Texas shore, Cabeza de Vaca and a shrinking number of companions by turns were rescued, enslaved, and sometimes executed by the various tribes that found them, likely bands of coastal hunter-gatherers known as the Karankawas. In their first several years in America the few Spaniards who survived lived among peoples who seemed regularly short of food. "For three months of the year" everyone ate nothing but oysters. Deer and other wildlife were so scarce that it seemed an accident that anyone possessed a deerskin. Many times, Cabeza de Vaca wrote, "I was three days without eating." The fat season was when the tunas of the prickly-pear cactus were ripe, but that season lasted only three months. This seemed a long way from a New World Eden.

Two developments allowed the final four surviving castaways to escape their lives of privation. First, "I set to trafficking," Cabeza de Vaca said, by which he meant that he became a trader, which exposed him to a larger world inland. And as a result of a classic first-contact misunderstanding— and tragic circumstances early Europeans unwittingly were about to release across America—all four Old Worlders discovered that in the eyes of the natives they had become sacred beings.

It happened this way. In Florida, some of the Spanish colonists had been ill. Not long after the final shipwreck, the Indians surrounding them began to die from mysterious afflictions. Half the members of a band that took in the castaways died suddenly of a "disease of the bowels." This group blamed the Spaniards for their plight. But as Indians died and the castaways didn't, word of that strange circumstance began to spread. The Native conclusion was that the Spaniards must be wizards and healers. As Cabeza de Vaca would later write it, "They wished to make us physicians without examination or inquiring for diplomas."

This saved them. Hearing from inland groups that there were Christians like them far to the west and the south, the four still alive—Cabeza de Vaca, Andrés Dorantes, Alonso Maldonado (whose father in Spain actually was a doctor), and the African, Esteban—were able to launch a transcontinental crossing of what is now the American Southwest. A common passage in the account reads this way: "We left there, and traveled through so many sorts of people, of such diverse languages, the memory fails to recall them." Cabeza de Vaca did recall that on their entering Native towns the residents rejoiced over the famous healers, hoisting them like heroes and carrying them "without letting us put our feet to the ground." The next day everyone in the towns would line up to be touched and cured. Word spread that they were "children of the Sun." (The tribes they traveled through were "all very fond of romance," Cabeza de Vaca wrote.) Eventually they moved through an increasingly desert landscape accompanied by a retinue of four thousand or more attentive to their every need.

What Cabeza de Vaca conveys in his remarkable first-contact account is the state of a Native America that had been lived in for at least fifteen thousand years. It's the story of a truly historic moment in time. In it America is not a "virgin" of any sort. The part of the continent Spaniards traveled through was brimming with people. As a result it was not always brimming with animal life.

As the healers and their followers moved westward they left the America most like Europe and entered the more Asia-like part of the continent. Here three kinds of "deer" (mule deer and elk now joined whitetails) appeared and became more numerous. One evening locals brought the healers five dressed deer each. There were also burly creatures the Spaniards called wild cattle, which their informants said sometimes migrated from a northerly direction all the way to Florida. Cabeza de Vaca saw these new and unfamiliar creatures only three times. They had small horns and "flocky" coats, he said. Some were tawny in coloring, others black. (The next Spaniard to write a description of buffalo, Pedro de Castañeda of the Coronado expedition, added intriguing additional details about this new animal: they had manes like lions, carried their tails over their backs "like scorpions," and were as numerous "as fishes in the sea.")

So populated was the America these Spaniards moved through that they stayed in "towns" virtually every night of their journey. But of the villages they visited, only among the "cattle" hunters—the most southerly peoples who lived among buffalo—did the Spaniards get a sense of American Natives who lived very well off a great surplus of wild animals. This "Cow Nation" seemed rich, with many skins to gift. "They had nothing they did not bestow." One of the Spaniards, Dorantes, traveled farther north and found himself among the farming masters of the Southwest, the Pueblo descendants of the Chacoan Empire. These, he said, lived in "the fixed dwellings of civilization." That line in Cabeza de Vaca's account would spin off events ultimately leading to the founding of the Spanish Southwest and luring a steady stream of incoming Old World settlers.

Eventually the Spaniards reached Mexico, where frontier troops rescued them. They had lived through one of the Contact period's most harrowing experiences. With the advantage of our perspective, though, what might appear a grand adventure was a tragedy on an operatic scale. At the heart of Cabeza de Vaca's account are stories of Native people growing sick and dying of contagions the Europeans unleashed among them. Yet everywhere these agents of the Old World traveled they were among Native people who treated them—the source of those very maladies—not as pariahs. Instead they were holy men.

THE NEW PEOPLE began appearing in this ancient world in the 1500s. By the early 1600s, when Bartholomew Gosnold visited Cape Cod and (in the wake of Cabeza de Vaca's and Francisco de Coronado's travels) Juan de Oñate arrived to colonize New Mexico, the Old Worlders were coming in such numbers it was clear this was not a momentary thing. Native America had hardly remained static, but change had been slow. Now the pace of history began to accelerate.

Half a century after Cabeza de Vaca's rescue—the year was now 1584— a thirty-four-year-old English adventurer named Arthur Barlowe became one of the first Old Worlders to witness the Atlantic shore and write about what he saw. Barlowe was along on the first voyage to Roanoke Island, off the North Carolina mainland. The report he made to Sir Walter Raleigh on the natural bounty their party experienced became the initial template for a powerful idea that resonated down the centuries: Virgin America.

Barlowe's own first contact began on a bright morning in early July. As the white sails of their ships filled with a gentle west wind, America announced her presence before a single British eye ever registered land: "We smelt so sweet, and so strong a smel, as if we had bene in the midst of some delicate garden," Barlowe wrote. That smell was the very first sensory impression in the English version of the Wild New World.

The next day, as they tacked off the islands fronting the mainland, the explorers had vast flocks of white cranes rise beneath their ships with a sound like an army shouting all at once. The woodlands they sailed past brimmed with deer, hares, and birds "in incredible abundance." An Indian they watched fish offshore filled his canoe almost to the point of sinking in half an hour. The Natives they met called the country "Wingandacoa" and related that years before, a large vessel similar to the ones now before them had wrecked on their coast. Their fathers had combed the wreckage for every metal nail and spike, which became some of their most treasured "instruments." Might these ships before them hold similar items? they inquired.

Indeed, they might. And what did the residents of Wingandacoa have in exchange for nails and other metal? the Englishmen asked. "Chamoys, Buffe, and Deere skinnes" was the answer, Barlowe said. A copper kettle brought fifty such skins. Hatchets, axes, and knives brought even more, and once the newcomers did a demonstration of sharpness and an ability to hold an edge, the Natives "would have given any thing for swordes." To the English the trade possibilities seemed limitless as the sky. As a wide-eyed Arthur Barlowe wrote, "I thinke in all the world the like abundance is not to be found."

The following year, 1585, a mathematician and astronomer named Thomas Hariot spent even more time in the land the British were now calling "Virginia." Only twenty-five, Hariot brought the excitement of youth and never-dreamed wonder to his effusive account of America's wildlife abundance. Yet from the start he had his eye on what this ecological diversity might mean for him. "All along the Sea coast," he wrote, "there are great store of Otters" that could not help but "yeelde good profite." America also appeared home to quite unbelievable numbers of whitetail deer. He marveled that "dressed after the manner of Chamoes or undressed [they] are to be had of the naturall inhabitants thousands yeerely by way of trafficke for trifles." There also were bears, wolves, lions, turkeys, and parrots. And (Hariot wrote), "I haue the names of eight & twenty seuerall sortes of beasts which I haue heard of to be here and there dispersed in the countrie." Who knows what animals he meant—bison? wolves? jaguars?— but bizarre creatures no one had ever heard of before, in either the Bible or any classical texts, hinted at the enormity of the exotic continent.

Without realizing it Hariot also left the beginnings of an explanation for why the wildlife of 1580s America appeared far more abundant than when Cabeza de Vaca had traveled the Gulf Coast sixty years before.

In every Native town the newcomers visited, he wrote, "within a few dayes after our departure . . . the people began to die very fast, and in short space." He went on: "This happened in no place that wee coulde learne but where wee had bene." To the English, this depopulation of potential adversaries was God's work. To the Native people it was a horrifying and inexplicable mystery. The affliction that killed them was "so strange, that they neither knew what it was, nor how to cure it; the like by report of the oldest men in the countrey neuer happened before, time out of minde." All anyone among either Natives or newcomers could tell for certain was that everywhere Europeans set foot in America, the local Natives died en masse almost within days.

While Virginia's Indigenous peoples marveled at new trade possibilities but saw their health collapse in the presence of Europeans, another momentous encounter unfolded even farther north. This one featured a classic, firsthand description of American wildlife in post-Contact times that further refines the view of America's transformation in the 1600s. Bostonian William Wood was fifty-four and the Puritan colony in Massachusetts only a decade old when his *New Englands Prospect* appeared in 1634. By this time the new arrivals were scattering across North America. Between 1565 and 1610 Europeans made settlements at St. Augustine in Florida, Jamestown in Virginia, Quebec on the St. Lawrence, and Santa Fe in New Mexico. As had happened in the Near Southwest and Virginia previously, everywhere they went a diverse Old World suite of diseases arrived with them to exploit what epidemiologists call "virgin soil" populations. Biologically isolated from the rest of humanity for many thousands of years, America's human population was now confronting diseases entirely new to them, to which they had almost no natural immunities. All the prerequisites were in place for a cataclysmic human population crash.

There was not only tragedy but irony in this incomprehensible loss of human life. As Native people died, America's wildlife underwent an explosive ecological release, the very phenomenon that produced the "Virgin America" mythology. By the time he was writing *New Englands Prospect* William Wood was documenting this old continent's new face as the Eden of Animals. The distilled source for John Locke's famous line, "In the beginning all the World was America," was at hand. But the true irony was that *this* America was not a survival of the primeval world, as

Locke and other Europeans assumed, but a place created by their own arrival and the biology that accompanied them.

In Wood's account, the English colonizers saw both the familiar and the inexplicable differences in many of the wild creatures they were seeing in America. As in England there were deer, but these were much larger than English deer and more brightly colored. Bears in America were "a great blacke kind of Beare." There were squirrels of three kinds, one of which actually "flew" from tree to tree. The hares were familiar. Not so "a beast called a Moose . . . as bigge as an Oxe," which put some overly optimistic settlers in mind of domestication. America's waterways held otters, martens, muskrats, and beavers. Of the latter, Wood wrote that "the wisedome and understanding of this Beast, will almost conclude him a reasonable creature." There were "terrible roarings" in the deep woods, which the English judged to be "either Devills or Lyons." Wood thought lions the better possibility. There were many, many birds, ranging in size from "one of the wonders of the Countrey," a minuscule bird "no bigger than a Hornet" yet as "glorious as the Raine-bow." These were called "Hum-birds," he told his readers, because the humming sound they made even when they hovered was a common music in America.

At the opposite end of avian size were very impressive eagles. One eagle seemed familiar, but the other "is something bigger with a great white head, and white tayle." There were wild turkeys of prodigious size, "much bigger than our English Turky." And the sensory impression these animals made on the European mind? Wood's words convey his main point: (deer) "there be a great many," (bears) "they be common," (squirrels) "there be the greatest plenty," (moose) "so fruitful . . . a great store of them," (wild turkeys) "forty, threescore, and a hundred of a flocke," (ducks, geese, and partridges) "in great abundance."

Then there were the pigeons, "something different from our Dove-house Pigeons in England." Wood knew that Old Worlders, who had devastated so much of their wildlife centuries before, would scarcely believe what he was about to say, but no better example of America's stupendous wildlife existed. In the "beginning of our Spring," he wrote, "I have seene them fly as if the Ayerie regiment had beene Pigeons; seeing neyther beginning nor ending, length, or breadth of these Millions of Millions." Nothing—the shouting of onlookers, the rattle of gunshots, nothing— could deter their flights, which sometimes continued without break for

Black bear.   Photograph by Dan Flores.

five hours or more. Where they nested, so dense were their gatherings that "the Sunne never sees the ground in that place."

In case any of his readers were especially slow-witted Old Worlders, Wood laid out in plain language what the possibilities might be. All this vast number of creatures could without asking permission of anyone be killed and harvested and turned to entertainment and profit. Passages like that fell like manna from heaven to Europeans, whose feudal system had reserved wildlife exclusively for the nobility. Many of the Native people lay dead "like rotten sheep" in their towns, he said, but to procure the skins of America's furbearers Europeans could turn to the Indians who survived, "whose time and experience fits them for that imployment."

Wood made one additional comment about a very specific animal that would have caught the attention of every European, for the abundance of this one in America gave all potential settlers pause. This was the wolf, which struck Old Worlders as a very special problem. True enough, these seemed somewhat different from wolves in Europe's fairy tales and memories. In America "it was never knowne yet that a Woolfe ever set upon a man or woman." Neither did wolves seem interested in English horses or cattle, although they did sometimes attack pigs, goats, and calves.

American beaver. Photograph courtesy Ben Goldfarb.

But wolves in such numbers were unexpected. From the very begin-ning Europeans fantasized about the volume of deer, moose, and bears they could exploit if only something could be done about the wolf. Wood may have been the first to express the sentiment: "It is not to be thought into what great multitudes" all these animals might increase "were it not for the common devourer." Innocent of future notions like coevolution, ecology, or keystone predators, Wood was just the first of many to imag-ine how wonderfully splendid a wolf-free America might be. But in 1630s New England he despaired. When it came to wolves, there just was "little hope of their utter destruction."

Wood's judgment about his fellow settlers' and their descendants' abil-ity to deal with wolves turned out to be a spectacular miss.

❧

THE WOLF, then, from the start of colonization became a special ani-mal for Europeans. Colonists who came from France or Spain still knew wolves firsthand, but England's last wolves hadn't endured beyond the 1400s. Virginians and New Englanders were living among wolves for the first time in their lives and as William Wood implied, they didn't like it in the least.

Eastern wolves (*Canis lycaon*), found in northern New England, and red wolves (*Canis rufus*), which ranged from Texas to southern New England, were the "common devourers" Wood warned about. Animals in the sixty-to-one-hundred-pound range, these Gulf/Atlantic wolves were ancient American canids distinct from the gray wolves farther west. Some were grayish, others cinnamon-buff, and others black. According to recent genomic science, black coats in America's wolves sprang from a hybridization event between wolves and domestic dogs in the northwest of the continent approximately thirteen thousand years ago, during Clovis times. That mutation had also conferred a fitness advantage, perhaps in disease resistance, that other wolves sensed. The visual clue of blackness then affected mating choice, allowing black wolves to greet Europeans on Atlantic shores thousands of years later.

For a people who at one time lived among wolves, Old World settlers seemed to know precious little about them. Their knowledge of predators came from their herding-culture religion, but handed-down folk knowledge also shaped their understanding. While Abenakis and Narragansetts admired wolves for their bravery, hunting skills, and devotion to mates and packs, Europeans saw the same wolves as degenerate cowards, the very definition of evil in nature. Folk stories of werewolves, memories of the human–wild animal therianthropes from the Paleolithic, still circulated in colonial times and may have fed a suspicion that wolves were avatars of a bestial nature in humans. All the folk stories, all the biblical passages about "ravening" wolves must have been confusing when America's wolves showed no aggression toward people. That didn't matter. Fresh encounters with wolves in America lent the canids a reputation for "cowardliness" but didn't quell the hatred. That hatred matters now, because it's difficult to look back on this history without feeling moral outrage about how the unsuspecting animals must have experienced colonialism.

Real wolves bore little resemblance to the animals in the stories Old Worlders brought to America. Devoted to social life, wolves spent their lives in family packs of related animals led by high-status breeders, or alphas. Wolves avoided breeding with close kin, so a pack's grown pups eventually moved on in search of mating opportunities. While they had individualistic personalities, like young humans wolf pups learned from their elders and were much influenced by pack culture. Wolves are emotional animals, strongly attached to one another. After absences they greeted by standing on their hind legs and "rallying," and they interacted with a remarkable range of body language and facial expressions. When they howled, with heads thrown back and muzzles elevated, they sang a timeless symphony of the continent. For wolves, howling was a way to express emotional states. Howling was also contagious and enabled them to recognize other wolves from the harmonic structure of their songs.

Wolf natural history and human natural history readily explain why tamed wolves became our first companion animals. Our social lives and ecological niches were similar. Wolf societies were configured much like hunter-gatherer bands. In both instances the leadership tended matriarchal. While the alpha female wolf directed the pack's movements, the larger males—especially those between about two and five years old—were the primary hunters. Wolves mated in February and bore four to five pups in April, and the pack, often including experienced wolves as old as eleven or twelve, raised and educated the alpha female's pups, plus any born to lower-ranking females. The population of wolves in a given region rested on food availability. That didn't just determine pup survival, it also meant that packs competed with one another for prime prey territories. Before Old Worlders arrived, in fact, the main mortality in wolves came from other wolves.

Europeans imagined America's wolves as vicious, efficient monsters of the kill, routinely murdering prey merely for fun. In the real world a very different process was playing out. Although they have strong jaw muscles, the geometry of wolves' long muzzles actually inhibits their bite force. And chasing down and neck-wrestling big animals armed with hooves and antlers is dangerous in the extreme. So wolves, like us, went for low-hanging fruit. They scavenged animals already dead when possible. Highly perceptive about cost-benefit, on the hunt they tried for fawns and young animals, or injured or old ones. Their strategy was to test prey in search of those least dangerous, the ones that offered the least resistance. Even then, among the whitetail deer wolves were primarily hunting in the eastern and southern forests of colonial America, their chase success could dip as low as 10 percent.

As for murdering for fun, sometimes wolves caught deer yarded up

Black color-phase wolf. Courtesy Shutterstock.

in winter snows or caribou on their calving grounds and killed several
in a flurry. But their practice was to return and feed until the remains
were gone. Nonetheless, stories of sport or surplus killing circulated
about wolves as something common and criminal enough for punishment. What the new settlers really wanted was for America's wolves to
disappear. Even before William Wood fantasized about a wolf-free America, the Massachusetts colony passed the first wildlife law in American
history. It was a one-penny bounty on wolves, the first extermination
attempt of a great many to come.

The aspiring naturalist John James Audubon left the future a chill-
ing account of how wolves experienced the new war Europeans were
about to level at them. This was in 1814, when the attitudes of Americans toward wolves had hardened into a rare viciousness. Spending the
night with a farmer on the Vincennes Trace, Audubon accompanied his
host to a capture pit that held three wolves. The wolves' sin? They had
attacked the farmer's loose stock in a country by then bled of almost
all its deer, bison, and elk. From his colonial forebears this farmer had
learned exactly how to respond. Climbing into the pit, he one by one sev-
ered the wolves' hamstrings with a knife, "exhibiting as little fear as if he

had been marking lambs," Audubon wrote. Then he dragged the wolves out so his dogs could tear them to pieces.

Audubon helped him pull up the largest, a black male wolf in the prime of life. Audubon described this beast of Old World horror stories as "motionless with fright, as if dead, its disabled legs swinging to and fro, its jaws wide open, and the gurgle in its throat alone indicating that it was alive."

Petrified and in shock, the black wolf offered no resistance. It took the dogs less than a minute to stop the gurgling and extinguish his life.

※

IN 1683 the Dutch scientist Anton van Leeuwenhoek looked through the lens of a microscope and beheld a teeming world of bacteria and viruses no human had ever suspected before. Two hundred years earlier, when the first Europeans saw America, the mystery of disease had helped make their arrival a success they could only attribute to God. The truth was that a biology of unseen forms became the architects of colonial disaster for one people and triumph for the other.

Today we know that the human genome preserves snippets of genetic material from hundreds of viral and bacterial contagions from across our evolutionary history. Brucellosis infected Neanderthals butchering wildlife kills and that surely was not the first. The sources of most human diseases come not just from our ancient ties with the animals we hunted, and later those we domesticated, but from our evolutionary kinship with them as fellow creatures ourselves, which is what makes us susceptible to "spillover." So five centuries ago, unseen and unimagined viruses and other pathogens, evolved in the Old World and entirely novel to the New, were about to lay waste to humanity in the Americas.

Neither Europeans nor Native Americans had a clue that invisible agents transferred through plumes of breath or simple touch could be death sentences. The new colonists from Europe represented thousands of generations winnowed by Old World diseases, herd immunities, and selection for survivors. Eurasia is the largest interconnected landmass on the planet. Any animal disease that someone in the Middle East, Southeast Asia, China, India, or Europe contracted eventually diffused to almost everyone else on that giant landmass. That connectivity was the same factor that had enabled the spread of cultural and technological ideas in the Old World. The Bronze Age, the Iron Age, the invention of gunpowder in China, sophisticated instruments of navigation from the deserts of the Middle East—all these breakthroughs rested on millions of minds sharing ideas that eventually spread across Eurasia. That's how Europeans arrived in America with iron and guns and all manner of new goods. Old Worlders also brought with them to America the original sources of many of Eurasia's pathogens. Their ships carried once-wild animal species their agricultural revolution had turned into domesticates. Sheep, goats, cattle, horses, hogs, chickens—all were now coming to America, too, bringing germs and viruses that were long ago spillovers from domestic animals to humans and long since mutated to transmit between humans. Now this invisible cargo was exploding among human populations with no immunities to any of it.

Almost nothing confirms our animal origins like our susceptibility to contagions from other species. In the twenty-first century COVID-19 is forcefully reminding us that 60 to 75 percent of our infectious diseases entered humans from other animals. Modern plagues like H1N1 influenzas come largely from hogs and chickens. HIV entered the human population from chimps. The coronavirus diseases—SARS, MERS, and COVID—jumped into us from bats, camels, and civets via (perhaps) pangolins. But four and five centuries ago horrifying disease epidemics were still indecipherable. You appealed to whatever god or gods you believed in and hoped against hope to survive.

Few accounts of infection in colonial America offer clear diagnoses of which disease or cluster of them was responsible. There were simply too many potential illnesses. The diseases Europeans inherited from animals charted out a domestication timeline. Humans carry twenty-seven diseases that originated in dogs, among them, worms of various kinds, salmonella, scabies, rabies. Those particular diseases would not have been novel to Native Americans, who'd brought dogs with them to America. And Native people had syphilis, which Europeans caught and took home. But cattle had infected Old Worlders with at least thirty-one diseases, including smallpox and tuberculosis, and those were brand-new to Native people. So were diseases from horses (thirty-one), from sheep (thirty), from goats (twenty-two), and from hogs (thirty-one). Domestic chickens and ducks contributed many of the crossover influenza viruses, which along with smallpox apparently were major killers of Native people in America. Diphtheria, typhoid, malaria, and eventually cholera were also in the mix.

All these were as novel to Native people in the Americas as COVID is to twenty-first-century humans. If our modern projections are accurate the mortality rate for America's Natives reached 90 percent across the first century after Europeans arrived. Up and down the Americas a death rate that high, according to the latest estimates, means as many as fifty-six million people perished in the "Great Dying." They died badly, and bewildered. Deaths on that scale represented a shocking 10 percent of the human population of the planet. In North America a Native population approaching five million shrank in a century to about nine hundred thousand. Every demographic group was susceptible, but in many cases it was the young and healthy—exactly those who drove Indian economies and reproduction—who were struck down. Overwhelmed, their immune systems were thrown into the shock epidemiologists today call a cytokine storm. Under assault by an invasion of viruses and pathogens exotic to them, their own immunity likely overreacted and destroyed them.

With so many people suddenly gone, along with the fire ecologies they practiced and the immense hunting pressure they exerted, the colossal scale of the Great Dying disaster transformed the hemisphere. A mega-effect may have been an alteration in the climate. One of Earth's infamous climate anomalies in the past thousand years is the cold spell known as the Little Ice Age. Recent climate modelers have pointed out that the timing of this odd climatic alteration—1550 to 1850—suspiciously times up with the American disease holocaust. They've speculated that a sudden drawdown in hemispheric airborne carbon when fifty-plus million people reliant on fire died within a century may have precipitated three centuries of cool, moist weather. Anthropogenic combustion ceased and villages and farms rapidly reverted to forest, potentially soaking up enough carbon dioxide to chill the climate.

Whatever caused the Little Ice Age, it helped usher in ecological alterations that exploded the ranges and numbers of many of America's wild animals. The rapid buildup in America's animal populations occurred in the years from 1600 to 1800, exactly during the heyday of European colonization. The legendary "Virgin America" mythology those settler colonists and early naturalists bequeathed to American history sprang from an ecological chain reaction, launched by one of humanity's most tragic biological disasters.

The human deaths didn't end magically in the colonial period, either. As they had done in Eurasia for millennia, measles, influenzas, smallpox, later diseases like cholera, continued to sweep across America generation after generation. A smallpox epidemic that began in Central America in 1779 took five years to reach all the way to Hudson's Bay, killing more hundreds of thousands. The folk-medicine creation of a smallpox vaccine from cowpox in 1798, when an English doctor made a critical observation about the immunity of milkmaids, couldn't prevent another smallpox epidemic in western America in 1837. That one nearly wiped out several entire nations, among them nationally famous Indian leaders and their families.

You should understand this about colonial America. The loss of human life and rebound of animal life set up much of our subsequent story. As Native populations collapsed and struggled to rebuild, and wildlife numbers soared in response, new peoples from distant shores were replacing the ancient inhabitants and becoming Americans. They saw all this freshly released abundance of wild creatures in terms of the main chance. Here was money to be made.

✵

JOINING THE OLD and poetic "invisible realities" of the continent's original peoples, Europeans brought their own understanding of animals to America. This understanding was ancient. Some elements of it may have originated in the Pleistocene, when Europeans were killing off their own great bestiary. Europe also possessed a human past three times older than America's. And humans in Eurasia had developed the art of transcribing human speech into writing, which increased by orders of magnitude a precision in the accumulation of human knowledge down the generations. By the time North America became a colonial target, writing had turned knowledge into an eruption of information reaching back to the Greeks. That gathered information went along with settlers to Jamestown, to Massachusetts Bay, to French settlements on the Mississippi, to Santa Fe, and to California on the Pacific.

I was introduced to one element of the European worldview about animals sometime around the age of five. Perhaps I was only four, because this is my oldest memory, and child-development experts say earliest memories are often those of four-year-olds. Whatever age I was, the outline and details have stuck with me, likely encoded into my neural chemistry when so much else was lost because this experience was an emotional one.

My parents lived in a small town in Louisiana only three miles from where my grandparents had a farm. Rural life and family indulgence allowed me to grow up surrounded by animals. I had puppies by five, who turned into an unceasing string of dog companions. At six I had a pet goat who followed my every footstep. I had horses by ten (a family photo shows me at that age sitting on my horse, Star, with my dog, Frito, as passenger). I don't recall any time in childhood when I didn't have animal companions. The first of them, though, was a little yellow chicken I called Chicky.

I didn't just provide water and feed that Chicky pecked from my stubby fingers. This chicken and I were playmates. Our primary game was chase, the high excitement of pursuit through home obstacles of tables, couches, sewing machines, tricycles. One day our chase game ended tragically. Somehow I miscalculated speed and moves and stepped on Chicky.

The death of my pet chicken was my introduction to a core of Old World beliefs Europeans brought to America. My mother and I, both of us heartbroken, gave Chicky a funeral in the backyard. With Chicky in the earth, between sobs I turned to my mom with one last pleading hope: "I at least get to have Chicky again in heaven, don't I, Mom?"

Heartbroken though she may have been, my mother was from flinty Midwestern Methodist stock and had a reputation all her life for delivering the unpainted version of things. Whether you were fifty or five (or four) didn't matter. "Why, no, Honey. Chickens don't go to heaven. They're different from you and me. They don't have souls. You were made in the image of God and have an everlasting soul, so you'll have a life after death and go to heaven. Animals don't get to do that. They just die."

No matter (as I would discover decades later) that the founder of Methodism, John Wesley, sometimes preached that animals possibly did have souls. Or that Joseph Smith and Brigham Young, who launched the Mormon religion my dad's family had embraced, held a similar idea. Mom was channeling the more traditional conviction in Judeo-Christian theology and Cartesian science that we are not animals, and animals are not us.

❧

FOUR CENTURIES AGO religion was the ultimate explanation of all things for almost all humans. Like Native peoples, Europeans in America generally understood animals in supernatural terms. But for Europeans the terms were their own. They did believe in invisible realities, but their worldview didn't include supernatural animals, didn't accord animals the ability to reincarnate, and included no provision for ceremonies that could cause animals to emerge in renewed numbers from the Earth. Our colonial ancestors most certainly didn't regard animals as close kin. Only humans were godlike and exceptional. But European religions did feature a supernatural creation for animals and included a couple of ideas Native peoples didn't find strange at all. Europeans believed all animals had a divine origin, which meant they had existed unchanged, exactly as they were in the present, since their moment of creation. Europeans also believed that because it was a god who had given animals the spark of life, no animal species had ever disappeared in the past nor could any species ever disappear, now or in the future.

The Bible was the primary source for settler ideas about the animals they found in America, but European views about animals went more deeply into the Old World past than Judeo-Christian books and teachings. It's hard to say just how far back. The Greeks are an obvious reference for Europeans of the colonial age, but it's difficult not to suspect that much of Greek knowledge may have come from preliterate times five thousand to ten thousand years ago, when inhabitants of Eurasia were starting to domesticate animals and herd them. Plato and Aristotle likely were codifying into written form ideas that many generations of earlier Eurasians had thought first. Nonetheless, Aristotle's *Historia animalium* is still our best origin source for many of the essential ideas about animals that Christian Europe incorporated, adjusted, and brought to America two thousand years later.

Plato and Aristotle began with an essential premise. There must be a deity, an invisible reality now missing in action, who had created the

Earth and everything on it. Plato investigated a critical distinction in this idea: that humans were earthy and animallike but clearly separate from other animals. The explanation for that separation must lie in a difference between us and them, ergo, an invisible and individual spirit in humans that permits us a connection to the deity. Looking at the orderliness and beauty around him, Aristotle sketched out that order into one of the most important intellectual ideas in Western thought. He called it the Great Chain of Being. At its pinnacle was the deity, accompanied on the immediate "chain space" directly below by spiritual assistants Europeans called "angels." All other divinely created life occupied the descending links in the Great Chain, with humans below the angels and other known life-forms arranged in descending order of "perfection." Perfection translated into how useful a particular species was to humans.

The Great Chain of Being was a model of the known world that struck Europeans as so self-evidently true, and so useful a blueprint, that it survived in books, conversation, and peoples' minds from three hundred years before Christ until the beginning of the 1800s. You can say that for two thousand years, in one part of the Earth, at least, this became a deeply internalized imagining of how the world worked. Thinkers and ordinary people alike embraced its ideas both of a hierarchy and a constant biology. A deity had created everything all at once, everything that existed had a proper place, and most things existed because they had a potential use for humans. The world was divine and perfect and everlasting. These were big, reassuring ideas.

The settlement of America coincided with an unquestioned acceptance of the Great Chain of Being as the world's design template. Primarily that was because of the success the religion of Christianity had in replacing the older pagan nature religions across the previous thousand years of Europe's history. The authors of Christianity's sacred books had folded Greek ideas about the world into their texts, which now reached an endless parade of generations through sermons and services. So the vast majority of Europeans who migrated to Virginia or New York in the 1600s brought with them a herding culture's book that answered any questions they had about their proper relationship with animals, wild or otherwise. At the beginning of the Old Testament (Genesis 1:28), God gives Adam, on behalf of humanity, dominion over everything that lives. Further along in the story, when Adam's "sin" produces his "Fall" from grace, some animals turn against Adam and his progeny. But in Genesis 9:2-3, the sacred book goes on to say: "The fear of you and the dread of you shall be upon every beast of the earth, and upon every fowl of the air. . . . Into your hand are they delivered." The next line—"Every moving thing that liveth shall be meat for you"—was Christianity's stamp of approval on self-interested human use of other animals. Two million years of human carnivory had found its justification.

The prevailing religion among colonists from Spain, France, and England didn't stop there. In the Judeo-Christian tradition, humans occupied an exalted place removed from other mere animals. Genesis (1:27) once again clarified things. God had made humans alone, no other creatures on Earth, in his own image. While animals ceased to exist when they died, humans had something that set us apart from all the rest of creation. That something Christians pronounced the "immortal soul" that promised life after death, secured by the culture hero Jesus's crucifixion and resurrection.

Europeans settled on "the soul" as the boundary separating humans from animals in deference to the Bible (and perhaps Plato), but also because other efforts to distinguish humans were unconvincing. We had the same internal organs as almost all other mammals. Standing upright was a doubtful badge of distinction when animals like bears could do the same, if more briefly. Speech was perhaps dubious, too. Animals did appear to have some form of communication among themselves. We were self-aware and aware of death, and other animals struck us as largely unaware, so we had that going for us. But our best bet at distinction seemed to be the soul. Unfortunately, souls turned out to be another invisible reality, difficult to locate or demonstrate. If we *had* emerged from the animal world—as some heretical thinkers believed— then either all animals had souls, which meant we weren't special, or we, too, lacked this key to immortality, producing the same conclusion. Europeans looked at one another and in the mirror and concluded that being made in the image of a deity and possessing an everlasting soul were what made us exceptional.

Starting in the 1630s, exactly the moment when Europeans were settling America, the French writer René Descartes guided Europeans to one additional step with respect to nonhuman life. On behalf of the new scientific method that would undergird the Age of Reason and the

Enlightenment, Descartes and his followers argued that animals didn't just lack souls, they were biological machines. They didn't appear to reason and didn't seem to be self-aware. More importantly, animals had no emotions and probably even lacked sensations. It was a way of thinking about animals that didn't threaten everything humans believed about themselves, including the immortal soul, which Descartes never questioned, and that subjected animals to exploitation without guilt. That wasn't the new empirical science's best opening move for exploring human-animal relations. Even many Europeans couldn't accept it.

Western civilization's premise of a benevolent creation struggled most with predators, which were on such widespread display in America. The Puritan rebel Roger Williams thought the American wilderness stood as "a clear resemblance of the world, where greedy and furious men persecute and devour the harmless and innocent as the wild beasts pursue and devour the hinds and roes." Pretty perceptive, actually. Greek philosopher Plotinus had long before contributed the idea that "amongst animals and amongst men a perpetual war" raged. But it took Christianity, building on that idea, to explain wolves and other predators as by-products of "the Fall." To the Greeks, predators and their victims completed the wholeness of creation. But to Christians, unhelpful animals were a curse one had to endure because of Adam's ancient transgression. When you'd herded domesticated animals for eight thousand years and religion was your way of understanding things, it wasn't a big leap to see wolves as a supernatural malediction. Since they could be destructive of human endeavors, predators clearly were evil. Adam's Fall was a likely and appropriate origin for them. Didn't wolves share the yellow eyes medieval illustration often gave to Satan?

Keeping themselves elevated above "mere animals" would become an all-encompassing project for Europeans arriving in America. Religious leaders especially saw a "savage" continent and the lure of reverting to "an earlier state" as perpetual threats to a civilization whose veneer seemed uncomfortably thin. Prim townspeople in colonies like Massachusetts Bay particularly frowned on the old human stories about therianthropes. They even refused to portray animals in their entertainment. Bestiality (although not incest, interestingly) became a capital offense in most colonies. Puritanical fears extended to the human body, whose lust was especially animallike, and those fears colored how some settlers thought of other peoples, even how men thought of women, since females of the species struck some as closer to the animal state. Accepting sex and giving birth both seemed untactfully bestial.

Despite all this, or maybe because of it, the lure of the wild in America became irresistible to many.

~

THEN THERE WAS the class issue. Ordinary people who settled the European colonies had resented the upper classes for generations over access to animals. That had spawned the legend of Robin Hood, a folk hero who became England's most famous deer poacher, chased unceasingly by the authorities (and nowadays in the movies). So in the colonies one of the first celebrated "freedoms" associated with America was the freedom to "take" wild animals. Some Native leaders attempted to persuade colonial authorities to recognize that wildlife was *their* property, that whitetail deer were "the Indian's cattle." But the vast majority of ordinary settlers resisted the idea that wildlife belonged to anyone, including the tribes. Their view was that without the Old World markings of property ownership such as fences and signage, or wardens like the Sheriff of Nottingham, the American landscape was open to roam and hunt and all animals were accessible to everyone, free for the taking. Killed or trapped, an animal became one's own property. For the middling and lower classes of colonists, for the French and the English particularly, this kind of unimaginable access to the wild world was downright euphoric. It reawakened old yearnings in human nature and produced a pattern of action that dominated the human-animal story in America for the next three hundred years.

That pattern began with a natural abundance that, from what we can determine now, truly was stunning to everyone who saw it. Europeans began their settlement of America at the very moment when the continent's creatures were at their fullest expression since the spread of agriculture grew Indian populations three thousand years before. Some animals were expanding their ranges into whole new habitats. Bison and elk, those ancient Asians from the West, migrated out of the Great Plains into parkland and canebrake country in the Midwest and sometimes beyond the Appalachians and Alleghenies. The Spanish exploring expedition led by

Hernando de Soto never saw live bison in the Deep South in the 1530s. A century later, Europeans were encountering bison herds from Louisiana to Georgia to Florida.

No one knows how many animals the settlers built their farms and towns among. The numbers would have been highly variable across time because of weather patterns, winter severity, and habitat changes. America was in a rapid state of change, with fewer Indian-set fires, and with forest clearing and new fences going up among settler private holdings. But a few brave ecologists have speculated. According to some of the most recent estimates, in the 1600s North America held as many as 62 million whitetail deer along with 5 to 13 million mule deer. Caribou existed in a hundred discrete herds that hovered around 3.5 million. Bison numbers were in the range of 22 to 30 million on the Great Plains, with another 5 million spilling eastward and westward. There were between 15 and 35 million pronghorn antelope on the plains in the 1600s and another 6 to 10 million west of the Rockies. Elk numbers, released and growing as a result of Indian depopulation, grew to 2 million from the Rockies westward and another 2 million from the plains eastward. Bighorn sheep numbered as many as 2 million. There were at least 50,000 grizzly bears in what would become the Lower 48. And wolf numbers—gray wolves in the West and eastern and red wolves in the East—exceeded 3 million animals across the same territory.

Those kinds of figures might be impressive, but standing alone they don't really convey a sense of what it must have felt like to be among numbers of animals like that. A century ago, though, one of America's pioneer ecologists, Victor Shelford, captured things considerably better with this estimate. He calculated that for settlers in Connecticut, New York, Virginia, or the Carolinas, in the early 1600s an average ten miles square of America harbored four hundred whitetail deer, fifty to two hundred wild turkeys, one to three wolves, three cougars, five black bears, and three hundred to six hundred beavers, the last depending on how well watered the ten-mile square happened to be.

How did people—the majority of them, at least—who had been cut off from nature and wildlife for generations, whose religion taught them that animals exist for human use, whose science told them animals have no emotions or sensations, and who started their whole trajectory as evolved carnivores, anyway, react to that kind of abundance and diversity? The answer to that is the spine of a story that defines humanity's encounter with America's animals for the next three hundred years.

As the newcomers settled in, some rules of thumb gradually formed. Rule one was that local wolves or beavers lasted only five years. Rule two was that whitetail deer survived at most for a decade. Rule three: kill as many of all these as you possibly could, get your share before everything was gone. From the Big History perspective, there's some clemency in the fact that this was simply another example of what self-interested hominin carnivores had been doing to wild animals all over the world for two million years. Some of the colonies tried to regulate the hunt. They couldn't. The Massachusetts Bay Colony attempted to shut down the deer hunt as early as 1694 but didn't hire a game warden until forty-five years later. In 1698 Connecticut tried to stop the common practice of killing deer in the spring and summer when does were gravid or raising fawns. The colony of New York issued closed-season orders on several birds—wild turkeys, grouse, heath hens, and quail—in 1708. A decade later, alarmed by the disappearance of whitetail deer, Massachusetts tried to end hunting for those once wildly numerous animals for three years. These laws failed to produce a deer recovery.

The colonial wildlife crisis became so dire by the onset of the Revolution that in 1776 the British Crown actually moved to ban all further deer hunting in all the colonies except for Georgia. This was the first time in American history for a national government to try to effect a wildlife edict. Predictably, Americans despised it as an overreach and an attempt to rein in their freedom. Across most of the colonies it was too late for the whitetail herds, anyway. There would not be another federal try at a general wildlife law for more than a century.

In the mid-1700s a ninety-year-old American colonist lamented to a visiting European about the orgy-like slaughter of Atlantic Seaboard wildlife he'd witnessed in his lifetime. No one would even accept blame. When the animals disappeared, everyone pointed fingers at someone else. It was the Indians' fault, the colonists claimed. Or the fault of the French, or the Spaniards. As for the Native peoples, they blamed the whites. Or other tribes. Why have none of your governments passed laws against such a thing? the European asked. The answer was the summation of an emerging and enduring American sensibility. The "spirit of freedom" in America, the old colonist told the visitor, would never brook such an infringement of individual action. Governments could pass all the animal laws they wanted, but his fellow citizens "would not suffer them to be obeyed."

≈

JAMES FENIMORE COOPER'S 1823 novel about the settling of upstate New York, *The Pioneers*, looked back on this settler destruction of wildlife from the distance of the nineteenth century and found it appalling. Describing something like a colonial Burning Man rave around the massacre of a passenger pigeon flock that "the eye cannot see the end of," Cooper's literary version had pioneers firing guns into the sky without bothering to aim, striking birds out of the air with poles, even blasting into the frightened, wheeling flocks with cannon fire. No one pretended to collect the thousands of downed, fluttering birds until the shooters finally paid young boys to crush the skulls of the wounded birds and pick up a few. Cooper has his famed literary hero from better breeding, Natty Bumppo—who for his own use had downed a single flying pigeon with a single rifle shot—supply a condemnation of the pioneers' almost bizarre commitment to destroy: "It's wicked to be shooting into the flocks in this wasty manner." But another pigeon hunter, Judge Temple, expressed what by Cooper's time was becoming a national remorse: the hunters just "purchased pleasure at the price of misery to others."

There are historians who argue that the pioneers had good reasons for laying waste to America's wildlife. In the initial years the settlers felt truly threatened by wolves and bears, even by pigeons or parakeets that could destroy their hard-won harvests, or so the argument goes. Some of the most outrageous of their massacres of animals thus were "moments of perverse but joyous revenge" spawned by the hellish effort it had taken to create farms out of eastern forests filled with wild creatures. The stories that subsequent generations heard from families that had endured this period perpetuated a settler animosity toward the wilderness and its creatures. And they had a religion that gave them every rationale for taking revenge on a natural world whose vexations were born of Adam's curse.

The reality was that fashioning a privatized replica of Europe, trying to create farms and introduce domesticated animals and an orderly environment, ran head-on into the presence of America's wild animals,

Shooting Wild Pigeons, *in* Illustrated Sporting and Dramatic News.   North Wind
Picture Archives/Alamy Stock Photo.

which ignored Old World–style property boundaries and saw livestock
with caution bred out of them as hapless sitting ducks. The diversity and
abundance of America's animals undermined the colonization enterprise
in another way, too. Teeming colonial-era animal populations not only
drew many surviving Indian groups away from farming back to the hunt,
they also sparked a kind of social de-evolution among the settlers themselves. Cooper's Natty Bumppo himself was an example. America seemed
to afford such people one last chance at living the life hominins had forever known.

The woodsman types became America's first folk heroes. The most
widely handed-down settler stories tended to be tales about those who
had slain the fiercest, most dangerous wolves, bears, and lions, or killed
the largest number of deer, pigeons, or beavers. These were the people,
always masculine, who became the local avatars of the colonial enterprise, the Daniel Boones and Davy Crocketts, of whom there were untold
thousands. Into the 1800s dozens of them left memoirs of their lives,
which if nothing else demonstrate today that they were very good at the
stoic assassination of animals. One hunter in New York claimed a lifetime

tally of 2,550 whitetail deer, 219 black bears, 214 wolves, and 77 cougars. Meshach Browning, a counterpart in Virginia and Maryland who liked to refer to his kills as "fights," estimated 1,800 to 2,000 whitetails, 300 to 400 black bears, 50 cougars, and "scores" of wolves. French writer Michel de Montaigne once wrote that hunting without killing was like having sex without an orgasm. Colonial America was his proof.

With the imprimatur of their religion and no restriction beyond Indian outrage and personal conscience about how far to push such a life, a vast number of young men from rural colonial towns did not deny the lure of it. The result across the English colonies from Georgia to New England, and among French settlers up and down the Mississippi, was inevitable. By the mid-1700s, as one pioneer wrote about the upper Hudson River Valley, "no deer, or other useful animal or next to none exist; and scarce a living creature is to be seen." Eastern forests that had swarmed with animals of every kind a century earlier were now "as still as death."

⁂

LIKE JAMES FENIMORE COOPER, Henry David Thoreau looked back on colonial history and felt personally injured. By the nineteenth century the list of creatures that had disappeared or drastically declined in Massachusetts since colonial times was shocking to anyone who paid attention. The Atlantic world's original penguins, the great auks, were entirely gone, driven to extinction. Whooping cranes and sandhill cranes were rarely if ever seen. The local inhabitants had pushed deer to scarcity and exterminated both wolves and wild turkeys. Heath hens, passenger pigeons, trumpeter swans, even pileated woodpeckers and ravens were rare and endangered. Reading accounts like William Wood's of the New England they both shared, but two centuries apart in time, Thoreau sat down to his journal one morning in March of 1857 and, as thought followed thought, not only compiled his own expanded list but realized his relationship to those losses with this stark line: "I am that citizen whom I pity."

"When I consider that the nobler animals have been exterminated here," he wrote, "the cougar, panther, lynx, wolverine, wolf, bear, moose, deer, the beaver, the turkey, I cannot but feel as if I lived in a tamed, and, as it were, emasculated country." He went on: "I take infinite pains to know all the phenomena of the spring, for instance, thinking that I have here the entire poem, and then, to my chagrin, I hear that it is but an imperfect copy that I possess and have read, that my ancestors have torn out many of the first leaves and grandest passages, and mutilated it in many places."

No one else had put colonial history in quite this way, and no one since has said it so well: "I should not like to think some demigod had come before me and picked out some of the best of the stars," Thoreau wrote. "I wish to know an entire heaven and an entire earth."

CHAPTER 5

THOU SHALT ACKNOWLEDGE THE WONDER

Once this scene was one of the most famous ones in American literature. Its author, in his late twenties when he lived this experience, became the rare American to capture the imaginations of literary lions like Wordsworth, Coleridge, the Shelleys, and Byron. The language enchanted readers with its unprecedented blend of romance and science. But so did the author's encounter with a near-chimerical creature in the wildest and most exotic place on Earth. The time was the 1770s and the book, published in 1791, was William Bartram's *Travels*.

Bartram, son of the colonies' King's Botanist, John Bartram, was on his personal walkabout in the tropics of colonial America. He'd come to a river called the St. Juan, today's St. Johns River in Florida, whose current languidly spools just inland from the Atlantic Coast northward through marshes and lakes toward Jacksonville. Bartram had been navigating the St. Juan in "a neat little sail-boat" with a companion from a local tribe. But the Native, conveying "a dislike to his situation," abandoned him, so the author was continuing an exploration of the sun-drenched tropics alone. Sailing one day beneath palm trees with fronds waving ninety feet overhead, the Philadelphian looked for a spot to camp for the night. He chose a slightly elevated perch under a magnolia tree surrounded by a world of waters.

About then the "crocodiles began to roar and appear in uncommon numbers." And not merely that. As Bartram watched, two huge specimens, with "waters like a cataract" spilling from their open mouths and "clouds of smoke" from their nostrils, entertained him with a reptilian struggle, the pair intertwined "in horrid wreaths." Everywhere Bartram looked there were alligators. They transformed his plan for dinner—to use his boat to cross to a lagoon to catch fish—into an anxiety-filled ordeal. He was halfway to the lagoon with the sun a red ball just above the waters when the alligators attacked. Rows of serrated white teeth clacked "so close to my ears, as almost to stun me." The biggest tried to breach his craft, but although slight, the young traveler flailed wildly and "at random" with a club until the giant, lizard-like reptiles withdrew.

Bartram made it back to camp with a string of bass, but his Florida alligator evening wasn't done. A twelve-footer followed him, waddling up the bank and looking him "in the face," he told his readers. That one the author dispatched with a gunshot to the head. In the clear water another suddenly emerged almost on top of him as he scaled his fish. That animal's "incredible boldness," he wrote, "disturbed me greatly." But the crowning moment came at dusk-dark when a flotilla of alligators—"floods of water and blood rushing out of their mouths, and the clouds of vapour issuing from their wide nostrils"—arranged themselves like a living dam across the river to intercept a surging fish migration. The scene was like a medieval dragon fantasy. Yet it was real and a "shocking and tremendous" thing to witness. Or to read.

Up to this point in his life, Billy Bartram's circumstances would sound familiar to plenty of twenty-first-century Americans. Mother, largely invisible, spread her attention among several children. Very successful father. Raised in the church. Good education, in his particular case at the Philadelphia Academy, whose instructors were excellent and pushed a great-books curriculum. Daydreamed a lot. No penchant for business, as he lacked the killer instinct. Spent idle time drawing, which seemed pointless, but he became quite proficient and received encouragement from distant mentors. His famous father, exasperated at his twenty-five-year-old's aimlessness, invited him along (the year was 1765) on a "business trip" that included natural history collecting along the St. Johns. Tropical Florida captivated Billy and he remained to try his hand at farming. A friend sent a report home months later. Billy had no wife, no friends, no neighbors. His house was a "hovel" and "not proof against the weather."

William Bartram, "Aligator of St. Johns." From Bartram's *Travels*.

And Billy was far too gentle and dreamy a soul to succeed at the hard labor of farming. A year later he moved back in with his parents.

And then he set off on his multiyear roam of the Southeast from Georgia to Florida to Louisiana and transformed himself into a naturalist. Like some character out of Chaucer he visited and traveled with a wild mélange of characters, from Cherokee "maidens" to distinguished fellow naturalists to a class of rural southerners that seems hardly to have changed over the centuries. His walkabout was right in the middle of the American Revolution, which Bartram never appeared to notice. He then wrote everything down in the oddest mix of scientific precision and poetic sensibility that had ever coexisted on a page. The result was a breakout book that dazzled both sides of the Atlantic. *Travels* made young William Bartram the most famous literary naturalist of all those who pursued, studied, painted, and wrote about the American bestiary.

Bartram's *Travels* was a timeless journey into a mysterious, unknown natural world and became the archetype for a very American genre of writing. It lives on today in the form of "Bartram's Trail" across the South and also as a kind of literary touchstone. Over the perfect late-night mix of beer and weed college students still tell one another the story of how Samuel Coleridge fell asleep reading *Travels* and smoking opium, then woke with every line and image of the poem "Kubla Khan" as clear in his mind as the fountains of the St. Johns River. As he feverishly scribbled down a "Xanadu" landscape straight out of colonial Florida, the poet was interrupted by a knock at the door. When he returned to his desk and took up his pen, he discovered he'd entirely forgotten lines that had been perfectly lucid only moments before. He never finished "Kubla Khan, or A Vision in a Dream."

However Bartram survives in the modern world, you can also say this about him and his fellow naturalists. They became the perfect colonial expression of D. H. Lawrence's dictum on the considered life: Thou Shalt Acknowledge the Wonder.

~

AMERICA TURNED OUT not to yield up griffins or sirens, mermaids or unicorns, those chimerical beings out of Old World fantasies. The occasional rumor of mermaids and unicorns did pop up in the journals of European observers, though. A British ship's captain said he saw "a siren of great beauty" in New England waters in 1614, and in 1720 the French explorer Bernard de la Harpe claimed a unicorn sighting in today's Oklahoma. But America's real flesh-and-blood creatures possessed an enduring fascination. Like Bartram's monster Florida lizards, the unexpected and puzzling almost seemed the norm.

To start there were the "hum-birds" that seemed to hover in midair. There was a bird that mimicked every other birdsong it ever heard. There was a mammal that carried its young in a stomach pouch, and a squirrel that flew like a bat. Rivers of wild pigeons flowed over towering trees at fantastical speeds. People told stories of wild cattle with humped backs and manes. Skeptics in Europe dismissed them, but passenger pigeons and buffalo were real. So were America's poisonous snakes. From the first landings, Europeans were chilled to find that America possessed deadly snakes, including one that telegraphed intention to strike with an angry, rattling warning. There were, as one account put it, "a thousand different kinds of birds and beasts of the forest which have never been known neither in shape nor name neither among the Latins nor Greeks nor any other nations of the world."

Throughout all our evolutionary past we had paid close attention to other animals and had attempted a rudimentary taxonomy, a first analytical classification. Human migrations around the globe made us aware that new geographies held never-before-seen creatures. Since our interest in animals was inherent, those encounters had ecological, economic, and philosophical elements. Inclinations bequeathed by our evolution—and our cultures—lead us to expose toddlers to different birds and animals and test them in identifying one from the other. That's the bedrock foundation of human cultural training. We were made by our past as a species to be naturalists and to begin our understanding of the world in just this way.

Native peoples had spent twenty-three thousand years studying and learning America's animals, a body of oral knowledge badly damaged in the Virgin Soil epidemics that took away so many Native intellectuals. But Europeans and Africans came from other, older parts of the globe. Almost everything about America was brand-new to them. Following their Greek forebears, when Europeans studied nature at least part of their intent (so they said) was the contemplation of their Judeo-Christian deity. Thus one specific philosophical goal of colonialism was to find patterns in nature to translate divine thought.

By the 1600s natural history was acquiring additional purposes. Europe's new Royal Societies, themselves creations of the colonial age, decided that it was natural history's task to determine whether all the new species emerging from the Americas and elsewhere held advantages for the colonial enterprise. This wasn't necessarily an escape from religion. The new scientific method rested on a critical assessment of evidence and conclusions from designed experiments that other, disinterested, researchers could confirm to discover the reality of the world. But science's supposed purpose was to enable humanity to reestablish the control over nature it had lost when the Judeo-Christian God expelled Adam and Eve from the Garden of Eden. No one suspected that a more profound understanding of nature's workings might one day remove the Fall entirely from the equation.

The scientific approach that accompanied the colonial age carried forward from Aristotle the idea that, based on their anatomy, birds and animals existed as "species," which Western religion taught had existed unchanged from the moment of their creation. Animals also fell into categories—domesticated animals as opposed to wild ones, wild ones humans could eat, or not, and inedible ones with otherwise useful qualities versus animals who hindered human projects. These divisions clarified things. Even when they weren't on the menu, some of the continent's wild creatures were desirable because of particular body parts—their fur, for instance. Even if their pelts had value, carnivores like wolves or big cats were inedible and big hindrances to the colonial project, thus double sinners. As one Massachusetts Puritan who had never seen a cougar or wolf before coming to America wrote, "They be evil to us and thirst after our blood." Naturally. Settlers everywhere were expected to take the lives of beasts of prey, or "vermin," at every opportunity.

Yet at that same moment in history, naturalists like Bartram, John Lawson, Mark Catesby, Antoine-Simon Le Page du Pratz, Jonathan Carver, Alexander Wilson, and John James Audubon sought out and engaged the unique wildlife of America's eastern forests in a far more inquisitive and appreciative way. Aided by the geniuses of the age—Newton, Linnaeus, Buffon, Humboldt—natural history showcased America's astonishing biological diversity and offered up some tantalizing speculations about what it all might possibly mean.

*

IN MY ANCESTORS' PAST there is a river called the Loire, which has its headwaters in upland hills south of Paris. After a long, coiling loop through western France, the Loire empties into the Atlantic midway down the French seacoast near a port city called La Rochelle. Like so many other colonial-age ports in Europe—Glasgow, Bristol, London, Amsterdam, Lisbon, Cádiz, Vladivostok—La Rochelle was one of the destinations for the natural resources of America's *pays sauvage*, its wildlands. A vast percentage of the wild-animal parts funneled to La Rochelle by a colonial enterprise called the "fur trade" came from the part of America where I would one day grow up. In other words, at the same time that naturalists were besotted with the beauty and strangeness of so many of the continent's animals, their contemporaries were arranging for the wholesale destruction of the same.

When the British and Dutch and Spanish and French came to America, they didn't merely claim and settle New England, Florida, California. They intended to ransack the place. America was the grand storehouse for all the things Europe was running short of five hundred years ago. Timber, iron, fish, whales, twenty times the amount of gold and silver in Europe—all of that and more Europeans cut, dug, mined, and caught in America, then shipped off to the depleted Old World. Those resources from America and elsewhere hauled back to the European "Metropolis" were the extracted wealth that made early capitalism work. This was a primary catalyst to the birth of a global economy, based on a premise unprecedented in human history: perpetual economic and demographic growth. Despite harboring less than 20 percent of the people on Earth at the time, Europe was in the process of making the whole planet its oyster.

America's wild creatures were one of the most lucrative of all these resources. No one can wrap a mind around figures like the following, the numbers representing the sacrifice of individual living beings. But consider. By 1700, from New England's shores, outer banks, and inland rivers, 10 million tons of fish annually were filling the holds of vessels plying the Atlantic. In another part of colonial America, from just one port in one year (it was 1706–7), British vessels shipped out of the Carolinas 121,000 skins peeled from the carcasses of southern whitetail deer. By 1750, the fur economy, which created America's first big business, had sent the pelts of 2 million beavers to Europe. This, everyone understood, was merely foretaste.

In one single year, 1743, the La Rochelle docks that received the spoils of my Louisiana homeland took in 127,000 beaver pelts, 30,300 marten furs, 12,400 river otter furs, 110,000 raccoon pelts, along with its big haul for that year, the stripped skins of 16,500 American black bears.

And what of those who fed this market? The answer points to one of the most insidious elements of this new trade. In Louisiana's records there is an account of a particular family that in the autumn of 1807 brought the skins of 400 whitetail deer and the glossy pelts and oil of 118 black bears to the Louisiana town of Natchitoches, where my European ancestors first settled. The family was Native, likely Caddo, from a prominent local tribe. That was one autumn's hunt and those bears came from the very parishes where I grew up two hundred years later. Not once in my youth did I ever see a black bear in the woods. As for that Indian family, it was like tens of thousands of others that got ensnared by the market that Europeans introduced to America.

A whitetail deer or a black bear giving its life to humans is not in principle an immoral thing. Humans have been taking life since we became carnivores in long-ago Africa. But now wildlife was in the sights of an economic system that was converting animals into commodities. In this system the life of an animal had no meaning beyond satisfying the desires of distant peoples who utilized animal body parts in industry or in making statements about fashion or status. This was an economy that made some who dealt in wild animals very wealthy and also supported a colonial working class—some of those workers the new Americans, but many of them Natives—who eked out an existence killing animals for "the trade."

However uneasily, natural history and the market existed side by side in the colonies and in the early United States. Discovering the biological diversity of America and ransacking it for wealth and status twinned into a Moebius loop that shaped a new, global destiny for the continent's wild animals.

IT WAS ACCIDENTAL, but an accident that shaped the initial views about America, that the most southerly of the European colonies first drew naturalists. Imperial Europe founded the Carolinas, Georgia, Florida, and Louisiana in the semitropics, and naturalists going there were revealing life endemic to dense forests, lazy coastal waterways, and tea-colored swamplands. This was where almost all America's early naturalists, including William Bartram, explored. The result was to press onto the global imagination throughout the eighteenth century an idea of a verdant Wild New World that actually was not that typical of the continent.

The naturalist who set this template was named John Lawson. A young Brit resolved to see the world, Lawson said that by serendipity he met a man in London in 1700 who told him "that Carolina was the best country I could go to." Landing in Charleston, Lawson performed a typical European maneuver. He became a developer, a job description that in eighteenth-century America meant both wide exploration to describe resources, and negotiations to extinguish the land claims of the Native people. In late December of 1700, with five English companions and a changing retinue of Native guides, Lawson commenced a 550-mile trek that described a semicircle from Charleston inland to present Charlotte, then returned to the coast in a northeasterly direction. Two months in the exotic continent had dazzled him enough to add a third ambition: writing a natural history of America. Aside from a slim volume by a self-proclaimed "Indian," actually Virginia native Robert Beverley, published while Lawson was in the interior, there was nothing in the field. By then it had been 120 years since those initial British chroniclers, Arthur Barlowe and Thomas Hariot, had written their cursory impressions of Virginia/Carolina shores, and an artist named John White had painted watercolor sketches to provide Europeans an initial mental image of the continent's eastern edge.

When Lawson's *A New Voyage to Carolina* appeared in print in 1709 it was clear the developer/naturalist would not address any of the great natural-philosophy questions of the day. What Lawson did instead was provide his readers with basic and often colorful descriptions of some of the South's singular creatures. He introduced Europeans to America's marsupial, the opossum, as "the wonder" of its land animals. He called the raccoon "the drunkenest creature living." Lawson's bald eagle was a foul-smelling "cowardly bird." And while America held no nightingales, a bird called the mockingbird sang "with the greatest diversity of notes that it is possible for a bird to change to." Passenger pigeons flew overhead in such numbers that "you might see Millions in a Flock; they sometimes split off the Limbs of stout Oaks."

Europeans were disappointed to learn that there were "endless Numbers of Panthers, Tygers, Wolves, and other Beasts of Prey" in America. (Some later biologists believe that reference to "Tygers" hints at jaguars in the South.) The largest animal of the eastern forests Lawson described this way: "The Buffelo is a wild Beast of America. . . . He seldom appears among the English inhabitants, his chief Haunt being in the Land of Messiasippi, which is for the most part, a plain Country; yet I have known some killed on the Hilly Part of Cape-Fair-River, they passing the Ledges of vast Mountains." Unlike Cabeza de Vaca (or Coronado) far to the west 150 years earlier, Lawson did not actually see live bison himself. But he continued on hearsay: "These Monsters are found to weigh (as I am informed by a Traveler of Credit) from 1600 to 2400 Weight."

Lawson planned to write a full American natural history, but unfortunately he failed to survive a second trip into the American wilds. Charged by the Tuscarora tribe as a stealer of Indian lands, which was true enough, he was tied up and impaled with hundreds of fine splinters of wood 'til

John Lawson's American bestiary. From Lawson's *A New Voyage to Carolina.*

he resembled a porcupine, then set afire in a grisly execution. Nonetheless, it was John Lawson who successfully pioneered an emergent American literary form, a style that featured first-person travel writing through the natural world. Jonathan Carver, Bartram, Lewis and Clark, Audubon, John Wesley Powell, Teddy Roosevelt, John Muir, John Burroughs, Joseph Wood Krutch, Rachel Carson, and Edward Abbey would all track this same path. Nature-travel writing has been the most popular nonfiction about America for a good deal of the country's history.

That may be the reason the name Mark Catesby doesn't readily come to mind. Once Catesby was "the colonial Audubon." But because he didn't write a travel narrative, few of us know his story now. He did write a book, *Natural History of Carolina, Florida, and the Bahama Islands*, which he published in 1747 after working on it for twenty years. But even then his reputation was largely in England and largely among scientists. Nonetheless, Mark Catesby was the first really serious naturalist-artist to work in

America, and much of what the wider literary/scientific world concluded about the continent's nature came from his *Natural History*.

Catesby was from a prominent East Anglia family, a status that rewarded him at the age of thirty with a trip to Williamsburg, in the century-old colony of Virginia. The year was 1712 and his connections brought him to the attention of wealthy landowner and colonial wit William Byrd II, eight years Catesby's senior. A fellow of the British Royal Society, Byrd possessed the largest library in the colonies. His *History of the Dividing Line* about a survey of the Virginia–North Carolina boundary was one of colonial America's first recognized literary works. Byrd had long wanted someone to publish a natural history of the southern colonies, and wilderness trips with Catesby over the next seven years left him convinced the young Englishman was his man.

Byrd's backing and Catesby's painting samples secured the naturalist Royal Society approval (although no money) for the project. He turned down a trip to Africa, then America's chief rival as an exotic wild destination, so he could return to Virginia in 1722. The field *was* wide open, the only competition coming from amateurs like Beverley and Lawson and a French soldier, Antoine-Simon Le Page du Pratz, whose *Histoire de la Louisiane* would appear with engraved drawings and a natural history list a few years later. Catesby spent years probing inland along rivers like the Savannah. His focus was mainly on plants and secondarily on birds (he believed eastern America's mammals were little different from Europe's). But because America's snakes filled Old Worlders with particular dread, he made a special study of reptiles. In *Natural History* he said he painted all the watercolors of his animals from life. That set a bar for others who followed.

In the early 1700s natural history was ambitious but far from a mature science. Holes were appearing in the Great Chain of Being that scientists were having a hard time ignoring, but as yet there was nothing to replace it. Catesby, like others, hoped to replicate in natural history the surety of Sir Isaac Newton's breakthrough 1687 book on physics. *Principia's* mathematics had demonstrated that the force that causes you to fall when you leap off the roof of a house is the self-same one that produces the tides and also the orbits of the planets in the solar system. Newton proved a tough act to follow. Finding general laws in the messiness of the natural world—especially in America's natural world, so unlike the perceived orderliness of nature in Europe—turned out to be maddeningly elusive.

Catesby's trips into the Southeast, set against experiences he was also having in the Caribbean, imparted some ideas he thought might be general. He saw patterns between latitude and species diversity: "Animals in general, and particularly Birds, diminish in number of species so much the nearer they approach the Pole [and] there is a Gradation of Increase at every Degree of Latitude approaching the Tropick." By describing America's bird migrations, he corrected the belief that birds disappeared in winter because they hibernated. He also thought hard about the origins of America's wild creatures. He inferred that at least some seemed Asian, which led him to a remarkable speculation for the 1730s. Was it possible Asia and America had once been geographically closer? Of all Catesby's ideas, though, he became most associated with seeking out what we would now call ecological connections. His paintings portraying birds and animals in relationship with settings and plants that were important to them was a novel new thing.

The impression many drew from the *Natural History* was that the Wild New World wasn't just disorderly, it was downright chaotic, which seemed the antipode to Newton's fixed gravity. At one point in his adventures Catesby witnessed a coastal hurricane. In its aftermath, he wrote, "Panthers, Bears, and Deer, were drowned, and found lodg'd on the Limbs of Trees." Many species were "beat in Pieces, and their Fragments (after the Waters fell) were seen in many Places to cover the Ground." In the storm's confusion he was shocked to see that "the voracious and larger Serpents were continually preying on the smaller, as well as those of their own Kind." Even when there were no hurricanes Catesby found southern swamps with their alligators, poisonous snakes, snapping turtles, and predatory fish to be settings of disturbing violence. Southern America seemed a kind of green hell. The violence or lack of morality in wild American nature even appeared to infect humanity. Transplanted Europeans struck him as indifferent to the natural beauty around them. As for the Native people, the whites had somehow managed to seduce them into a perverse participation in the market that was reducing their sacred animals to body parts with commercial value. Catesby himself gloried in the beauty. He found all the rest distasteful.

Catesby's two-volume *Natural History* featured 240 painted plates with descriptive text. Flip through the volumes now and the plates strike you as oddly composed grab bags of insects, birds, snakes, and vegetation, often assembled without a common perspective. Among the vertebrates, he portrayed 109 American birds, 20 snakes, and half a dozen mammals, modest numbers, although Jefferson would use his birds as a rejoinder to Buffon about America's lack of diversity. One of the "patterns" he reported ended up shoplifted by the Prussian naturalist Alexander von Humboldt, whose "isothermal zodiac" was a map of latitude lines encircling the globe showing increasing natural richness toward the equator. The Englishman's *Natural History* does present an America of rich color and beauty. It also struck many as an America too wild, chaotic, and violent for Old Worlders ever to tame.

It was also dangerous. One of Catesby's best paintings and most notable passages dealt with the "Rattle-Snake." America had turned out to harbor many poisonous snakes—copperheads and coral snakes in the hills, water moccasins in the swamps—but the continent's star billing went to its many species of rattlesnakes. Since human primate ancestry left us special neurons in the brain's medial pulvinar for snake detection, we come by serpent titillation naturally. Add to the serpentine visual an aural *buzzzz* of danger and it's easy to see why colonists couldn't resist lurid rattler stories: "The Rattlesnake is reckoned by the *Ab-origines*, to be the most terrible of all Snakes, and the Master of the Serpent-kind," one New Englander wrote. Even Indians wouldn't travel through the woods when it was raining, he said, as moisture silenced the rattles and there was a "fear of being among these Snakes before they are aware." He'd seen an eastern diamondback a "full five Foot and a half long, and as big as the Calf of a Man's Leg."

A friend of Benjamin Franklin's told what it was like to be bitten by a big rattler. The bitten hand "swelled, grew black and stunk." Spots shaped like snake scales speckled the arm from wrist to shoulder. The bite victim's dreams grew frightful. After four months "my arm swelled, gather'd and burst. So away went the poison spots and all. Heaven be thanked."

Catesby couldn't resist. After describing a rattler that had crawled into his bed and, once discovered, struck at everyone who entered the room, he added this story of "the largest I ever saw." It was an eight footer weighing nine pounds. "This Monster was gliding into the House of Colonel *Blake* of *Carolina* [while] the Domestick Animals alarmed the Family with their repeated Outcries; the Hogs, Dogs and Poultry united in their

Mark Catesby, Vipera caudisona. From Catesby's *Natural History.*

Hatred to him," Catesby told his readers, "while he, regardless of their
Threats, glided slowly along." Intended to chill the blood by illustrating
a rattler's casual arrogance, the passage likely misinterpreted the scene.
Rattlesnakes sometimes go weeks between feeding, and when they do
eat their metabolic rate can rise by 700 percent. My experience with rattlesnake behavior from years of living in a West Texas canyon was that a
snake that "glided through" a house paying no attention to people or animals had almost certainly struck a mouse or rat that fled into the house.
The apparent arrogance of Catesby's snake was likely a single-minded
focus on pursuing the walking dead. But no one who read Catesby's passage forgot the image of a gliding, imperturbable monster.

THE MAN MOST RESPONSIBLE for entrenching the American economy
into a free-market form was a Scotsman, born in the coastal town of
Kirkcaldy, near Edinburgh, in 1723. Adam Smith entered the world the
year after Catesby returned to America to begin fieldwork for his *Natural*

History. Smith was twenty-four and studying at Oxford when Catesby's book came out, but all he seemed to gather from it was that America held "objects of vulgar wonder and curiosity." Smith taught for a few years in Glasgow, then—still unmarried—returned home to live with his mother. And to think and write.

Adam Smith's masterpiece was of course *The Wealth of Nations*, which appeared in 1776. It's one of the delicious convergences of history that the "Father of Capitalism" published his call for an economy based on human nature at the very moment the United States saw birth with the Declaration of Independence. But to understand his mind, you need to know that in 1759, exactly one century before Darwin would stun the world with his own breakthrough, Smith had written a predecessor book. Since the famous premise of *Wealth* is the Scotsman's insistence that self-interest is the prime directive of human nature, and therefore an economy based on the freedom to be selfish is "natural," it's intriguing to know about this earlier book. Smith called it *The Theory of Moral Sentiments*. It argued that for psychologically healthy humans, selfishness was softened by an empathy for others. And human empathy, he said, sprang from our deep history as social creatures.

Which makes you wonder. If Adam Smith had been more interested in the natural world might he have grasped that we could well extend empathy to species other than our own? But other than writing that the actual wealth of nations rested on their "animal and vegetable products," in his most famous book the philosopher of the market (and empathy) had no interest in nature. Instead he focused on how self-interest could create a "natural economy" where everyone "is left perfectly free to pursue his own interest in his own way, and to bring both his industry and his capital into competition with any other man." Here, according to Adam Smith, lay true freedom, real liberty. Here, obviously, lay the American future. And justification for what had been happening on the continent for 150 years. Smith's natural economy thus blessed our conversion of nonhuman animals into mere commodities. In Adam Smith's mind it must have seemed inevitable that all those "vulgar" American animals had been waiting across the centuries for Old World humans to show up and start organizing them into wealth. Unfortunately, in the wake of Darwin's insights Adam Smith economics now seems an immature, green-fruit science.

Make no mistake, though, killing wild animals and turning their body parts into an economy didn't begin in America. Across the previous eight hundred thousand years the sewn pelts of wild animals had enabled human migrations out of Africa and around the world. Egyptians, Phoenicians, Greeks, and Romans had all turned the furs of wild animals into trade goods, as had scores of Native American cultures from the moment humans first settled North America. Europeans had initiated a systematic sweep of their own landscapes for beavers, otters, wolves, and foxes, whose tanned pelts and dyed leathers had become clothing and bedding. In the process they had extinguished populations of favored fur animals across much of western Europe. In the decades prior to the discovery of America, Russia had become Europe's latest hunting ground.

There was a method to this madness for fur, and it, too, was ancient. Wearing high-quality pelts or feathers from certain favored species was a key way for humans to demonstrate another directive of human nature: showing status. With animal populations dwindling in the Old World, European monarchs and nobility—and soon, the merchant class—began to restrict the wearing of certain furs to themselves alone. What better way to flaunt one's status, class membership, or au courant fashion than to wear an item of animal clothing the hoi polloi could only eye with envy?

During those first-contact encounters in America, Europeans were endlessly on the lookout both for animal possibilities and for Native willingness to supply furs. Englishman Henry Hudson, sailing for the Dutch in 1609, had returned to Europe with a report that brought a focus to the fur possibilities in America. Ascending a "fine river" some 150 miles into the interior and encountering "polite" Indians willing to barter deerskins and furs, Hudson had set in motion the first resources rush in eastern America. The Spaniards may have found real gold from Mexico to Peru, but farther north America's riches didn't reside in dead metals pulled from inert geology. There wealth came from living creatures.

IT IS EARLY OCTOBER of 2019 and in the picturesque little village of Norfolk, Connecticut, I am having a conversation with the beaver historian and advocate Ben Goldfarb, author of a recent prize-winning book on beavers, called *Eager*. Ben and I are part of Norfolk's fall book festival, and I am soaking up as much as I can from him. Tall, lanky, well spoken, sporting a ponytail and wearing a "Beaver Believer" T-shirt, the former Yalie is telling the audience that their area of New England, whose local tribes were brought into the market for furs by Dutch traders, was one of the first places in North America to lose its beavers. "As it adapted to North America across eight million years," Ben is saying, "the beaver became a perfect animal, an ecosystem engineer." Beavers are partisans of slack water, and from the Pliocene down to Contact, the streams their dams converted into ponds had made much of America an aqueous paradise, a vast world of water storage and enhanced humidity. "The destruction of the watery world created by America's unmolested beaver populations," Ben continues, "was akin to seminal ecological disasters in America like deforestation, hydraulic and dredge gold mining, and the Dust Bowl."

With my wife and old friends who were hosting us in Norfolk, I'd spent the previous day ascending Bear Mountain, at 2,316 feet Connecticut's loftiest peak. We'd looked for beaver sign in every creek we passed, spending most of the day on waters that flow off these verdant hills down to Long Island. So I was arrested by Ben's comment that "the moment of the beaver's disappearance from this part of New England is plainly evident in the composition of geologic core samples taken in Long Island Sound." Those core samples demonstrate the pattern. With beavers gone, almost everything about the hydrology of America changed overnight. Formerly trickling streams now flashed their waters off in torrents that flushed more sediment out to sea. Accumulated sediment in the cores signaled the eradication of beavers upstream.

The beavers that were so central to the animal market were found across almost all of the continent except the swamps of the Southeast and the far deserts of the Southwest. But beavers grew their warmest— therefore most valuable—fur coats above about thirty-eight degrees of latitude. Beavers are rodents, of course, which would imply easy population recovery, but they are giants of the rodent world, reaching lengths of four feet if you count their elongated, paddle-shaped tails. Adults reach fifty pounds and some are twice that size, but for all that they are not prolific breeders. Never mind natural history details, though, because lying against the skin inside their silver-tipped guard hairs, beavers grow a woolly undercoat with an astonishing thirty thousand to sixty thousand barbed hairs per square inch. Pressed and intertwined into a felt, that undercoat created the fabric for the legendary beaver hat, the most desirable, high-status headgear everywhere across Europe and America. As long as the beavers that supplied the raw product lasted.

As they dammed up flowing waters and went about their inoffensive lives, beavers that had been largely unmolested for millennia suddenly found themselves target number one in America's wild-animal trade. They weren't alone. The animal market wasn't especially picky. While streams that produced the resounding *clap!* of beaver tails enthralled the Dutch, the Brit New Englanders, and the French in the upper Mississippi Valley, farther south it was the skins of America's timid whitetail deer—utilized not for fur but for leather—that drew market attention. Whitetails were far and away the most numerous and common big forest mammal in eastern America, a relied-upon source of wild food for both Natives and settlers. Until, unbelievable as it sounds given their abundance, whitetails, too, would all but disappear.

Black bears were another target of the market. The truth is that capitalists devised a way to exploit almost every American animal. Beyond the Appalachian chain, where small herds of bison and elk roamed colonial America, everyone of every nationality shot and killed those big animals, too, although it would be in another setting and decades later that there were cost-efficient ways to get heavy bison robes and elk skins to market. But any animal that had a skin or fur drew market attention. Moose, wolves, bobcats, raccoons, red and gray foxes, river otters, skunks, opossums, muskrats, all the various kinds of American weasels, even eight-ounce tree squirrels, all lost their lives to satisfy some human's profit margin and some other human's sense of style and status.

A measure of just how important the animal trade was in early America is evident from looking at big-picture geopolitics. Every one of the colonizing nations used its fur traders to gain Indian allies and establish land claims against other colonial powers. The British famously used the animal trade to develop an alliance with the Iroquois, or Haudenosaunee, Confederacy, promoting that confederacy's expansion westward in search of more animals to trade, which precipitated their so-called Beaver Wars with the Hurons (Wendats), trade allies of the French. The government in Versailles sent La Salle to explore the Mississippi in 1682 so it could claim the vast lands watered by that river for the purpose of capturing the continental interior's trade in animals. Earth's first great world war—the "French-Indian War," the English colonists styled it—was a geopolitical struggle for that same landscape, again because of its enormous wildlife population. The first of King George III's edicts that put Americans in a revolutionary mood, the Proclamation of 1763, had one primary purpose: to leave Indians in the Mississippi Valley unmolested so they could funnel all that animal wealth to England. Once those animals were gone the tribes would confront the sobering realization that the only thing they had left to trade for Old World goods was their lands. But while they lasted, America's animals were the favored extractive resource.

The whole story begs an answer to one big question. How did the Native people become central players in a war of animal attrition? Why, as an early Dutch writer would put it, did "the Indians, without labor and exertion on our part, provide us with a handsome and considerable peltry trade that can be assessed at several tons of gold annually"?

NATURALISTS IN THE eighteenth century were exploring America in the midst of a biological revolution. In far-off Sweden, another intellectual giant was turning the natural sciences inside out. Surveying an Earth bigger than any prior generation had known, where naturalists were collecting exotic new species nowhere found in the Great Chain of Being, Carolus Linnaeus was clear-eyed enough to see what the moment needed.

Born in 1707, when John Lawson was exploring Carolina, Linnaeus received a college degree at Sweden's Uppsala University and went on to advanced training in natural history in the Netherlands. In that age this meant a medical degree, since most cures came from the *materia medica* of plants. The breakthrough that made Dr. Linnaeus a name for the ages came with the 1735 publication of his book *Systema naturae*. With reports of seemingly endless numbers of new mammals, birds, reptiles, and plants coming out of America and elsewhere, many of them oddly familiar yet different from the life-forms of the classical world, science found a universal method of classifying and naming species essential. Linnaeus concluded he was the man to do what no one since the incep-

THOU SHALT ACKNOWLEDGE THE WONDER 171

tion of the Great Chain of Being had succeeded in pulling off. But rather than spiritual relationship to a deity as organizing principle, the Swedish big thinker's new model of nature rested on seven layers of kinship that all Earth's life-forms shared with one another.

He called those Kingdom, Phylum, Class, Order, Family, Genus, and Species. It was a system for a new way of apprehending reality through science. Its genius was that it offered naturalists from all over the world a way to place new discoveries into a logical template. (A biology professor my freshman year of college imparted a tip for memorizing it. "Just mutter to yourself—*Kingdom, Phylum, Class, Order, Family, Genus, Species*—every time you swig a beer." I've never forgotten the list.) The new bison from America, for example, could now join other creatures with which it was similar. It worked this way: The bison's Linnaean classification fell in the broad kingdom called Animalia, in the phylum Chordata (animals with backbones), in the class Mammalia (animals with mammaries), in the order Artiodactyla (cloven-hoofed), and in the family Bovidae (bovine, or cattle-like). Linnaeus named its genus of closely related animals *Bison*, and since the American animal seemed the characteristic one in its genus, the Swede called its species *bison* as well. So the American buffalo's official name in Linnaeus's new system, using Latin "binomials"—genus name plus species name—was *Bison bison*. Every naturalist around the world now had one precise name for an American buffalo.

As one new edition of *Systema naturae* followed another, Linnaeus eventually classified and gave new names to 4,400 animals and 7,700 plants. To his eternal credit (there were many who objected), Linnaeus did not blanch at how to treat humans in his system. We remained in the same kingdom, phylum, and class as bison. But our order was Primates, our family Hominidae, our genus *Homo*, our species *sapiens*. Hence the binomial, *Homo sapiens*. Wise, but still an animal, just like all the others.

Linnaeus was a narcissist as well as a genius. The first recorded creature of any species is called the "type" animal, and Linnaeus reckoned he was the perfect "type" specimen for humans. He gave himself credit for the discovery of thousands of long-known species he folded into his system, tended to refer to himself in the third person (always and forever a red flag), and repeatedly wrote his own autobiography. Linnaeus did have going for him his reliance on sexual characteristics to classify plants, which outraged the pious, for whom this somehow implied a lewd, licentious Creator rather than a puritanical one. But his adherence to the old religious idea that species had existed unchanged since Genesis was an indication of where the next big intellectual battle over natural sciences would take place.

For American naturalists, the Linnaean revolution took a while to catch on. Jonathan Carver's Lawson-inspired wilderness travelogue of the Great Lakes country, published in 1778 as *Travels through the Interior Parts of North America, in the Years 1766, 1767, and 1768*, at least focused natural history on some other American geography than the tropical South. Carver, a native of Massachusetts, was the first to talk about a mountain range—the "Shining Mountains"—west of the Appalachians. And he included an eighty-five-page appendix to his book that listed the animals, birds, and reptiles he'd seen. Another natural history travelogue, his book sold well. But his amateurish work was innocent of Linnaean taxonomy.

The more professionally trained naturalists working in America didn't rush to embrace Linnaeus, either. Although *Systema naturae* had been in print for a decade when Catesby published, and he corresponded with Linnaeus, Catesby used his own system of classification. Almost inexplicably, so did William Bartram. For all his science couched as poetry—he could write of a murmurating flight of sandhill cranes as "they all rise and fall together as one bird . . . ascending aloft in spiral circles"—Bartram's species names were his own inventions.

As the nineteenth century dawned, natural history still struggled to make sense of America and its creatures. What did American nature have to offer beyond raw materials for an emerging worldwide market? The great French naturalist, the Comte de Buffon, author of the nature classic, *Histoire naturelle, generale et particuliere*, thought he knew. A man whom the wife of one of France's ministers described this way—"M. de Buffon has never spoken to me of the marvels of the earth without inspiring in me the thought that he himself was one of them"—Buffon had read Catesby, du Pratz, and Bartram, which meant he was reading about the Carolinas, Louisiana, and Florida. Knowing little else about America, Buffon assumed the semitropics stood for the continent as a whole.

Thus Buffon described an America where "the air and the earth [are] overloaded with humid and noxious vapors [that] are unable either to purify themselves or to profit by the influence of the sun, who darts in vain." As in all tropical regions, insects and snakes swarmed Buffon's

America. Yet paradoxically, the winters were much colder than Europe's, another explanation for why America (he asserted) had little of the Old World's biological diversity. The continent held nothing remotely like Africa's megafauna, no elephants, no lions or anything even close. America's animals, in fact, were small and timid: "Very few [are] ferocious and none formidable." It led the French naturalist to a theory that America was likely the last of the continents created, by which time life's energies had run soupy-thin.

The French and the Americans have long specialized in insulting each other, but at least Buffon set naturalists a goal. The next group of them would be motivated to observe, write about, and paint American creatures that would put the lie to Buffon's assessment.

WEST OF THE GREAT LAKES, on the edge of the prairies, Siouan-speaking performers assembled bands of their people, gathering them around big fires that sent smoke and fiery embers spiraling into the cold winter nights, and told this story.

Coyote had spied a beautiful young chief's daughter he badly wanted to bed. As beautiful chief's daughters tend to do, she ignored him. But hearing that along the eastern seashore there were now white people who possessed many wonderful things, Coyote used magic to go there and returned with four objects no Indian had ever seen. Then he set up a lodge in the girl's camp and over the next four nights pounded and banged away as if he were a mad inventor. The first morning he showed the girl a choker of brightly colored glass beads he'd acquired from the whites. As Coyote guessed, she was dazzled and offered him a kiss in exchange. The next day he produced an iron pot, better for cooking than anything she had ever seen, for which she let him fondle a breast. The third day it was a red wool blanket with stripes in several colors, and to possess that she let Coyote feel her buttocks. Finally, on the fourth day, Coyote produced a beautiful mirror. After observing herself in it for several long moments, the chief's daughter so desired it that she let Coyote look between her legs.

Upon which Coyote frowned, "Too bad, too bad that you've been made upside down. That really should be fixed." The beautiful chief's daughter took her mirror home but thought long and hard about what Coyote had said. If her sex truly did need remaking, who else to do it but Coyote, who'd made so many magical new things?

"Go, and fetch Coyote and do it quickly," she told her girlfriend.

However ribald, or by modern standards casually sexist—and Coyote stories can be both—this was a Coyote tale that spoke to tribal people about a major question of the time. Why *were* Native peoples willing to trap, kill, and skin the animals Europeans and later Americans seemed so desperate to possess? Various writers since have offered various answers, one even arguing that pelts came from a Native "war against the animals," prosecuted for religious reasons because beavers and other animals had allowed diseases to sweep through Native villages. The evidence—including from Native people in the form of stories like this one—more realistically demonstrates that despite cultural differences, tribal peoples shared the same motivations all humans do. When traders approached them with beautiful new goods that could serve as markers of status, or metalwares infinitely sharper and longer lasting than flint tools, eventually with firearms to confer advantage or protect yourself—and ultimately with the one trade good, alcohol, for which there was a bottomless psychological demand—how else to respond than by offering in trade the one thing you have that they seem to want?

The outcome was that Native people became one more labor force for colonizing America. Not an enslaved one like Africans working colonial crops, but laborers ensnared by the market. There was also this difference: Indians were also consumers, and in an America where traders from different countries competed for tribal business, that gave Native people power. It wasn't a power that allowed them to escape a spiraling dependence on the goods of the industrial revolution, but it gave them preferences for French gunpowder and French generosity, for the quality and low prices of English goods, and for the willingness of Americans to trade guns and alcohol. Consumerism allowed them to play Frenchmen against Englishmen, or Americans against Spaniards and Russians.

Native consumer power didn't rescue America's animals, though. While the details varied from place to place and tribe to tribe, figures like the following were common. For a single English blanket, Cherokees and other southern nations trading in Charleston in 1716 killed and prepped the skins of 16 whitetail deer or beavers. Five animals lost their lives for a shirt, three for a hatchet, eight for a yard of cloth, 20 for a pistol, and

30 for a long gun. Prices had dropped some by 1750. A long gun cost the Choctaws only 16 deerskins by then, and blankets only asked for the lives of eight whitetail deer. Prices for small products like scissors or knives or strands of beads required only a single deer to die. Native people probably regarded this trade as gift exchanges rather than global-market trade. But a little simple multiplication conveys the biological cost. In the year 1750 the Choctaws bought 1,700 blankets, along with 150 muskets, 2 tons of gunpowder, 18,000 flints, and 2,500 trade shirts. And 200 pounds of colored drinking glasses. For the blankets and muskets alone they sacrificed the lives of sixteen thousand whitetail deer.

Historians believe that by the 1780s the Choctaws—only one tribal group among scores—were killing one hundred thousand whitetails for the market every year. Even prolific animals like whitetail deer could not take that kind of pressure. When the herds began to thin, the Choctaws did just as the Native beaver hunters did farther north. They began moving westward in search of country where deer weren't yet hunted out. Humans abandoning hunted-out country and searching out places with unexploited animals was, of course, a reprise of what hominins had been doing over the previous eight hundred thousand years.

As the animals-for-industrial-goods market evolved, in American fashion the desired items drifted more and more in the direction of luxury or entertainment items. By the time the Choctaws had hunted deer to local extirpation, Native people in the Ohio Valley were trading the beavers, otters, wolves, and deer they killed for things like sugar, tea, and chocolate, for brass candlesticks, trunks, tablecloths, flannel, Irish linen, ruffed shirts, silver brooches, harmonicas, and petticoats. And there's no skirting it: among people who valued mind-altering experiences, liquor became the clinching argument in the animal trade. As one Choctaw leader explained, liquor was a lot like a desirable woman: "When a man wanted her—and saw her—He must have her." American consumption was assuming a recognizable form.

It's a mistake to imagine this colonial tragedy without understanding that there was as well a working-class set of settlers, of several nationalities, who were engaging the cash economy by doing their own best to destroy the continent's animal life. Many of them were doing so as if they were players in a colonial killing contest, a race of extermination with prizes at the end. When Americans pushed beyond the Appalachians into

Kentucky and the Ohio Valley, they not only had deer to pursue, they had finally reached the range of significant numbers of elk and bison. Animals that large, moving in herds, no less, excited a wild lust to kill. As Richard Henderson, a companion of Daniel Boone's in founding Boonesborough, wrote: "We found it very difficult at first and indeed yet, to stop the great waste in killing meat." Some among them, "of wicked and wanton dispositions, would kill three, four, five, or ½ dozen buffaloes, and not take half a horse load from them all."

Boone himself introduced a law in their assembly to stop this "wanton destruction of game." As was the case elsewhere, however, Americans in Kentucky regarded restrictions like Boone's law as a violation of their freedom. Within a mere two years the region around Boonesborough was destitute of the big animals that had drawn settlers there in the first place.

There was always a repeated refrain, that waterfall of excuses: someone else was to blame. Beyond the Appalachians the advancing Americans were far enough west to hear stories of French hunters plying the Mississippi, the product of their hunts bound for the port of New Orleans. Surely it was the *French* who were wiping out wildlife west of the Appalachians. Father Jacques Marquette, descending the big river to Illinois in 1673, had reported that wild-animal numbers in the valley were phenomenal: "cows, stags, does and turkeys are found there in much greater number than elsewhere," he wrote. Father Louis Hennepin had said that in Louisiana, prior to the market the tribes had never been able to exterminate bison "for however much they hunt them these beasts multiply." The French thought buffalo and whitetails alone would yield every year a return of more than 2.5 million *livres*.

One New Year's Day in upper Louisiana a French force toasted their officer this way: "In the name of the King, all you cats, bears, wolves, buffalo, deer, and other animals . . . shall recognize our commander as your governor, and you shall obey and serve him as he commands you." His orders, presumably, were that they surrender and die forthwith. But everyone of every nation understood: there was much more of America toward the sunset. Who knew what kinds of animals, how many, what kind of wealth from them might be out there?

MUCH IN THE MANNER of Linnaeus giving Latin binomials to thousands of plants and animals known since antiquity, then crediting himself with their discovery, naturalists who described and named America's creatures were appropriating an ecology Native people had known for ten thousand years and more. Just as Indians commonly knew every animal and bird where they lived, Enlightenment naturalists hoped to do the same. The two most famous of them called on the artist's brush and palette to set a complete class of American creatures before the world. They were the country's bird painters, one famous, one whose name you'll puzzle to recognize.

The little-known one was the first in time, which reverses the usual order of things in science. Alexander Wilson emigrated from Scotland at the age of twenty-eight with no particular purpose. A dour, lanky loner, Wilson found his calling when he realized he had come to a place "filled with strange birds." Almost none were familiar to him, and all were "much richer in color" than the birds adapted to monotone Scotland. The year was 1794, and soon finding himself in Philadelphia, Wilson managed to coax one of Charles Willson Peale's sons, Rubens, to give him pointers on painting. The Peale family had founded the major museum of natural history in the young United States, and Wilson's connection with them opened doors. One of those doors landed him a friendship with William Bartram, by then in his sixties and largely confined to the family estate outside Philadelphia. By 1803, at Bartram's urging, Wilson decided that he was "about to make a collection of all our finest birds."

He had the Peale brothers, several of them painters, critique his efforts, and he brought to his quest the energy for travel and the will to work. Alexander Wilson now embarked on a project that consumed the rest of his life. He called it the "American Ornithology," and by 1807 he had its master plan in his head. He would do ten volumes, each including a dozen color plates with several different birds on each plate. To raise the money to prepare the publication, he committed to traveling the country selling subscriptions for the entire book. And the painter himself would also supply the text. Wilson's text has become the best thing about his book. He wrote stories about bird species that were numerous and stood as iconic symbols of America when he roamed the eastern woods, shooting and painting them. Now many are lost and observations like Wilson's are our only time machine for resurrecting them.

Alexander Wilson, Passenger Pigeon, Blue-mountain Warbler, and Hemlock Warbler. From Wilson's *American Ornithology*.

He was traveling near Wilmington, North Carolina, when he shot the adult ivory-bill he painted for *American Ornithology*, breaking the giant woodpecker's wing and rendering it flightless. On being seized, Wilson's captive "uttered a loudly reiterated, and most piteous note, exactly resembling the violent crying of a young child." These gigantic black-and-white woodpeckers, with two-and-a-half-foot wingspans, red mohawk topknots (on males), and the ability to rip limbs bare of bark with their slate-white bills, astonished everyone who saw them. It wasn't just their size, it was their calls, which ranged from percussive *kent*s to ear-splitting shrieks. Ivory-bills sounded like "toy trumpets in the forest," some said. The Caddo people called them "horned screamers." The only recording of an ivory-bill's calls, from 1935 Louisiana, confirms all these descriptions. Along with their vocals, their double *bam-BAM* bass-drum pounding for grubs in the bark of the Southeast's old-growth forests was one of the signature woodland sounds of early America. The naturalist had wounded one of the continent's most impressive ornithological wonders.

Wilson left the wounded bird in his hotel room and returned to find daylight shining into the room, the bed littered with chunks of plaster from a hole the size of a fist the ivory-bill had hammered through the wall. Securing his captive to a mahogany table with a string on its foot was no improvement. The woodpecker turned the mahogany table into toothpicks. Wilson spent three days with this ivory-bill, offering him various foods, all of which he resolutely refused. His cries grew weaker as Wilson painted him. "I witnessed his death with regret," he tells us. He also tells us that it took days for his arms to recover from all the stab wounds he'd taken from his frightened giant.

Another of North America's most dramatic lost birds traveled with Wilson for weeks. Outside Cincinnati he had fired into a flock of Carolina parakeets. We twenty-first-century Americans are habituated to there being no wild parrots flitting through our trees, but then extinction can do that. So reorient yourself about what you should in fact be seeing in southern, midwestern, and some western forests if we'd had the presence of mind to save these brilliantly colored birds.

Yellow and green in plumage, traveling in bright, noisy flocks, America's native parrots measured more than a foot from head to tail, with wingspans of nearly two feet. Adapted to tough fare—cockleburs were a favorite food—the big parrots took a liking to maize as soon as Native people began growing it. So their switch to eating corn from colonial farms could have been a natural move, although we're not at all certain they actually did so. Ivory-bills were birds of ancient old-growth forests and bayou cypress, and logging those trees put one kind of bull's-eye on them. Carolina parakeets somehow struck settlers as too familiar. Their bull's-eye was more literal and direct.

Wilson's luck with captive birds was not good. He did get the parrot he'd winged, whom he named "Poll," to eat the cockleburs he offered, so there was that. Carried in a handkerchief in his coat's pocket, Poll was with Wilson for a thousand miles. She was there for his fabled meeting in Kentucky with John James Audubon, when Wilson first learned that someone else was painting America's birds. She was in his coat when the naturalist visited the inn where Meriwether Lewis had committed suicide the year before. Once he put another Carolina parakeet he'd wounded in Poll's cage and the two "nestled as close as possible to each other." When the new bird died Poll was inconsolable until Wilson put a mirror in her cage, whereupon "she was completely deceived." Maybe.

Poll got her revenge "by cutting and almost disabling several of my fingers with [her] sharp and powerful bill," Wilson wrote. Even with her damaged wing, she still tried for jailbreaks at every opportunity. Sailing through the Gulf of Mexico for home, after Poll had traveled with him for weeks, Wilson woke one morning to find her floating facedown, wings outspread, in the Gulf's murky waters. She had drowned herself attempting a final, fatal lunge to freedom.

Wilson never had a captive passenger pigeon to observe close-up, but he shot many for his paintings and near Lexington visited a roost and watched a pigeon flight for nearly five hours, at the end of which "the living torrent above my head seemed as numerous and as extensive as ever." He guessed he had seen two billion pigeons between one thirty and six that afternoon. He likely could not have imagined a future in his adopted country when every last one would be gone.

Wilson was a superb observer but he never quite mastered painting. Eventually he became skilled at getting down the markings and nuances that distinguished one species from another. But the truth is that the "Father of American Ornithology" did full-body bird portraits that rarely rose above field-guide quality. The details were precise but he never was able to snag from nature the vibrancy that so thrilled him. Despite setting down a wealth of information few others possessed, a big picture about birds also eluded Wilson. He came closest to something profound when he realized that the wood thrush was too similar to the European thrush to imply separate creations. The question was why the two thrushes, and other American birds that had near-but-not-exact twins in the Old World, were different. Maybe "at some time after the creation," Wilson speculated, the European thrush flew across the Atlantic and in America "became degenerated by change of food and climate" until it somehow turned into our bird.

But what mechanism in nature could enable a species to change like that?

⁓

OF ALL AMERICA'S early naturalists, it was John James Audubon who best answered the Old World's fantasy imagination of what a European-become-American nature man should be. With wavy, shoulder-length chestnut hair, gray eyes, and the easy grace of an athlete, Audubon remained lean and cut a striking figure all his life. Speaking English with an accent—a visitor once asked him, "You are a Frenchman, sire? You look like a Frenchman and you speak like one"—Audubon was the very definition of Romantic charisma, a rough-hewn New World Byron. As his own brother-in-law said of him, the man was not just a painter of nature's subjects, or noteworthy for the elegance of his figure, he was also an expert with guns, an excellent swimmer, and a fine fencer and dancer who had a way with dogs and horses. And with women, like the New Orleans beauty who asked to pose nude for him, or women of high station all over Europe. Like Byron, Audubon actually shared many traits with the Native American Coyote deity. He was talented and mesmerizing but, like Coyote, also vain, jealous, and too often ungenerous.

Eventually one of early America's celebrity imports/exports, John James Audubon was the out-of-wedlock son of a wealthy Frenchman, Captain Jean Audubon. Audubon spent his entire life hiding the actual facts of his birth, but he came into the world in the Caribbean, his mother the captain's young mistress, Jeanne Rabin, who died soon after giving birth. Audubon always denied his mother, claiming he was the offspring of a Spanish woman of good breeding, not the result of his father's fling with a peasant chambermaid. Raised in Nantes along with another of Captain Audubon's illegitimate children, he dodged Napoleon's draft at age eighteen and fled to America in 1803. By the time he arrived in Pennsylvania he had anglicized his name and promptly fell in love with a well-educated young neighbor named Lucy Bakewell.

Audubon was captivated by drawing and painting as a child. Picturing nature was his first love, but for all his later mastery he was self-taught. He claimed to have studied under Jacques-Louis David in France, but there exists no evidence of it. In 1808, when he quit his father's farm and fled with new bride Lucy to Kentucky and then Louisiana, painting seems to have hovered in his mind like a beckoning planet in the night sky. The couple had two sons—Victor and John Woodhouse—by 1812. So to support them Audubon tried farming, then business. He lost everything the family had accumulated in the great financial crash of 1819, and that disaster led him to try, at age thirty-four, to become a full-time painter. It was the literal fulfillment of the notion of art as an act of desperation.

Audubon's art interest had always been painting birds, but therein lay a problem. By 1819 all of the volumes of Alexander Wilson's book were out. Wilson had passed away, but his work was widely respected. Of their meeting when the Scotsman was in Kentucky in 1810, Audubon's version had it that after Wilson proudly showed his work, Audubon had laid some of his own paintings on the store counter, stunning Wilson into dismayed silence. Wilson's account barely mentioned Audubon. No friendship ever developed between them.

Audubon intended his opus, *The Birds of America*, to be a comprehensive book that would portray every bird in the United States life-size. That meant the naturalist had to find, observe, collect, study, and paint every species, which brings up an uncomfortable topic. All these early naturalists were shootists. To capture in paint the iridescent color shadings of a hummingbird's wing, the bird had to be in hand. For the bird to be in hand, it had to be dead. Audubon certainly observed and wrote about living birds, but to paint he needed specimens, which he wired in lifelike poses and tried to render rapidly with either hand, and sometimes both, before death glazed their eyes and dulled the vibrancy of their coloring.

After all the fieldwork, then he had to find an engraver and a team of printmakers. Then enroll subscribers while he was writing the text that would become the final book. When no publisher in either Philadelphia or New York would take on such a project, in 1826 Audubon hauled four hundred of his bird drawings to England. Here, at least, the handsome American was an immediate sensation, a "long-haired Achaean," someone wrote, with locks spilling down his back. Once the subscribers tore their eyes from Audubon, his gorgeous paintings—with the birds in dramatic, animated poses, and every one set in "a landscape wholly American, trees, flowers, grass, even the tints of the sky and the waters"— worked their magic. He found his publisher in London and even merited advance praise by the French, the Parisians scarcely expecting such genius from America.

Somehow Audubon completed all these tasks on both sides of the Atlantic in just twelve years. When it finally appeared in 1839, *The Birds of America* didn't just present beautiful birds shown in their habitats. Audubon had painted them life-size, a showstopper. The book was stunning, at thirty by forty inches the size of a small house window. Even so, the biggest birds, like whooping cranes, bald eagles, and great blue herons, had to strike unusual poses to fit the page without cropping legs or wings. The final version contained 435 plates and, with the 85 western birds whose

John James Audubon, Ivory-billed woodpecker. From Audubon's *The Birds of America.*

skins he acquired from other naturalists, enumerated 489 species of American birds. The world, especially the European world, where Audubon traveled, dined, partied, and sold subscriptions, was utterly entranced. No less than famous Parisian naturalist Georges Cuvier called *The Birds of America* "the greatest monument ever erected by art to nature."

ALL THE WHILE Catesby and Bartram, Wilson and Audubon were leaving their monuments to American nature, the snapping of metal traps, the hiss of netting, and the rattle of gunfire never let up. The United States was free of England now and was acquiring the French empire in the Mississippi Valley and soon would rip away all of Spain's original settlements in the West. But the animal slaughter didn't stop, it just migrated. America's naturalists had shown by example a different, appreciative way of relating to the new country's wildlife, but as Alexis de Tocqueville observed of Americans, most were too busy tearing up the continent to notice. Among the eastern Indians, shamans like Neolin of the Delawares and the Shawnee prophet Tenskwatawa, the twin brother of the great diplomatic leader Tecumseh, begged their followers to give up the market hunt for animals, to forsake white trade goods and relish their own productions. As with most messages preaching anti-consumerism, that was a hard sell. The majority of tribal peoples just moved farther west, where there were still huntable animal populations.

In the introduction to his famous *Travels* William Bartram had written about a pursuit of two black bears with a companion. When his companion shot one of the animals dead, Bartram was shocked to see that the other bear "approached the dead body, smelled, and pawed it, and appearing in agony, fell to weeping, and looking upwards, then towards us, and cried out like a child." Bartram was moved by compassion to beg the hunter not to inflict a second "cruel murder." But his entreaties to his companion were to no avail, since "by habit he had become insensible to compassion towards the brute creation."

Late in his life, in an essay he never published, perhaps thinking of this bear story America's first great native-born naturalist wondered whether, unrestrained, we humans wouldn't destroy "the whole Animal creation." West of the Mississippi River in the nineteenth century, we were about to find out.

CHAPTER 6

THE NATURAL WEST

On a gray Tuesday in November of 1805, with a chill wind scattering autumn leaves into the puddles of Washington's muddy streets, White House staff admitted a caller there for a private dinner with the president of the United States, Virginia planter Thomas Jefferson. That November 16 evening Thomas Freeman must have felt his future was made. Jefferson was about to offer him a plum appointment, the leadership of one of his prized explorations into the Louisiana Purchase, one of the most fascinating parts of the globe for scientific study. With Meriwether Lewis and his party already on the shores of the Pacific, Jefferson was turning to Freeman, a civilian astronomer who'd emigrated from Ireland, to lead an exploring party into southwestern America.

Jefferson called this new probe the "Grand Expedition" and he was aiming it at the Red River, which natural history titan Alexander von Humboldt assured the president would take American explorers into vast deserts and the southerly ranges of the "Shining Mountains." Jefferson had canvassed a number of scientists, including Humboldt, who'd been gathering information about the Southwest and he was fascinated. The young United States had several geopolitical reasons for exploring Louisiana, but at heart Jefferson was a naturalist who had dug fossils and written his own Lawson-type book about Virginia. His informants told him "wonderful stories" about volcanoes and tigers and herds of wild horses among innumerable buffalo and wolves. He knew that camels, "the llama or paca of Peru," still existed in similar country in South America. And since there was already evidence of elephants in America—by then Charles Willson Peale had laboriously (and badly) reassembled the skeleton of one for his museum—elephants might still be in the West, too.

Meriwether Lewis had shipped enough reports and specimens back from the Missouri River that science was already buzzing about animals and birds never seen in the eastern states, so Jefferson's hope was that William Bartram himself would accompany Freeman. In his late sixties, though, Bartram instead promoted Alexander Wilson as Freeman's naturalist. Reluctant to man his expedition with two immigrants, Jefferson decided to choose a young Virginian whose family he knew well. Thus did a Penn medical student named Peter Custis become the first scientist trained in an American university to win a posting as a naturalist to the West. Congress had come up with twice the funding for this expedition as it had for Lewis and Clark's party, so when Freeman stepped into the Washington night holding seven pages of exploring instructions written in Jefferson's clear handwriting, he was ready—as he wrote a friend—to "stick or go through" wherever the president pointed him.

Jefferson's instructions, which Freeman must have scanned repeatedly, still exist in the Library of Congress. They included three crucial lines that also appeared in the exploring instructions the president had given Meriwether Lewis. They were simple and direct. "The following objects in the Country adjacent to the rivers along which you will pass will be worthy of notice . . . the animals of the Country generally and especially those not known in the maritime states. [And] the remains and accounts of any which may be deemed extinct."

The western half of North America was the country, and the 1800s was the century, that ultimately answered many fundamental questions about the distinctive biological life of America, and how U.S. citizens would respond to the abundance and diversity they found there. With the Louisiana Purchase Jefferson's administration had effected a continental future for the country. Like a stone rolling down a mountain, the U.S. in the 1800s would claim and buy and seize much more of the continent, eventually incorporating everything from southwestern deserts to Alaskan tundra. What had been and could have remained Native America, or French, British, or Mexican territory, became part of the United States.

For the West's wild animals, that implied an extension of the exploitation already unfolding in the East and South. The rapidly growing nation was about to answer William Bartram's question when he'd watched his companion gun down those multiple black bears "by habit": How would western America's animals fare under an economic system of self-interest built in part on the commodification of dead animals?

Then there was that final line in Jefferson's natural history instructions. Was extinction real? Were camels and elephants still out there, or had they somehow vanished from the continent? Could living creatures disappear entirely? If so, what force or forces could possibly push bird and animal species to wink out like dying embers in the night? For all their promise, Thomas Freeman and Peter Custis unfortunately didn't lend much assistance with those questions. In 1806 they did go west and they explored up the Red River for several months. But their expedition reached a point only 650 miles upriver, short of the great deserts and with the Rockies still out of sight beyond the curve of the earth. A Spanish army suspicious of American designs on its colonial territories intercepted them and turned them back short of Jefferson's goals. Freeman neither stuck nor went through. In fact he bounced, right out of American memory.

॥

URGING WESTERN EXPLORERS to look for the remains of extinct creatures was proof that Jefferson was a man of science, willing to change his mind when the evidence called for it. Extinction was a new idea in the early 1800s, challenging to the dominant view of the world. Most Americans yet held to the Great Chain of Being with all its premises, among them the impossibility of extinction. But confronting the excavated remains of bizarre creatures no human had ever seen alive, scientists like Benjamin Smith Barton in Philadelphia (who trained Peter Custis at Penn) came to the conclusion that extinction was real and that the Great Chain possessed a fatal flaw. Once Jefferson had vigorously resisted extinction. By the time he wrote his exploring instructions for Lewis and Freeman he'd obviously decided he'd been wrong. Nonetheless he hedged his bets, hoping that mastodons or camels were still somewhere out beyond the Mississippi and that Spanish, French, and British travelers had just missed them.

The possibility of extinction raised one other issue for natural history. If creatures that once lived on Earth had become extinct, however mysteriously that had happened, might species still *alive* disappear? The emerging answer to that question was going to push American wild animals into the public eye for the next two centuries. But in the Enlightenment Age, despite the obvious decline of animals like deer and beavers, few people initially worried about losing living animals entirely. Likely no one in early American history witnessed the destruction of wild creatures on the scale that John James Audubon did. One day on Louisiana's Lake St. John, Audubon watched as two hundred hunters wiped out what he estimated were 48,000 black-billed plovers. The best shooter destroyed, by actual count, 756 birds. But Audubon never speculated about the impact days like this had on shorebird populations. Even after watching all the wanton, wasteful methods Americans utilized to slaughter passenger pigeons, the painter still contended those birds *could* never become extinct. "Persons unacquainted with these birds might naturally conclude that such dreadful havoc would soon put an end to the species," he wrote. "But I have satisfied myself, by long observation, that nothing but the gradual diminution of our forests can accomplish their decrease, as they not infrequently quadruple their numbers yearly, and always at least double it." Even in an age of declining animal populations and escalating scientific understanding, Audubon hewed to the old notion of "the astonishing bounty of the Great Author of Nature in providing for the wants of His creatures."

As the nineteenth century proceeded, Audubon's idea seemed valid based on reports from the West. The Natural West—the sunset half of the continent the United States was incorporating—was a country of unending oceans of grass, of towering mountain ranges draped like exposed skeletal backbones over sere deserts. On its farthest edge lay a thousand miles of crashing seacoasts that merged into a rainforest in the continent's northwest corner. Ecologically the West offered a bounty of wild animals of diversity and numbers no Old Worlder had seen in ten thousand years. Although it was a very old country humans had long lived in, at the start of the 1800s America's West seemed a planetary touchstone of everything wild and exotic, and who even knew what unimagined mysteries it held?

The first of the European nations to feel that enchantment was Spain. Although Spain's scientific revolution had lagged initially, King Charles

IV spent more on scientific investigation than any monarch of his time. In 1789 Charles's government launched a major, six-year exploration across half the globe. Alessandro Malaspina commanded this maritime expedition, with a distinguished cast of scientists in tow to examine the coasts, ports, and missions of most of Spain's possessions, including those from California up the coast to Alaska.

By Malaspina's time Franciscan missions and associated ranchos already sprinkled California from Los Angeles to San Francisco. In a reprise of what happened on the Atlantic side of the continent, though, an apocalypse of Old World diseases was in the process of wiping out some 75 percent of California's Native population. All hunter-gatherers, Native people had been numerous across California when the first Spanish settlers arrived in the 1770s to found their missions and presidios. As with Cabeza de Vaca's experiences along the Gulf Coast 250 years earlier, the initial Spanish settlers in California frequently faced starvation for want of wildlife. But as the great continental smallpox epidemic of the 1780s swept through California, the collapse of the Indian population sent animal numbers soaring.

By the time his expedition visited Carmel Mission on Monterey Bay, Malaspina's scientists were seeing a California rebounding in wildlife of every kind. Giant brown bears ranged from mountaintops to the seacoast, where they joined huge black vultures in demolishing whale carcasses washed up on the beaches. There were stories of lions and of wolves of a small kind, the *coyotl*, that attacked Spanish sheep and goats. The explorers also heard how rapidly horses and cattle brought up from Mexico ran wild in the rolling golden hills inland. Northward along the coast to the Tlingit and Haida villages the Spanish scientists couldn't miss the staggering numbers of elk on the Marin headlands, or the vast numbers of sea otters and fur seals that were already becoming targets of the global market.

It was a gravid moment for Old Worlders. A newly found unknown world was at hand.

⁄⁄⁄⁄

A YEAR BEFORE Thomas Freeman had his private dinner with Jefferson, his predecessor explorers were on the verge of big new insights. It was the early fall in 1804 and for four months twenty-eight-year-old Meriwether Lewis and thirty-two-year-old William Clark had been leading their "Corps of Discovery" up the Missouri through a setting and among animals similar to those of Virginia, where both grew up. They had found whitetail deer, black bears, and elk "numerous" on the lower river. But late that summer—geographically they were roughly where Nebraska, South Dakota, and Iowa now meet—the country began to change. Trees were shorter, horizons were more distant, and on the hills above the river they no longer had views of continuous forests. What their French hunters called "prairies" were replacing views of trees. The air was growing drier and the splines and pegs on their keelboat started falling out of their fittings.

In the next three weeks Lewis and Clark passed into the American version of Asia, from which so many of the West's animals had come, or Africa, whose plains those horizontal grasslands resembled. The transformation happened between about ninety-seven and ninety-nine degrees west longitude, some two hundred miles due west of the Mississippi River. If our modern cities had existed then, the zone of change demarcating Appalachia America from Laramidia America would have run, roughly, through Austin, Fort Worth, Oklahoma City, Wichita, Fargo. Today some of those towns are more intriguing than others, but one thing they all share is a sense of being on an edge. I grew up a little farther east but I still recall the powerful feeling of some undefined but profound alteration looming just beyond my western horizons. The feeling was always there and it excited me enough that as soon as I could drive a car I at once went west to see.

Lewis and Clark got to experience the change entire. Instructed by Jefferson to seek out any and all new life-forms, on August 23 the party killed and dined on the first bison most of them had ever seen. By September 7 they were among their first prairie dogs, colony squirrels that made nests in burrows in the ground rather than in trees. A week later there was an even more remarkable encounter, a "Buck Goat of this countrey . . . more like the Antilope or Gazella of Africa than any other Species of Goat." The pronghorn was one of America's original contributions to evolution, the striped thoroughbred of the West. Three days later the explorers encountered "a curious kind of Deer of a Dark Gray Colr . . . the ears large & long" with a strange, pogo-stick gait. They'd seen a mule deer, familiar to Spanish settlers in New Mexico and California but unlike any deer seen by Americans from east of the Big River. This came on the same day they saw "a remarkable bird of the Spicis of Corvus." It was their first sighting of a black-billed magpie, "a butifull thing," Clark wrote.

The day after that, in the vicinity of present Chamberlain, South Dakota, the explorers encountered another American original. All that September they'd been seeing what they assumed was some new kind of fox, and on the eighteenth William Clark finally shot one. The sleek canine was no fox, but it wasn't the eastern wolf the Americans knew, either. They decided to call it a "prairie wolff." Decades later naturalists would discover that this was the same animal many Native people and Spanish settlers knew as the "coyotl," the avatar of the western Indian deity Coyote and another special contribution from North American evolution.

Lewis and Clark had passed through the portal into North America's version of the Serengeti. The analogy isn't specious. Despite Jefferson's hopes, mastodons, mammoths, and camels no longer roamed the West. But its historic bestiary preserved poetry and spectacle, with thronging masses of bison playing a role similar to that of East Africa's migrating wildebeests, pronghorns resembling nothing less than antelopes or gazelles, gray wolves filling the niche of wild dogs, coyotes doing an almost exact impression of jackals, and an ancient American animal that was now fast returning—escaped wild horses—functioning like zebra herds. Africa had retained its lions and elephants, its hyenas and cheetahs while America had not. But America's Serengeti had another king of beasts, the grizzly bear. As Lewis and Clark were about to discover, this formidable bear played a lion-like, almost godlike, role in the West.

The open country was filled with a cacophony of sound. Prairie dog towns loaded aural space with chirruping and trilling. Red-tailed hawks screamed overhead, bull elk with racks heavy enough to affect how they moved whistled haunting challenges. Bison muttered and bellowed with a sound that resembled faraway continuous thunder. Coyotes yipped and wolves howled the continent's original national anthem morning and evening and throughout the nights. And grizzly bears—well, grizzly bears had a repertoire of sounds, some hair-raising beyond all experience. These Americans hadn't heard them yet. When they did they would never forget.

They got their first intimation they were in grizzly country on October 7, 1804, when they saw bear tracks three times the size of their footprints. But the fascinating animal of the moment, in part because it was so unprecedented, was the pronghorn. It struck them as an almost inexplicable creature. Having already shot a buck and now hoping to collect a female for science, Lewis stalked a harem of seven only to have them whirl away and disappear. In another few minutes he saw the same band three miles distant. "I had this day an opportunity of witnessing the agility and superior fleetness of this animal which was to me really astonishing," he wrote in his journal. "When I beheld the rapidity of their flight along the ridge before me it appeared reather the rappid flight of birds than the motion of quadrupeds."

The Americans had no way of knowing that in pronghorns they were seeing one of the best extant expressions of deep continental evolution. Pronghorns are Pleistocene ghosts. When Meriwether Lewis stood amazed at their speed, their predators were gray wolves and coyotes, neither of which can run much more than forty miles an hour. Yet with their light bones, broad nostrils, and windpipes delivering turbocharged oxygen to outsize lungs and hearts, 120-pound buck pronghorns can top fifty-five miles per hour and slighter does can hit sixty-five to seventy miles per hour, giving us a good idea just how fast American cheetahs could run ten thousand years ago. Selecting males to sire their fawns, female pronghorns still picked the fastest runners, then bore twin fawns—an heir and a spare—on the evolutionary assumption that something would get one of them. In the manner of gazelles in Africa, they even engaged in a social behavior called "selfish herding," crowding lower-status members to the outside in case a predator picked somebody off. To detect danger at great distances they had gigantic eyes. Yet the predators of adult pronghorns had all vanished in the Pleistocene extinctions! No wonder naturalists were dazzled.

On October 20, a week shy of the Mandan villages where they planned to spend the winter, and in the midst of "great numbers of Buffalow Elk & Deer, Goats," the Americans got their first look at a "white bear." In a premonition of the relationship they would soon forge with grizzlies, they shot it. But the bear seemed to shrug off the hit and escaped. That winter of 1804–5 they heard many grizzly stories from their Mandan hosts, whose awe of and respect for the giant bears had the Americans snickering in private.

Heading up the Missouri in the spring of 1805 the Americans got their

American pronghorns. Photograph courtesy Ed Breitinger.

first sense of grizzly natural history in what is now North Dakota, about halfway between today's Minot and Williston. Notice, first, that these cities are far out on the Great Plains, which in fact was a primary grizzly range in the West. The party found drowned bison washed up on the riverbanks as well as "many tracks of the white bear of enormous size." They were seeing proof that grizzlies scavenged buffalo casualties. Over the next two weeks, as the expedition pushed upriver toward the present Montana border, bear tracks in the river mud increased, but the only bears they glimpsed "are at a great distance generally runing from us. . . . The Indian account of them dose not corrispond with our experience so far," Lewis confided to his journal. That was about to change.

On the morning of April 29, walking along the shore, Lewis encountered two grizzlies and shot both. One escaped but the other, a young male, at once came after the explorer, pursuing Lewis until more shots downed him. This young animal was the first grizzly bear Lewis was able to examine close-up. "The legs of this bear are somewhat longer than those of the black, as are it's tallons and tusks incomparably larger and longer. . . . It's colour is yellowish brown, the eyes small, black, and piercing. . . . The fur is finger thicker and deeper than that of the black bear."

While he conceded that "it is a much more furious and formidable ana-mal, and will frequently pursue the hunter when wounded," Lewis also concluded that against their heavy rifles, grizzlies were "by no means as formidable or dangerous as they have been represented."

That was a foolhardy arrogance about American technology, and a few days later, on May 5, an encounter with a fully grown bear would sow the first seeds of doubt. This time they came up against "a most tremendious looking anamal, and extremely hard to kill notwithstanding he had five balls through his lungs and five others in various parts." They also began to perceive the individuality of grizzly personalities. This bear did not attack but instead made "the most tremendous roaring from the moment he was shot," then swam to a sandbar and took twenty minutes to die. Doing a kind of field autopsy Lewis found the bear's heart to be the size of that of a large ox. The maw, ten times the size of a black bear's, was filled with flesh and the fish he had been catching when he had been unfor-tunate enough to fall under the gaze of the American travelers. In his journal William Clark wrote that this bear was "the largest of the Carniv-orous kind I ever Saw." Around the campfire that night several members of the party decided their curiosity about grizzly bears "is pretty well satisfyed."

But this was a group of men straight out of the colonial experience. They were used to shooting virtually every animal they saw. With griz-zlies, in almost every case they were firing on an unsuspecting animal minding its own business. Unless they're guarding a carcass, are sur-prised at close range, or are females protecting their young, grizzly bears don't normally charge people. But they don't react well to being attacked. As this rodeo repeated again and again the bears increasingly assumed the role, in the explorers' journal accounts, of "monsters." In Lewis's almost classic line, "These bear being so hard to die reather intimedates us all," there was an obvious solution. He could have just told his men to stop shooting every grizzly they encountered. But that was not the American way.

I know what it is like to be charged by a grizzly bear. In twenty years living in Montana with grizzly bears in the mountains I never had a bad bear encounter. But I have had the experience of seeing a grizzly coming on at full gallop, so I struggle to understand intentionally provoking a grizzly bear—like some frat-party dare—by shooting it in the ribs with a

Grizzly bear in the West.   Photograph by Dan Flores.

muzzleloader. My own thirty or so seconds with a charging grizzly was in Alaska. Our group saw a grizzly up ahead, climbing a riverbank. As we approached the spot my companions and I were suddenly presented with the heart-stopping vision of a chestnut-blond grizzly launching at full speed directly at us. All the eye could register were detached details— rippling fur in the sunlight, tiny and focused eyes, glimpses of curved white fangs, and an irresistible onward motion closing the distance too fast to register.

Then our guide loosed a piercing whistle that skidded the bear to a stop barely sixty feet away. Handsome, symmetrical face and upright ears scanning, the bear's black eyes suddenly locked. A split-second WTF crossed his face, followed by a studied, almost nonchalant turn to his right . . . and then like a quarter-horse under quirt he was bounding and crashing through the dwarf willows as hard as he could run. *Away* from us. After a few moments of *the-gods-we're-still-alive!* confusion, we realized what had happened. The bear had heard our voices bouncing off a distant cutbank, thought that's where we were, and had fled in exactly the wrong direction. The "charge" we'd just experienced was actually a grizzly bear running for its life to escape us.

Meriwether Lewis didn't tell his men that science had been satisfied, so they just kept tempting fate. At one point six of them approached to within forty yards of a grizzly grazing quietly on spring grass in the open prairie. Four men shot him while two reserved their fire. "In an instant this monster ran at them with open mouth," Lewis scribbled that night. They finally perforated him with eight balls, each hit only serving to direct the furious animal to the shooter, until several of them had to dive off a twenty-foot cliff into the river. In a rage the bear plunged off the bank right after them before finally expiring. The stories of these needless assaults on grizzly bears raised no eyebrows back East when the Lewis and Clark journals went into print, but they did manage to lay the blame on the bears, giving the grizzly its Linnaean name: *Ursus horribilis*.

Lewis and Clark ended up encountering (usually confronting) thirty-seven grizzly bears during the course of their expedition. By following rivers they were conducting what field biology calls line-transect sampling. Bear ecologists believe Lewis and Clark's experiences on the Missouri in 1804–6 give us a good feel for how many grizzly bears were in the West when Americans first arrived. The explorers saw all their grizzlies in a thousand-mile stretch in the High Plains east of the main chain of the Rockies. They saw no grizzlies in the depths of the mountains or in the Pacific Northwest, and their line transect obviously missed grizzlies farther south—in California, Colorado, the desert Southwest. Nonetheless, thirty-seven bears in a thousand miles means they were seeing a bear roughly every twenty-five miles. As ecology infers population demographics, that translates to nearly four grizzly bears in every block of ground ten miles by ten miles. In the one and a half million square miles of the grizzly bear's range in the Lower 48, from California to the western edges of Texas, Oklahoma, Missouri, and Iowa, bear ecologists estimate that in the early 1800s some fifty-six thousand grizzly bears ranged across the West.

The Americans' line transect missed a great deal of the bird and animal riches of the West, which is why Jefferson planned other scientific explorations and why the rest of the nineteenth century featured new U.S. expeditions into the West every subsequent decade. Though tragically depressive, Meriwether Lewis was a kind of self-taught American Humboldt, a brilliant field naturalist for whom evidence-based science appealed more than religious or supernatural explanations. So it wasn't just grizzly bears or pronghorns, coyotes or magpies the party introduced to the new scientific understanding of nature. Among their many other discoveries were bighorn sheep ("These animals bound from rock to rock and stand apparently in the most careless manner on the sides of precipices of many hundred feet. They are very shy and are quick of both scent and sight. The horns occupy the crown of the head almost entirely"), plains gray wolves ("the shepherds of the buffalo herds"), black-tailed deer (a version of mule deer), Roosevelt's elk (like blacktails, a Pacific Northwest variation), mountain goats, white-tailed jackrabbits, swift foxes, western badgers, numerous species of ground squirrels.

And beyond the mammals, these: cutthroat and steelhead trout, white sturgeon, two new species of rattlesnakes, horned toads. As for birds: prairie chickens; four new corvids, including western ravens; sage grouse and five other new species of grouse; three new geese, including the lesser Canadian; five new species of woodpeckers, among them a western pileated and a woodpecker named after Meriwether Lewis; three new jays, including the piñon jay; a new nighthawk and a new poor-will; the western meadowlark; the western tanager; the western mourning dove; the long-billed curlew; three new gulls; and two new terns. More prairie birds, the prairie horned lark and McCown's longspur. And among several others, the whistling swan. Native people had known about all these species for more than ten thousand years, of course. But now they belonged to the world.

When they returned with their specimens and their voluminous notes, and despite their having seen wild horses "fat as seals" once again grazing the Columbia Gorge, all hope that the West might hold living mastodons or camels faded. The West offered up a completely different America than the East, which now seemed partly tropical, partly European. But with no elephants anywhere, extinction appeared to be final, even in the West.

∿

A BOYS' WORLD in the American West was now at hand and it would not be a pretty thing.

When Lewis and Clark spent their winter at the mouth of the Columbia River, the Western economy's unquenchable appetite for America's animals was already a presence on the Pacific Coast. While beavers and bears and deer had taken the brunt of the onslaught in the East and South in the 1700s, on the West Coast the targets were different. The havoc was familiar.

For hard-bitten men who knew how to kill animals for money, this wild new world became a destination for the ages. So you went with whoever could get you there. Vitus Bering was a Dane but he sailed for Russian Czar Peter the Great. Pushing out of Siberia he made landfall in North America in the summer of 1741. German naturalist Georg Steller was in his crew, hoping for discoveries, but market concerns took over. Bering found several target animals, including the northern fur seal, already on the decline in Siberia. But the prize, the "Pacific beaver," was the sea otter. Otters frequented the shorelines from Japan to the Aleutians and down the Pacific Coast to Baja California in numbers that seemed large, although there were probably fewer than three hundred thousand across their range. As ships from Boston and New York began showing up on the West Coast, word spread among sailors of several nations that prime sea otter pelts had sold for $120 apiece in China. One American trader said the fur of the sea otter, so luxurious in the hand (2.6 million hairs per square inch!), "excepting a beautiful woman and a lovely infant" was the most extraordinary thing on Earth.

Like humans, like wolves, sea otters are keystone carnivores. Ancient ecosystems had formed around sea otter predation, and so long as the six-foot-long otters were present the ecosystems held together. Impossibly appealing, these hundred-pound members of the family Mustilidae, which includes wolverines, badgers, and weasels, evolved as hunters of fish and sea urchins in shallow, shoreline kelp beds. Otters kept those kelp forests healthy by devouring as many as a thousand sea urchins a day.

Just as we are, sea otters are tool users. At some point in their evolution otters developed a culture utilizing rocks of a certain size and shape to break open the shells of their prey. There were even variations handed down among regional populations, some otters making do with a single rock, others using two at once. Tools were critical. As hunters of cold Pacific shorelines, sea otters need to consume as much as 25 percent of their weight in food daily to stay warm. Important to what befell them is that they do not become sexually mature until they are several years old, bear only a single pup at a time, and sometimes spend a year without producing offspring. In prime feeding grounds, undisturbed otter colonies can increase their numbers by 20 percent a year, no more. By the turn of the nineteenth century they were undergoing an extreme disturbance.

Once drawn to the Pacific Coast's animal possibilities, the boys involved in the market went as fast as they could before other boys beat them to it. In 1778 the global traveler, English explorer James Cook, sailed the shores of Oregon and Washington to Vancouver Island, finding Native peoples who rushed to his ships with otter skins, hoping to trade for any kind of metal, even nails. The next year, after islanders killed Cook in a shallow bay of the Big Island in Hawai'i, his men sold twenty of those pelts for $40 apiece in Canton, China. That wasn't $120 but it was good enough. The American Robert Gray happened on the mouth of the Columbia River and traded for otter and seal furs up and down the West Coast the very next year. After circumnavigating the planet Gray would return to Boston in 1790 having sold his haul in China for an astonishing $21,000.

At that point the great otter/fur seal rush was on, a destruction of nature contemplated today with profound unease, although it was clearly conducted without any sentiment whatsoever at the time. It took the Russian Gerasim Pribilof just two years, 1786 and 1787, to kill seven thousand otters and obliterate every last one on the islands now named for him. That was made possible by biological first contact. Most otters and fur seals had never seen a human before. They were trusting and tame, and with no empathy for living creatures the hunters violated their innocence. When Russia's professional fur hunters, the *promyshleniki*, descended on America in the wake of these reports, they added the next horrifying step: the forced conscription of the Aleuts and other Native peoples into an animal-killing labor force. As had happened in the East, on the mainland a lucrative exchange of furs for metal technology could seduce Native people into killing animals for the market. But Russian traders lacked goods of sufficient quality to pull that off, so they resorted to subterfuge, sometimes kidnapping family members to force Native men to pursue otters for them.

If in 1800 naturalists or American presidents doubted extinction was possible, by 1820 the fallacy of that position was becoming all too clear. First in the East and South, now on the Pacific Coast, North America was losing its animals at a frightening rate. Hunters from several nations wiped out otter and fur seal colonies with a speed no one could hold in the mind. Demand in China seemed insatiable and while otters lasted American ship captains unloaded twenty thousand skins a year there. In search of laborers to harvest the otters farther south, in 1812 American ship captains invited Alexander Baranov of the Russian American Company to send Aleut hunters down the California coast, with Americans hauling the take to China and splitting the profits with the *promyshleniki.*

In a next act, Yankee sealers slaughtered more than seventy-three thousand fur seals on the Farallon Islands off San Francisco Bay. That got the Russians' attention and led them to establish their famous trading post and fort at Bodega Bay, from which their conscripted Native laborers killed eighty thousand animals in just one season. But the seals were so tame and numerous that no Native labor force was really necessary. Using clubs or knives, European and American seal hunters murdered the animals themselves, stripped off the pelts and left the discarded carcasses to seabirds, condors, coyotes, and bears, then sold the pelts for a dollar apiece in China.

Less desirable and more numerous than otters, fur seals lasted along the Pacific Coast into the 1840s. But otters were so pursued, and their colonies so devastated, that by the 1820s there weren't enough left for hunters to justify chasing down the final few. And that saved them. One of Georg Steller's few discoveries, the Pleistocene giant now known as the Steller's sea cow, had the unhappy distinction of being the first of these Pacific creatures the hunters pushed into total extinction. Sea cows were gone by 1768, and if the agents of the market could easily have located them their other targets would have followed suit. But tiny remnants of otters and fur seals at least remained alive in a few hidden, inconvenient spots the hunters missed.

Now that same pattern, rooted in our predatory evolution and released afresh by market self-interest, was about to play out in the inland West.

※

AS RUSSIANS AND AMERICANS plied coastal waters with sea otter assassination on their minds, a young southerner was traveling the interior West in search of an entirely different animal. Barely into his twenties, Philip Nolan stood out everywhere he went. An acquaintance in New Orleans described him as "an extraordinary Character . . . whom Nature seems to have formed for Enterprises of which the rest of Mankind are incapable." There's little doubt about his self-confidence. More than a decade before Lewis and Clark, Nolan was venturing deep into the Southern Plains to live with Indians like the Wichitas, Comanches, and Kiowas, who also were impressed. It was not their lifestyle he sought. Nolan lived with the tribes but said he was never able "to Indianfy my heart." What he really sought were their horses, and if Native people held their horseflesh too dear, then he went for *wild* horses in their country.

This is not widely known, but at the beginning of the market's exploitation of western animals there was the horse. Science would eventually discover a truth horse-catchers like Philip Nolan never knew, that horses were ancient *Americans* Europeans had unwittingly returned to their evolutionary home. Preadapted to thrive in America, when they escaped Spanish missions in California and Texas or fled into the wild when the Pueblo Revolt drove the colonizers out of New Mexico, horses reinserted themselves among elk, bison, and pronghorn herds in a biological instant. By 1800 wild horses on the Great Plains were approaching two million animals and increasing. As they spread north and south they were changing the American Serengeti, lowering the carrying capacity for bison by cropping the same grasses and drinking from the same water sources. Wolves and lions preyed on them once again. Horses became the American Pleistocene reasserting itself.

In the 1790s all Philip Nolan knew was that wild-horse herds were all over the western prairies and their abundance meant money in his pockets if he could possess them. By Nolan's time horses were so numerous in places that distant views across the prairies sometimes made the landscape appear to be in motion "with long undulations, like the waves of an ocean." Experience made him realize: those undulating mirages were bands of running horses. The sensory feast didn't stop there. If he got close enough "the trampling of their hooves sounded like the roar of the surf on a rocky coast." What Nolan also knew was that if he could catch them alive, then be lucky enough to get them back to the settlements—to New Orleans or St. Louis or Louisville—he could sell them for $50 to $150 on the advancing American frontier. Southern Natives like the Seminoles and Creeks wanted them. So did the military, and soon enough, so would overland emigrants. Wild horses were one more animal commodity that made the West seem like a paradise, its wealth free for the taking. Horses were walking, nickering gold pieces.

From 1789 to 1797 Philip Nolan made four horse-catching expeditions into the West. They attracted enough attention in taverns and in letters that Thomas Jefferson heard of him and sent a request for an audience with the man who, at "the only moment in the age of the world" when such was possible, had witnessed horses in their original state as wild animals. Nolan actually departed for Monticello in May of 1800 with good intentions and a beautiful paint stallion for Jefferson, but somehow neither horse-catcher nor horse ever made it to Virginia. Instead Nolan stood up the next president of the United States, turned back for one more expedition, and got himself killed when a Spanish cavalry discovered his contraband capture pens south of today's Fort Worth.

Among all the American animals now being pursued, trapped, shot, and clubbed to death in the process of building the country, horses were unique. Everything about horse evolution fit them for American conditions, yet it had been a hundred centuries since wild horses had raced across the West. The fossils of those final Pleistocene horses revealed animals virtually identical to the horses that Old Worlders returned here. Spanish horses were descended from animals the Moors had ridden into Europe a thousand years before. These "Arabs" and "Barbs" were not only desert adapted, they preserved old genetics and antique morphologies. Like the fossil horses in American sites, their common coloring was deer-dun and freckled gray. And—we know these wonderful details because an escaped herd of original Spanish horses hid themselves away in the Pryor Mountains in Montana—they had characteristics common to ancient members of *Equus*, striped legs and black dorsal lines running from mane to tail. Like deer-colored zebras.

One other thing made horses unique. Humans had long ago domesticated them, which meant that the primary economic value of wild ones wasn't for their pelts or hides or rendered oil but for their potential to be domesticated again. That meant live capture. But the task wasn't easy, because horses born on the prairies were as wild as any pronghorn or gray wolf. Unsurprisingly it was a naturalist, the Swiss Jean-Louis Berlandier working in Texas for the brand-new Mexican government in the 1820s, who left us the best account anyone wrote of the nineteenth-century technique of live capture. When Spain had owned Texas it had gone so far as to proclaim the horse herds the property of the government and had imposed a tax on captured ones. When Mexico became independent, though, it followed the lead of the United States with wildlife and made wild horses the property of whoever caught them.

Berlandier closely watched how the Hispanic *mesteñeros* did it. After sit-ing and building immense capture pens they located near water sources, the crews then constructed half-mile-long funnel-shaped "wings" that could direct a running herd into the pen. The riders then divided themselves into three groups. Once they had reconnoitered a likely herd, one group on good horses, the *aventadores*, had the task of startling the wild ones into flight and pushing them toward the capture pen. At a crucial point the most skilled riders, called the *puestos*, rode in to aim at the funnel "that dreadful mass of living beings by riding full gallop along the flanks and gathering there, in the midst of suffocating dust, the partial herds which sometimes unite at the sound of the terror of a large herd." Finally, as the frightened horses realized they were being trapped, a third group of riders, the *encerradores*, dashed in to close the gate before the animals could dart back to liberty.

The Swiss scientist left an account of the aftermath, too, and it lingers in the mind. "When these animals find themselves enclosed," he wrote, "the first to enter fruitlessly search for exits and those in the rear . . . trample over the first. It is rare that in one of these chases a large part of the horses thus trapped do not kill one another in their efforts to escape. It has happened," he went on, "that the *mesteñeros* have trapped at one swoop more than one thousand horses, of which not a fifth remained."

Every group of animal hunters in the nineteenth century had a specialized vocabulary that described particular scenes of shooting, trapping, or capturing animals. In wild-horse capture the vocabulary did not shy. Of the horses that survived trampling, many died from *despecho*, nervous rage at being captured. Others fell from *sentimiento*, or brokenheartedness over capture. Finally there was the term *hediondo*, stinking. It designated a pen horse-catchers could no longer use because it harbored the scent of panicked, dying animals. That was a smell horses taught one another to avoid for decades to come.

🐎

LIKE MOST OF US, I live in the valley of a river. It's called the Rio Galisteo, a stream that runs (sometimes) and seeps into the sands (most of the time) just south of Santa Fe, New Mexico. The Galisteo is no grand watercourse but it does string a cottonwood corridor through the High Desert, and the celebrated American naturalist Aldo Leopold once used it as an example of how waterways were ruined when Europeans brought their domestic stock to the West. Leopold told the story of a drunken emigrant who in 1849 was able to walk across the Rio Galisteo successfully on a 20-foot board plank. Yet in the twentieth century, he wrote, the Galisteo had sliced its streambed into so many eroded gullies, known as *arroyos*, that a drunk wouldn't stand a snowball's chance of making the far bank. In many places a twentieth-century plank across the Rio Galisteo would have to span 250 feet of torn-up streambed.

I've little doubt the cow and the sheep made their contributions to this ecological set piece, but I also know the naturalist arrived too late in New Mexico to see what else had happened here in an earlier time. Heading on the flanks of Thompson Peak in the Southern Rockies, the Galisteo was one of the streams the western trappers we know as "mountain men" picked clean of beavers in the 1820s. Drawn to the southern mountains in the wake of Mexico's success throwing off Spanish rule, then the Republic of Mexico's opening of the Southwest to outside trade, trappers from the States began to operate out of Santa Fe and Taos shortly after 1821. They fanned out across the mountains all the way to the high parks of Colorado, and they made astonishingly quick work of every beaver colony they could find. One party of trappers cashed in $50,000 of New Mexico furs in St. Louis in 1831. Local authorities tried to control the carnage with a law banning nonresident trapping. The Americans ignored it. By 1832 trappers were even scouring the nearby High Plains in their rush to de-beaver every last trickle of water.

Set aside for a moment the predictability of it all, the deaths of wild animals in return for a brief few years of profits, and take the long view. As it had done in the East, beaver removal in the West abruptly terminated millennia of hydraulic engineering. A world where beavers had turned western rivers into ribbons of dammed ponds and year-round water storage now yielded to flashing runoffs that cut gullies and arroyos in places like New Mexico. Leopold may have blamed sheep and cows, but the destruction of the West's beavers also appears visible on the land two centuries later. As he later wrote, "Somehow the watercourse is to a dry country what the face is to human beauty. Mutilate it and the whole is gone."

Along with the extinction of bison in the wild, the extirpation of millions of beavers in just three short decades is at least an event modern

Americans recall from the West's slaughterhouse century, 1820 to 1920. The western beaver hunt was merely an extension of beaver mania that gathered momentum from the time of Henry Hudson. But by the time the beaver trade moved west, the United States was already giving it a peculiarly American cast. In neighboring Canada the British Crown planned and regulated the fur trade, granting a government-sanctioned monopoly to the Hudson's Bay Company. Right next door to the United States, this offered lessons in efficiency, decent treatment of the labor force, and even some conservation of the target animals. But back in 1763 King George's similar attempts to regulate wild animals in the American colonies had infuriated everyone from Virginia to New York. The American approach to market capitalism was too freewheeling and anti-regulation to be patient with a setup like Canada's. The United States did, briefly, try a so-called federal "factory" system, with official trading posts, licensed traders, sanctioned trade goods, and set rates of exchange. The factory system didn't survive a decade, although it did come up with the idea of extending Indians credit, based on behind-the-hand whispers that credit inevitably led the tribes into debt. And fur-trade debt produced land cessions.

But ultimately America went for no market planning beyond the natural laws of Adam Smith capitalism. In that clear space the country's first big business enterprise, immigrant John Jacob Astor's American Fur Company, made a heroic effort to dominate the fur trade by outcompeting everyone else. The son of a butcher in the German town of Walldorf, Johann Jakob arrived in America in 1783 after first spending a stretch in London to master English. In 1808 Astor founded the American Fur Company and began his quest to control the fur market. He built and supplied trading posts first in the Great Lakes country, then in the West. Vying with Astor's fur behemoth was a myriad of small private enterprises. They were eager and could wreak havoc on animals, but weren't always professionally run. One of the most notable was Manuel Lisa's Missouri Fur Company.

The wilderness trading posts that Lisa pioneered in the Missouri River country as early as 1808 were the Walmarts and Targets of their day. They provided the Native labor force handy box stores for European technology as well as warehouses for the gathered furs. River barges ferried the body parts from beavers, river otters, and muskrats downriver to stateside markets. After Astor's wildly ambitious attempt to monopolize the Pacific Coast trade collapsed (the British seized his "Astoria" post at the mouth of the Columbia in the War of 1812), the American Fur Company refocused on the interior West. Soon enough it put its chips on a new technological innovation, the steamboat, which could haul more trade goods upriver, as well as heavy quantities of furs, now including even bulky bison robes, back to civilization, where demand seemed insatiable.

Astor's corporate design even extended to health care for his labor force. Since Indian trade partners laid low by disease would never be able to generate profitable product, in the 1830s Astor called on the government's brand-new Bureau of Indian Affairs to vaccinate Native people against smallpox. Tragically, the 1830s was too late for that. One of Astor's steamboats making an annual resupply run in 1837, the *St. Peter's*, discovered on the way upriver that it had passengers falling ill with the dreaded pox. Instead of doing the moral thing and turning back to St. Louis, the *St. Peter's* continued its run, attempting to mitigate the danger by warning Native people at every stop that a contagious disease was onboard. The Arikaras and Mandans were unimpressed. The Assiniboines thought the announcement a hoax to preserve trade goods for someone else. The Blackfeet had always refused to kill beavers for the whites because Beaver was one of their deities. Plus they valued beaver ponds as critical sources of water on the dry prairie. So they compensated by killing wolves for the trade, and had plenty of wolf pelts and bison robes on hand. The Blackfeet had suffered a bout with smallpox in the 1780s, but this band said their historians had never heard of such a disease.

By the time the epidemic of 1837 had run its course nearly twenty thousand Missouri River Indians had died disfiguring, horrible deaths. One of the most famous Indians in the West, interviewed and painted by western travelers across the previous decade, was the Mandan headman Mah-to-toh-pa (Four Bears). A strikingly handsome, middle-aged war leader, Mah-to-toh-pa impressed observers as "free, generous, elegant, and gentlemanly in his deportment." He had survived a warrior's life, but the invisible virus struck him down without the slightest care for his bravery or grace. Mah-to-toh-pa blamed the traders. "I do Not fear Death," he is supposed to have said, "but to *die* With My face rotton, that even the Wolves will shrink With horror at seeing Me . . ."

Jacob Halsey, who oversaw the American Fur Company's Fort Union post, put the moment in terms Astor and the AFC could best appreciate.

The losses, he wrote, would be "incalculable, as our most profitable Indians have died."

The most remembered part of the beaver story in the West centers on the Rocky Mountain Fur Company of William Ashley and Andrew Henry, whose reassessment of how to attack beavers would have won a business school's prize for thinking outside the box had one existed. Posts and steamboats were expensive. Like the Blackfeet, several other Plains tribes refused to kill beavers because of their ecological importance. But since colonial times America had been full of men who fled towns and farms and marriages and desired nothing so much as to spend their lives camping and hunting, a life that seemed "natural." So why not sidestep the Indian labor force altogether and have such men trap for beavers and otters themselves, as Kit Carson and others were doing out of Santa Fe and Taos? They could remain in the West year-round and the new company would use overland wagons to supply them at an annual "rendezvous" site somewhere out West, then wagon the accumulated loot back to St. Louis.

It worked. At least it worked for a few years, as long as the animals lasted. Although barely aware of it, Ashley and Henry's mountain men became players in animal geopolitics. Because they could clear streams of beavers so quickly, the American trappers were endlessly pushing onto new streams farther west, which threatened British colonial claims. So in 1824, at London's request, the Hudson's Bay Company sent trapping brigades into the Rockies with instructions to "ruin the country," to create a fur desert that would turn these Americans back.

For millions of years the streams that ran snowmelt through the canyons of ranges like the Bitterroots, the Lemhis, and the Wasatch had known beaver colonies at roughly half-mile intervals, twenty-five colonies/dams/ponds for every fifteen miles of a stream and its side creeks. But British brigade leaders Peter Skene Ogden, John Work, and Alexander Ross, with fifteen or so trappers plus Native wives or girlfriends to do the cooking and pelt preparation, were easily up to the task of obliterating all the West Slope beavers. Ross described how. His brigade of twenty with 212 traps would scatter their sets up the length of a mountain stream in an afternoon. On a typical creek in the Bitterroots they'd catch ninety-five beavers the following morning and another sixty that afternoon. That got every animal in the drainage. Then, on to the next canyon and

Beaver dam in the Bitterroot Mountains. Photograph courtesy Ben Goldfarb.

repeat. Especially during the spring, when female beavers were pregnant or already had kits, that "ruined" things proper.

This wasn't the all-boys' world of the seal/otter hunt, as the women who went on these pursuits were major players, with the skills to dress pelts and keep everyone clothed and fed and pointed in the right direction. With that kind of female assistance, from 1823 to 1841 the British brigades destroyed thirty-five thousand western beavers, draining an estimated six thousand beaver ponds.

By now everyone knew what the end of this looked like. There was no chance beavers and river otters could last any longer than whitetail deer had in the East. What they wouldn't have understood, but we do, is how the extraction of sea otters and beavers was devastating finely balanced American ecologies hundreds of thousands of years old. Without otters to hold them in check, sea urchin populations exploded, then mowed down whole kelp forests, whose loss in turn threatened the red-algae reefs that grew those waving stands of kelp. Obliterating every beaver on stream after stream didn't just deprive the Native people of traditional camps. It entirely remade drainage systems all over the continent, altering growth patterns for willows and cottonwoods, destroying wetlands favored by waterfowl, raccoons, moose, in effect drying out America.

But the stories we tell ourselves! These mountain men with their rendezvous gatherings—combination hardware stores and all-night raves, except in leather and at the foot of the Wind River Mountains—became American working-class heroes. Back East Daniel Boone's biographers had already devised a romantic take on the masculine American hunter. Now the West broadcast the Boone model as bigger-than-life figures: Kit Carson, Jim Bridger, Buffalo Bill Cody. Literary types like James Fenimore Cooper and Washington Irving became their biographers (as did Herman Melville of the whale hunters). In one of his first books Teddy Roosevelt claimed these men were "the first to become Americans." Even the historian Frederick Jackson Turner echoed that. Turner's "frontier thesis" made a Darwinian claim that it was wilderness life (with all its dead animals) that turned Europeans into Americans.

These heroes out of the adventureland of early America in fact were detached, stoic killers. Some of them, yes, were expectant capitalists. But most were not even that. They were killing animals because it was an ancient human skill, and many of them were doing so because apparently they didn't know how to do much else. The truest, most unvarnished characterization of them came from one of their own, a peer—albeit a literary one—named George Frederick Ruxton.

A British adventurer and novelist who trapped with the mountain men out of Taos, Ruxton knew his colleagues firsthand. They made up "a genus" of men distilled into a "primitive" state, he wrote, whose personalities assumed "a most singular cast of simplicity mingled with ferocity." The western hunters he knew "rival the beasts of prey" and "destroy human as well as animal life with as little scruple and as freely as they expose their own." So they were *brave* stoic killers. And true, looking for animals they examined every nook and cranny of the continent and "paved the way for the settlement of the western country."

Romance about all this was misdirected, though. The West afforded the mountain men freedom from all restraint. Their response to freedom, as Ruxton saw it, was "ransacking" the place.

THE RANSACKING CONTINUED, and naturalists and artists raced against market forces in a frantic effort to record the animal life Old Worlders inherited in America. With even common animals and birds—whitetail deer, black bears, moose, sandhill cranes, wild turkeys, and wolves—locally extirpated in much of the East, and sea otters, fur seals, and beavers following rapidly in the West, the test for extinction in the modern world seemed at hand. But it was an Atlantic shore-and-island bird, a Northern Hemisphere penguin known as the great auk, that provided all the proof skeptics needed that humans under sway of markets, who believed wild animals were mere commodity resources, could actually reduce living species to nothing but fossils and skins in collections.

If some genetic miracle were able to bring a great auk back to life today, we would be convinced we were looking at a giant, heavy-beaked penguin. A flightless black-and-white swimmer that stood two and a half feet tall, the great auk thrived on coastlines and islands all across the North Atlantic and down the coast of North America. Great auks frequented American shorelines as far south as Massachusetts Bay and there's good evidence they were features of shore life all the way down to Florida. But in the face of seagoing Old Worlders, great auks presented two fatal flaws. One was that they nested on only a few islands and laid only a single egg per season. The other was that like most of the penguins (family Alcidae, genus *Pinguinus*) they evinced a classic tameness toward humans. A British sailor named Richard Whitbourne wrote of them in 1618 that "men drive them . . . upon a boord into their Boates by hundreds at a time; as if God had made the innocencie of so poor a creature to become an admirable instrument for the sustenation of man."

New genomic research indicates that while great auks had gone through a winnowing in the Pleistocene, they reached modern times genetically healthy and unstressed. Then, around 1500, European sailors started killing the otherwise long-lived birds (they had a natural life span of 25 years) for their meat, oil, feathers, and eggs. Recent work estimates their population in 1500 at 2 million. With an annual loss of 10.5 percent of the adults (210,000 of them) to human predators and an egg harvest of 5 percent, great auks would not have lasted more than 350 years. And that is exactly what happened. The English naturalist Thomas Nuttall, who came to America in 1807 and ended up teaching at Harvard, was working on the second volume of his *Manual of the Ornithology of the United States and of Canada* in the early 1830s when he became the last scientist to see great auks alive in North America, off the coast of Canada. When Audubon looked for that colony a few years later the birds had disappeared. Egg hunters killed the last nesting pair of great auks left on Earth in 1844.

Few modern Americans have even heard of the great auk, but in the nineteenth century the fate of this giant penguin spoke to an increasingly critical question. If great auks could disappear, then all the comfort humans once drew from the Great Chain of Being's denial of change in the world now was withdrawn. Modern living animals could disappear after all, never to be seen again. Losing the great auk meant staring into both the past, at the remains of mammoths and ground sloths, and into the future and wondering what might be next. The heath hen, perhaps, an eastern prairie chicken already sensibly diminishing? The passenger pigeon? The American buffalo? Of course not those. No force in nature, no human action, could destroy creatures as numerous as those.

☙

WHILE EASTERNERS WORRIED about their losses, in the West the world still seemed dewy, fresh, an American Eden. The naturalists who went there were ecstatic. In 1819 and 1820 a government expedition led by Stephen Long explored the Colorado Front Range and returned across the plains thinking they were examining Jefferson's old target, the Red River. Its naturalists included Edwin James, Thomas Say (John Bartram's grandson), and painters Samuel Seymour and Titian Peale. Crossing today's Texas and Oklahoma plains (actually on the Canadian River instead of the Red) they traveled through "inconceivable numbers of herbivorous animals, and . . . innumerable birds and beasts of prey." Herds of wild horses five hundred strong surrounded them, and with disease epidemics still suppressing Native populations, the animals were all so tame they "appeared wholly unaccustomed to the sight of men. The bisons and wolves moved slowly off to right and left, leaving a lane for the party to pass." Long's official report claimed that the western plains "seems peculiarly adapted as a range for buffaloes, wild goats, and other wild game," and was best left just as they had found it. Seymour's and Peale's paintings, portraying the West as a paradise of animals, were the first visual representations of the inland West the world saw.

The accolades kept coming. Englishman John Bradbury, a romantic nature lover who accompanied fur-company parties for safety, had never seen the like: "The buffaloes, elks, and antelopes had made paths which were covered with grass and flowers. I have never seen a place, however embellished by art, equal to this in beauty." American John Kirk Townsend, who accompanied Nuttall with a fur-trading party up the Platte in 1834, thought the pronghorn "one of the most beautiful animals I ever saw. . . . As it bounds over the plain, it seems scarcely to touch the ground, so exceedingly light and agile are its motions." On crossing South Pass over the Rockies, Townsend wrote: "I never before saw so great a variety of birds in the same space. All were beautiful and many new to me." He sent his collections of western birds (like the Townsend's solitaire) and animal skins to the Academy of Natural Sciences of Philadelphia.

It had become clear that the task of describing and adding all these new species to Western science was going to take naturalists decades. And the global public's appetite for this version of the wild new world was insatiable. That included imagery. In Philadelphia a self-taught artist named George Catlin, who was to become known as a portraitist of American Indians, went west with fur parties on the Missouri and with a military expedition into the Southern Plains in the 1830s. Catlin focused on Native life, but his rapid-fire watercolors captured scores of scenes of bison, elk, and wolves. His best animal painting from the Southern Plains was his nearly three-dimensional portrayal of wild horses. In the foreground are many duns and grays, the classic colors of horse evolution. Additional bands guarded by their stallions recede into the mists of distance toward the Wichita Mountains.

Catlin's horse painting is a visual tribute to the re-forming Pleistocene West. But a pair of watercolors from the Missouri the artist called *White Wolves Attacking Buffalo* may be his best natural history narratives, anticipating modern ecologists who insist on the perils of the predator's life. Catlin's bison was old, sick, and solitary, but even though "his eyes were entirely eaten out of his head—the grizzle of his nose was mostly gone— his tongue was half eaten off, and the skin and flesh of his legs torn almost literally into strings," he administered "death by wholesale, to his canine assailants," tossing them feet into the air with a swipe of his head, or stomping them to pulp under his feet. Numerous wolves from the pack, Catlin wrote, ended up "crushed to death by the feet or horns of the bull."

George Catlin, Wild Horses at Play. From Catlin, *Letters and Notes on the Manners, Customs, and Condition of the North American Indians,* volume 2.

Catlin wasn't alone in showing the world how nature looked in the West. One of Humboldt's students, Prince Maximilian of the German province of Wied, did natural history on the Missouri in the 1830s and brought along a young wildlife painter named Karl Bodmer. Some of this Swiss artist's works were exquisite animal portraits, among them close-in profiles of a coyote, a porcupine, a vulture, a whooping crane, a pronghorn, a bighorn sheep, and the prize of the group, a gorgeous quartering-portrait of a bull bison. Bodmer became a famous wildlife artist in Europe, but before he left America he also did a full-body painting of a bull elk, landscapes speckled with herds of bison, and a dramatic scene of hunters stalking two adult grizzlies on a carcass. The Swiss painter was a true talent, but his works ended up in Europe. Few Americans knew of them until the twentieth century, and then primarily for Bodmer's extraordinary portraits of Native people rather than his natural history art.

One final but poignant story from early natural history in the West featured a last starring role by the country's most famous naturalist. In the summer of 1843 John James Audubon made it to the West at long last. As

Karl Bodmer (Swiss, 1809–1893), Head of a Crane, *1833, watercolor and graphite on paper.* Joslyn Art Museum, Omaha, Nebraska, Gift of the Enron Art Foundation, 1986.49.337. Photograph © Bruce M. White, 2019.

The Birds of America gathered admirers, Audubon launched an ambitious new project. Assisted by his sons, Victor and John Woodhouse, and their able father-in-law, John Bachman, Audubon's team had published the first volume of the new work the previous year with a less catchy title than the bird book: *Viviparous Quadrupeds of North America.* Now the time had come to see America beyond the Mississippi. That lure finally brought one of the world's premier nature artists, aboard the steam vessel *Omega,* to witness the fabled bestiary of western America on the "grand and Last Journey I intend to make as a Naturalist."

This was a world apart from anything Audubon had experienced before. The man had spent almost his entire life in nature. He had witnessed what he estimated were billions of passenger pigeons in flight, waterfowl in unimaginable numbers migrating down the Mississippi Flyway, captive wolves terrified by the human hand. Yet his journal makes clear that he was in no way prepared for what unfolded in front of them as the *Omega* chugged up the narrowing river, their destination Fort Union at the confluence of the Yellowstone and the Missouri.

What Audubon saw stunned him. In the East the woods were alive with birdsong but mammals were often secretive and hard to see. But in this wide-open country animals were in sight almost constantly and their diversity and strangeness were breathtaking. Two weeks before they arrived at Fort Union, not far from the eastern border of today's Montana, he set down these scenes in his journal. I've excised some of his other observations here so we can get to the pure experience of what he was seeing and feeling:

> We passed some beautiful scenery and almost opposite had the plea-
> sure of seeing five Mountain Rams, or Bighorns, on the summit of a
> hill. We saw what we supposed to be three Grizzly Bears, but could
> not be sure. We saw a Wolf attempting to climb a very steep bank of
> clay. On the opposite shore another Wolf was lying down on a sand-
> bar, like a dog. I forgot to say that last evening we saw a large herd of
> Buffaloes, with many calves among them; they were grazing quietly
> on a fine bit of prairie. They stared, and then started at a handsome
> canter producing a beautiful picturesque view. We have seen many
> Elks swimming the river. These animals are abundant beyond belief
> hereabouts. If ever there was a country where Wolves are surpass-
> ingly abundant, it is the one we now are in.

"In fact," Audubon wrote, "it is *impossible to describe or even conceive* the vast multitudes of these animals that exist even now, and feed on these ocean-like prairies." Face-to-face with a spectacle equaled only by the Serengeti or the Masai Mara, this marvelous aggregation of big graz-ers and their predators, all visible in the bright light and vast spaces, had America's nature artist reeling. He closed a letter to his wife that summer with this sentence: "My head is actually swimming with excitement, and I cannot write any more."

Audubon portrayed himself as hale and hearty, although by the time he went west his hair had gone white and many of his teeth were missing. But the eight-month journey wore on his energy. He declined a buffalo hunt because he was "too near seventy" (actually he had just turned fifty-eight). He took on painting animals much as he had birds, by shooting them and wiring them into active poses, their eyes fixed on the viewer. But of the 150 plates in the *Quadrupeds'* three volumes, he would paint only half of them.

John Woodhouse Audubon, John James Audubon as he appeared in 1843, just back from his western trip. Courtesy the American Museum of Natural History.

His son John Woodhouse, far less talented, did the rest. Audubon had once described himself as "a two-legged monster with a gun." But in the West he soured on watching animals die, writing that "thousands multiplied by thousands of Buffaloes are murdered in senseless play. . . . What a terrible destruction of life as it were for nothing or next to it."

Audubon's sensory experience in the West was rich but the scientific returns meager. Meriwether Lewis, Catlin, and Bodmer had already been in this country. Of the twenty-seven mammals Audubon's party collected on the Missouri, only the black-footed ferret, a primary predator in prairie dog ecology, turned out to be a new discovery. They did find fourteen new birds. To approach the thoroughness of *Birds of America* John Woodhouse had to make a separate trip to Texas, where he collected an ocelot, a "Red Texan Wolf," and heard stories of fearsome jaguars, which seemed to range over most of the state. What naturalists already knew of the animal life of the southwestern deserts and the West Coast, the Audubons painted from specimens. The book was a heroic effort, but not nearly the cultural triumph of *Birds*, partly because Audubon himself could not complete it, partly because, as one reviewer put the difference, birds were "exalting" and spiritual while mammals, somehow, seemed "earthy" and base.

Because Audubon set it down in his journal, his lingering memory of how he experienced the West remains. His last impression of America's version of Africa as they headed downriver: "Wolves howling and bulls roaring, just like the long continued roll of a hundred drums. . . . Thousands upon thousands of Buffaloes . . . the roaring can be heard for miles." Four years after Audubon wrote that, his collaborator John Bachman visited him in his home on the Upper West Side of Manhattan and found "his noble mind is all in ruins." He was only sixty-two. John James Audubon passed away three years later.

⁓

THE CHANCE TO LEARN more about mammals, birds, and reptiles in the West came after the Mexican War and after the Far Southwest became part of the United States. Several expeditions that looked more closely at missed mountain ranges, the Pacific Coast, and America's great deserts included scientists but were led by the Army Corps of Topographical Engineers, or else were parties surveying routes for a rail line to the Pacific Coast. Some of their leaders—like John Charles Fremont, who became such a celebrity that in 1856 he garnered the new Republican Party's nomination to be its first presidential candidate—became cultural stars from these expeditions. But as is the fate of many scientists, history has forgotten most of the naturalists who wrote about America's animals when they were novel to discovery.

The most notable of these many expeditions was the Mexican Boundary Survey, which laid out a new U.S. boundary after the treaty that ended the war forced Mexico to cede its northern territories to the United States. Exploring that line fell to William Emory, one of the founders of the American Association for the Advancement of Science, with a young New Yorker named Spencer Baird as his assistant. Soon to be a naturalist of legendary proportions, Baird as a teen had wanted to go west with Audubon but his parents had demurred because of the dangers. Emory was himself an excellent field naturalist to send to the Southwest. During the war he had accompanied the American army's push from New Mexico to California and added several species of desert cacti to science (including giant saguaros) along with visiting ruins like Chaco, whose antiquity he thought must reach into America's deep past.

But this was the 1850s. Red-whiskered Emory was also a Maryland slaveholder whose closest boyhood friend was Jefferson Davis, soon to be president of the Confederacy. So when he got west his time and background led him to an odd lack of appreciation of how Spanish settlers interacted with western wildlife. That wild animals were still abundant around towns in the Southwest appeared certain proof to Emory of Hispanic "indolence and incapacity." Evidently it said something laudatory that back home Americans would never suffer animals within reach to survive like that.

A half century after Thomas Jefferson had aimed his second major exploration at the Southwest, Emory's survey from Texas to California at last illuminated the animals and birds of the western deserts. This was an exotic part of North America, with a mix of familiar creatures in combination with several Central American species at the northern limits of their ranges. Whitetail deer inhabited the Southwest in vast numbers, and the mule deer Lewis and Clark had found also ranged widely across this region, becoming especially numerous in California. Beavers yet dammed up the streams and "shy and comparatively rare" desert bighorn sheep looked down from the canyon rims. Prairie dogs and pronghorns populated the flats, and coyotes (Emory had already borrowed the southwestern name) became their camp followers, stealing food and snatching gear. Grizzlies were present but Emory couldn't decide whether those in the interior were the same as the bears on the coast, which were much larger. There were two different wolves, the "Red Texan wolf" eastward and farther west a gray wolf. This was a new subspecies, the Mexican wolf, but scientists wouldn't recognize that for another fifty years.

The new, unfamiliar creatures were especially intriguing. In Berlandier's collections in Texas they found the skull of a "yaguarundi" (jaguarundi), a small, long-tailed cat that even then was rare. Far more impressive and numerous was the huge cat Audubon's book had mentioned. As Baird described it, a "vast number of pumas and jaguars" were preying on wildlife and on the immense herds of wild horses and wild cattle. Known to Spanish settlers as *el tigre* and to most Americans as "tigers" or "leopards," jaguars hunted the jungles of the Americas as far south as Argentina. But their northern range clearly extended into open country in North America. Coronado's early 1540s exploration into the Southwest had mentioned "leopards," and fur trapper Rufus Sage claimed to have seen such an animal on the headwaters of the North Platte, in today's Colorado. Jaguars apparently were common predators of the deer and wild horses of Texas, as a German naturalist, Ferdinand von Roemer of San Antonio, reported several of the cats killed in the area in the 1840s, where their pelts sold for around twenty dollars. The Comanches and other tribes used jaguar hides decoratively. In 1853 there were accounts of a "large tiger" as far north as the Canadian River.

In Emory's report jaguars were also on the outskirts of Santa Fe. At some point as he worked on his survey he discovered a literary source describing a frightening morning in 1825 in what he assumed was New Mexico's capital. A massive flood in the nearby "Rio Bravo" had driven out wildlife, and on opening the church a lay brother found himself "face to face with a jaguar (tigre) of very extraordinary size." The big cat killed four clerics before survivors drilled a hole through a church door large enough for a rifle barrel. Jaguars would turn up in the Southern Rockies for decades to come, so their presence in the area was not amiss. But the story Emory included actually referred to an incident in Santa Fe, Argentina, on the Río Paraná Bravo.

Other creatures reflected a different America from East, South, or North. Emory's people saw swine-like collared peccaries, the desert's native javelinas. They collected a bizarrely armored South American emigrant called the armadillo, exclusive to the Americas and distantly related to anteaters and ground sloths. (This was a nine-banded armadillo. The story that they found it run over by an Indian travois has to be a twenty-first-century rumor.) They reported a remarkable variety of reptiles, particularly snakes and lizards, of which the most impressive were chuckwallas. They acquired the first specimen (and included an illustration in their report) of the remarkable Gila monster, although taxonomists in Washington misidentified it and didn't realize it had a poisonous bite. The bird life was prodigious. Among the more intriguing were "chaparral cocks," the big cuckoos known as roadrunners. The fierce, fleet roadrunner was a "great enemy of the rattlesnake," taking them on in "pitched battles" it usually won. And they found the "Mexican eagle," the outsize falcon known as the caracara, "extremely numerous" from the Rio Grande to the Sierra Madre. By the time all the identifications were in, the Mexican Boundary Survey added a total of 311 new mammals, birds, and reptiles to America's lists of native animal life.

As was standard by midcentury, it was Spencer Baird and other naturalists of the new Smithsonian in DC who classified all these new creatures. Baird had field experience but increasingly came to represent the new laboratory naturalist who did his work studying specimens. When he finally published his monumental *Mammals of North America* in 1859 and *Birds of North America* in 1860, based on the work of Emory, Fremont, and the other Pacific Survey naturalists of the 1850s, Baird increased the number of known American mammals by 70 and upped Audubon's final bird count from 506 to 738.

Baird's books were classics of the kind of science American naturalists were then doing. By that point all were Linnaeans, for whom natural history meant adding new species and classifying them via minute morphological descriptions of their physical characteristics. The science of ecology did not yet exist, genetic and genomic analysis were 150 years in the future, and the American approach showed little interest even in a close study of bison or beaver natural history, or origins. No one knew how to interrogate the shocking success of the wild horse in America, or why one would do so.

American naturalists were adding their pebbles to the Linnaean pile, and that pile was becoming very impressive. But by the 1850s there was a vague sense among many naturalists that something was missing. Classification pointed at kinships—red wolves and gray wolves in the Southwest obviously were related—but except for some mysterious supernatural cause, natural history still lacked an explanation for the existence of such similar animals. And what was natural history to make of the fossil remains of unknown animals western expeditions were turning up in profusion? Government explorers in the badlands of South Dakota that decade were digging up fossils of elephants, other remains that looked like camels, and sometimes what appeared to be the remnants of horses mixed in with elephant bones. When had creatures like these—*camels? horses?*—been in America? And if they had been here for some reason, why weren't they *still* here?

An answer wasn't long in coming. And like the laws of motion Newton had used to revolutionize physics and astronomy, this was a new idea that was going to turn the world people had long agreed upon upside down.

CHAPTER 7

SILENCE AND EMPTINESS

Naturalist Spencer Baird had a remarkable career at the Smithsonian. In 1883 the U.S. Government Printing Office issued a descriptive list of his published works that ran to a whopping 293 pages. It required an additional 69 pages to cover all the various species Baird worked on. But Baird had the misfortune to publish his two big books, on American mammals and American birds, at precisely the wrong moment. For all the work he put into them, Baird's capstone volumes were washed into back eddies by the flood of attention that focused on a British naturalist and a new volume titled *On the Origin of Species*. The epic insight about nature that Catesby, Humboldt, and many others had called for, had searched for, at last was at hand.

The turning point for Charles Darwin had come in 1831. At the time he was just twenty-two and studying for the ministry. At Oxford he'd become fascinated with natural history, though, and as a result of a professor's intervention he lucked into an appointment as naturalist on an upcoming five-year global expedition. The simple act of adding Darwin to the crew of the HMS *Beagle* turned into one of those small decisions that forever altered human history. When Darwin returned to England in 1836, then a mere twenty-seven, he understood something no one else on Earth did, or ever had. When he finally revealed just what this blockbuster secret was, he wrote a friend that "it is like confessing a murder."

On the Origin of Species was actually Darwin's second book. He'd published one of the great natural history travelogues, *The Voyage of the Beagle*, in 1845. In that volume, particularly in his account of the weeks he spent on an island chain with the odd name of Galápagos, Darwin laid out all the disjointed nuggets of information that would lead him to his eureka moment. The dozen islands some five hundred miles off the coast of South America had looked uninviting and smelled worse. But nature there struck the young naturalist as "eminently curious." For one, the island's birds and reptiles were so unafraid Darwin was able to collect his specimens with a switch or a noose, as if an upright human biped literally was invisible. But more, each island seemed "inhabited by a different set of beings," including three different kinds of mockingbirds—"a form highly characteristic of America"—along with six almost identical finch species. Almost identical except they had beaks ranging from warbler to giant nutcracker size. Darwin was astonished: "One might really fancy that from an original paucity of birds in this archipelago, one species had been taken and modified for different ends."

But modified by what? That passive-voice sentence in *The Voyage of the Beagle* left a cause hanging in the air.

Ancient humans, the Greeks, even Darwin's nineteenth-century contemporaries had asked questions about creatures that were similar but slightly different, and most had yielded to a supernatural explanation: God must have modified an original template. But on these rocky Pacific Ocean pinpricks—and Darwin was in the Galápagos a mere two hundred years ago, remember—the veil obscuring the diversity of life on Earth suddenly lifted. That revelatory moment in the mind came when Darwin witnessed the odd differentiation of familiar American creatures on the Galápagos. But his epic insight would never have been possible without the new scientific breakthroughs that prepared him to understand what he was seeing.

The larger context that set up Charles Darwin and the rest of humanity for the discovery of evolution began with all those fossils of vanished animals. Those puzzling bones in the ground had caused naturalists like Jean-Baptiste Lamarck to abandon the timeworn belief that Earth's life-forms had existed unchanged since creation. Studying ancient fossils, Lamarck understood that change over time occurred in a variety of species. A generation older than Darwin, Lamarck famously argued for an explanation based on the idea that "acquired characters were inheritable." The common example he used was that of a giraffe forced to browse higher in the treetops, who stretched its neck and then passed that trait on to offspring. Eventually Lamarck had decided that extinction didn't actually happen. Creatures simply "inherited" their way out of one form and into another.

By the early 1830s other scientific discoveries were preparing the ground for a more nuanced explanation. You first had to read widely enough to encounter Thomas Malthus's essay arguing that populations of most species (including humans) tended to outgrow their resources, which led to competition (or migrations) to survive. As a good naturalist you also read the blockbuster of the day, Sir Charles Lyell's 1832 *Principles of Geology*, with its game-changing assertion that the Earth was vastly older than the six-thousand-year biblical chronology. With those ideas uploaded, if you had the wild luck to go on a tour of the world and see Earth's diversity close-up, an open mind was all you needed to pick the lock to one of the oldest questions since the dawn of human consciousness. A similar preparation is why explorer-naturalist Alfred Russel Wallace separately came to the same answer Darwin did, spurring Darwin finally to publish. And why, when he read Darwin, pioneer anthropologist Thomas Huxley would respond: "How extremely stupid of me not to have thought of that!"

What Darwin styled "natural selection" was the answer to the question of life's diversity. Wonderment about why the world was brimming with so many remarkable creatures must extend back to the first curious hominin mind. The answer had been there all along, but it had taken millions of years of accumulated human culture to reveal it. Natural selection meant that individual life competed for survival in a world of limits. Whatever differences existed, however small, that conferred some survival advantage to particular individuals, natural selection invariably chose them. However they originated, advantageous differences then passed to subsequent generations. This, Darwin realized, was what sexual choice was all about. Over time and space, the advantages natural selection chose could accumulate to alter a species into something new. Let natural selection play out over oceans of time, on an endlessly changing Earth, and the result was not just differentiated mockingbirds and finches on the Galápagos Islands but the remarkable diversity of life-forms across the whole planet. Darwin even concluded that natural selection probably was a cause of many of the extinctions that nineteenth-century science had been unable to ignore. Sometimes traits evolved for one situation set species up to fail in brand-new circumstances.

One of the admirable qualities of Charles Darwin's career was that he never shied from his ideas even though he knew full well what they implied. He called *Origin* an "abstract" (it was a 311-page one) and in it he offered a teaser: "Light will be thrown on the origin of man and his history." That light came a decade later with two books that appeared almost at the same time, *The Descent of Man* in 1871 and *The Expression of the Emotions in Man and Animals* in 1872. For readers whose minds were nimble enough to see the whole animate world bathed in brilliant new light, these books fully melded humans and other animals, those same animals we had so long regarded as separate from us, beneath us, there for us to exploit.

"The sole object of this work," Darwin said in *Descent*, "is to consider, firstly, whether man, like every other species, is descended from some pre-existing form." This concept was even bolder in 1871 than you might think. The first Neanderthal sites in Europe had just come to light. Beyond those, fossil evidence for any "pre-existing form" was entirely lacking. But in *Descent* and *Emotions* Darwin nonetheless took on and demolished almost all the classic arguments for human exceptionalism. Exhibit A, humans were built on exactly the same physical model as other mammals, with all the same map of organs, the same skeletal makeup. Exhibit B, human physiology was so similar to that of other animals that we were easily infected by their diseases. Exhibit C, our reproductive strategies were "strikingly the same" as theirs. Exhibit D (a subject on which he was devoting an entire book), "animals, like man, manifestly feel pleasure and pain, happiness and misery." Exhibit E, "there is no fundamental difference between man and the higher animals in their mental faculties." Differences such as language and tool-using were *only in degree, not kind*. And finally, Exhibit F, while human *morality* might at first seem exceptional, morality actually had its origins in our evolution as a social species, and "any animal whatever, endowed with well-marked social instincts would inevitably acquire a moral sense."

As for "souls," since human physiology was built on exactly the model of other mammals, and since science had no evidence to present for souls, Darwin's books ignored the matter. But there was an implication. If we had indeed emerged from the animal world, then either we were biological and thus joined to the rest of earthly life by lacking this key to immortality, or else animals must also have souls.

AD (After Darwin), the newly anxious in both Europe and the United States founded churches to work at once on "saving animals' souls."

❧

FIVE YEARS BEFORE *Origin* rocked the world, back in America a fellow member of Darwin's class and country, Oxford-educated Sir St. George Gore, had arrived on the banks of Wyoming's Powder River in search of world-class "field sports." Having already tried East Africa, Gore was now embarking on a two-year safari in the American West. For centuries hunting had been a kind of masculine cult pastime among European elites. In America the working classes may have killed animals for money, but highbrow elites viewed hunting as a special imperial privilege. It was a way for men with means to express the power they had, and not just over other people. As they saw things, they were seeking adventure in the last wild places of the Earth. But the adrenaline rush of taking on a big animal fighting for its life was a subconscious connection with our deep past. This ancient power of life and death over animals had brought Sir St. George Gore to darkest America.

Gore was the 8th Baronet of Manor Gore in County Donegal, Ireland. Sitting by the fire in their castles, he and others like him read Lewis and Clark, Ruxton, and James Fenimore Cooper and developed an obsession. Buffalo, grizzly bears, and elk were larger, more charismatic, and more dangerous than foxes or game birds on the moors. And in America there were no hunt restrictions beyond avoiding getting mauled, stampeded, or captured by the Native people. An American safari implied complete freedom, a rare opportunity to indulge human nature. Gore and those like him called their western fascination Prairie Fever because it was the grasslands east of the Rockies with their teeming herds of big animals that drew them.

Gore wasn't the first. British adventurer Sir William Drummond Stewart had commenced luxurious, guided field sports in the West at the same time (the early 1830s) that William Cornwallis Harris was swashbuckling through his own hunts in South Africa. Although the Swahili word *safari* didn't enter English until the 1850s as a result of Sir Richard Burton's books, all the elements are recognizable to anyone who has seen the film *Out of Africa* or read Hemingway's "The Snows of Kilimanjaro." Robert Redford's "white hunter," Denys Finch Hatton in *Out of Africa*, rested on an old reality. Local guides were essential. You also needed to travel in a party large enough to deter whatever groups of Native people you encountered. On his initial hunt in 1833–35 Stewart had traveled with various parties of the Rocky Mountain Fur Company, shooting animals from Wyoming to Colorado and back again with William Sublette and Kit Carson as his guides.

With that experience, in 1836 Stewart had launched his own expedition and established a template. He brought two wagonloads of rare liquors, fine cigars, wines, brandy, whiskey, along with a companion at least one observer referred to as a handsome "young English blood" (he was actually a wealthy German adventurer named Sillen). Stewart and Sillen tented together and had all their needs met by servants, who cooked their meals and served their drinks, broke their camps, cleaned their firearms, and caped out their trophies. The next year, 1837, Stewart invited the young New Orleans painter Alfred Jacob Miller along to record their adventures. This time he and his companion camped in an expansive, rug-strewn, green-striped tent the size of a yurt. The writer Bernard DeVoto once characterized Stewart's few surviving letters as brimming with "mysterious longings and melancholies, romantic passions, unhappiness and frustration." He passed off Stewart's oddities to upbringing and a longing for military action. Some of Stewart's hunting companions were not so sure that's what was going on.

A decade before Gore's Powder River safari, Stewart had made one more western trip. In 1843, now forty-seven years old, he journeyed once more to the Wind River Mountains, although most of the hunting took place on the return trip along the North Platte River. On one day they saw what Stewart estimated were fifty thousand buffalo, which they shot into from daybreak 'til nightfall. That was just the advance swarm. A few days later a mass Stewart estimated at a million animals descended on them and completely enveloped them. The din and dust were beyond all experience. At night they had to deflect the immense living mass from their camp with bonfires. Stewart's party was in the midst of this incredible expression of western life for two full days. Although Stewart never did

American bison in the West. Photograph courtesy Ed Breitinger.

so, several members of the trip later expressed remorse over "the many murders we had committed among the poor brutes of the prairie," what one called "a tumbling ocean of buffalo blood."

So here was Sir St. George Gore, heading for the Powder River ten years later, on a trip that cost him $100,000 in an age when entire states had budgets barely that size. In Gore's case, his entourage included a staff of fifty—"secretaries, stewards, cooks, flymakers, dog tenders, hunters, servants"—along with imported foods and wines and an extensive library. En route to the Powder, the caravan of six wagons, twenty-one carts, a dozen yokes of oxen, fifty dogs, forty mules, and 112 fine hunting horses would string out for well over two miles across the plains. This was glamping at its most indulgent. Gore's personal wagons held a brass bed, a steel bathtub, and more pointedly, seventy-five firearms and three tons of ammunition.

Having alienated white hunter Henry Chatillon so thoroughly that he abandoned the expedition at Fort Laramie, Gore found a new guide in mountain man Jim Bridger. For the most part the Indian and white guides seemed to like Gore well enough, or at least were intrigued by his obsession, but they did find themselves sitting around camp until nearly noon every day while he slept in, then had to listen to him read from the classics by firelight at night.

Of course what made a safari, either the African or American version, was not just servants, haute cuisine on fine china, or companions who could at least tolerate Shakespeare. Safaris were blood sports. Success was measured in body counts, and almost no one in recorded history (George Gordon-Cumming, on safari in South Africa in the 1840s, is a possible rival) matched Gore's bloodlust. On the Powder River in 1855 Gore ended animals' existence as if they, their pain, their lives were meaningless except as convenient live targets. Some of the animals fed the party, true enough, although Gore could have taken his entire entourage to dinner in Denver every evening for less money.

This was mass death administered for pleasure, and it bore every resemblance to the mass hyena slaughter of gazelles that biologist Hans Kruuk would come upon more than a century later. Gore was indulging an ancient expression of surplus killing. It had never been moral, and there's no perspective from which his indulgence looks even partially defensible. He religiously kept totals: 4,000 buffalo and 105 bears, most of those grizzlies. He also tallied in his "game book" all the pronghorns, elk, mule deer, bighorn sheep, and wolves Bridger could find for him, the totals running into the thousands. Coyotes he regarded as merely targets for practice, so generously spared them a count.

Ultimately Gore's safari slaughtered so many animals that the Crows and Lakotas in the region grew resentful. They complained to their agents, who warned Gore's party that their bloodbath might invite attack. Even in the 1850s this kind of pointless destruction was politically incorrect, and more than just the threat of Native fury finally ended it. The Lakotas confronted Gore's party, demanding all their guns and clothes. In a small measure of justice, Sir St. George Gore, 8th Baronet of Manor Gore, ended up walking naked out of the Black Hills. He had to be rescued by a band of Indians friendly to the whites.

※

INTELLECTUALS, MINISTERS, and journalists spilled vast amounts of ink and heated words on Darwinian evolution, some of it even witty. One critic demanded to be shown a bed of oysters morphing into the British parliament. A reviewer wrote that "if we are so little above the dog, we may as well make out the dog to be as fine a fellow as possible." But in the American West there was still money to be made from wild animals and that trumped any philosophical debate elites might be having about humanity's animal origins. Working-class westerners saw things in more practical terms. Even if we were more like animals than we'd thought, animals hadn't invented guns. We had.

For wildlife in western America, guns and traps meant every decade in the nineteenth century was worse than the last. The 1820s had seen the near wipeout of sea otters. Beaver colonies were so devastated and beavers so scarce by the end of the 1830s that silk headgear replaced beaver hats in the market. The 1840s saw former New Englanders who were now Mormons arrive at the western foot of the Rocky Mountains and resume the old Puritan war against "wasters and destroyers" (wolves and cougars). Mining strikes from the late 1840s through the 1870s dropped tens of thousands of hungry miners into the outback mountains of California, Nevada, Colorado, Idaho, and Montana. And the newest transportation technology—the railroads—pushing across the West brought in huge crews of workers from the 1860s to the 1880s. Market hunters, some of them Natives, who supplied all these mining and track-laying camps unleashed episode after episode of regional wildlife destruction. California's experiences were typical. When miners first arrived they were gobsmacked by the wildlife they saw: forty grizzlies in sight at once from a hilltop in what would be Humboldt County, thousands of elk scattered over the grassy sweeps of future Sonoma and Napa Counties. There were four million acres of wetlands in the Central Valley, habitat for an estimated half million Tule elk and in excess of a million and a half tricolored blackbirds, whose flocks set up a din audible for miles. Then came the strikes and the miners. By 1895 a single Tule elk herd of twenty-eight animals remained alive.

Some of the new territories and states in the West grew alarmed enough that they tried to legislate a modicum of animal protection. California attempted a closed season on both pronghorns and elk in 1852, four years into the Gold Rush. Nevada banned the killing of mountain sheep and goats, and the territory of Idaho tried to stop the destruction of all major species—buffalo, elk, bighorns, goats, pronghorns, and deer—during their spring birthing seasons. Their intentions were good but the

West was an awfully big place for any enforcement. As a correspondent to the magazine *American Farmer* put it, "The right to hunt wild animals, is held by the great body of the people . . . as one of their franchises." Whenever governments tried to limit that franchise, he added, the country should expect its citizens to react "with the worst possible grace."

⁓

IT'S THE STORY of America's most iconic mammal, the buffalo, that stands as shorthand for the country's environmental history of the past five centuries, our epic of decline and loss. For Americans who are thoughtful about who we are and the kind of country we should have, the buffalo's fate is one of the stories we ought to understand and—as with climate change in our time—really get right. Yet for a century Americans have failed to come to terms with what really happened to our National Mammal.

Our historical miss about the buffalo story has an interesting origin that rested largely on a very effective misdirection of blame. In the first decade of the twentieth century Americans were still grappling with the shocking demise of the most numerous mammal on the continent. At a time when the term "conservation" was a new notion in the national consciousness, many were outraged by the stories of orgiastic slaughter by men who had destroyed millions of American animals in the previous three decades. Descendants of some of those hunters went so far as to destroy family papers and records to hide their shame.

Then, in 1907, a former buffalo hunter named John Cook published a memoir he titled *The Border and the Buffalo*. Appearing at the very moment conservationists were desperately trying to save the last few bison left, Cook's memoir introduced a new way to think about the bison story. Among the frontier adventures he wrote about was a startling claim: destroying America's bison had in fact been the goal of a secret conspiracy by the federal government and the U.S. military. Its purpose had been to end the Indian wars and force Native people in the West onto designated reservations. In Cook's telling, former buffalo hunters like him were no villains. They were unsung national heroes, the executioners of a secret plan of social engineering. The military, Cook averred, had not only encouraged the hunters to exterminate bison but had handed out free ammunition to expedite the animals' demise. There were government officials who secretly understood: market hunters like Cook deserved the thanks of the nation and medals of honor for what they had done!

Cook wrote that he had a "smoking gun," an actual public speech delivered by a high-ranking military official, who had unfortunately passed away a few years before. Over time that "speech" has become one of the best-known documents in America's bison story, endlessly quoted in books and articles since 1907, still available on T-shirts in the twenty-first century. Its origins, Cook wrote, lay in 1875, when Texas was on the verge of becoming the only state in America to pass a law to protect buffalo. At which point, the former buffalo hunter claimed, no less than Civil War hero General Philip Sheridan journeyed to the legislature in Austin to bring Texas in line with the secret policy.

Cook didn't hesitate to put words in Sheridan's mouth:

> These [buffalo hunters] have done more in the last two years and will do more in the next year to settle the vexed Indian question, than the entire regular army has done in the last thirty years. They are destroying the Indian's commissary, and it is a well-known fact that an army losing its base of supplies is placed at a great disadvantage. Send them powder and lead, if you will; for the sake of lasting peace, let them kill, skin, and sell until the buffaloes are exterminated. Then your prairies can be covered with speckled cattle, and the festive cowboy, who follows the hunter as a second forerunner of an advanced civilization.

"Let them kill, skin, and sell" comes close to rivaling George C. Scott's memorable delivery in front of the American flag as the war film *Patton* opens. You can easily imagine an American flag (maybe a Confederate one, too) waving behind Sheridan as he delivers the words Cook has him say.

There is one problem with Cook's piece of conspiracy evidence. It was in fact a complete fabrication. There was the hint of invention in Cook's story all along. He introduced Sheridan's supposed address with the telling, passive-voice phrase "It is said." Scoundrels by the score have gotten off scot-free via the passive voice, and so, too, does the perpetrator of this story, since Cook never identified a source, if there really was one.

But there is no corroborating evidence that any of what Cook described happened. There is no record that Texas ever introduced or debated a bill to outlaw killing bison, and there is no evidence that Philip Sheridan ever testified before the Texas legislature. During those same years the U.S. Congress *was* considering laws to protect all big mammals in the western territories, buffalo cows specifically. Those bills met with derision from members of the Texas delegation, a reaction that speaks to how Texas politicians in the 1870s thought about laws to protect animals. They dismissed them as misplaced sentimentalism.

But Cook's story recast buffalo history and it worked for three reasons. All the world knew of the intimate linkage between buffalo and Indians. Americans also knew that Union victory in the Civil War had featured military destruction of the South's farms and resources. And a bison story that blamed the animals' demise on "the government" didn't just make heroes of those who'd shot down buffalo for money. Its misdirection obscured the role of America's hallowed free market in destroying a world-famous animal. Much like the twentieth-century rewriting of the cause of the Civil War that many Americans embraced—that the war that nearly tore the country apart had not been about the enslavement of African Americans but about states' rights and preserving an agrarian "Southern way of life"—Cook's version of what had happened to the buffalo made Americans feel better about themselves.

※

THE REALITY OF the country's buffalo story was that it was part of a horrific history of wildlife slaughter in America, a destruction that has no rival anywhere else in world history. The story did involve Indian people, and it did feature a transformation of a hundred centuries of Native America into a re-creation of Old World civilization on the new continent. The real bison story included elements like climate change and new ecological upheavals that few remembered later. But most people did know that the story had a big historical context. Understanding that context begins in an unlikely place, with an anthropologist named Franz Boas.

In the early twentieth century Boas developed an argument he would send graduate students Ruth Benedict and Margaret Mead around the world to test. Boas believed that human cultural practices were so variable and idiosyncratic that there was no consistent future toward which everyone was aimed. This was the beginning of what we now call "multiculturalism" and it was very different from the way nineteenth-century thinkers imagined human history. To them, humans everywhere were on the same "ladder of progress." Hunters would eventually become herders, then farmers who built cities, wrote constitutions, and founded capitalist republics that looked remarkably like France or the United States. It was no doubt comforting to think that everybody else in the world wanted nothing so much as to become just like you. No one was supposed to retreat back down the ladder. In nineteenth-century America this was the organizing principle behind an Indian policy of converting tribes to agriculture, the ownership of private property, and eventually "assimilation."

But for many Native people the America of the 1600s through the late 1800s seemed the perfect opportunity to descend the ladder, not climb it. During those centuries an unusual number of Native groups determined that times were propitious for becoming hunters again. This had been an old colonial fear for Europeans about their own people. How would their outposts of enlightenment survive the enticements of wilderness that lured young men away from farms to hunt and trap? But it wasn't just Europeans, either. Native peoples all across the country began to abandon their cornfields and village lives and move west to hunt again.

Aside from the market, for Native people an animal revolution that centered on the acquisition of the horse drove this. That revolution created a grand historical moment that captured the imagination of the world. For at most ten human generations, conditions were perfect for fashioning a legendary American scene: the horse-mounted Indian as hunter of buffalo and other western animals. The climate anomaly called the Little Ice Age assisted by producing more than two centuries of cool, moist weather in the West, watering grasses and growing bumper crops of buffalo. This Native Eden emerged around 1650 and lasted until the early 1880s. Out on the continent's great grasslands buffalo numbers soared, to probably as high as thirty million animals in good years. History, the market, and climate set the stage for a legendary time for Native people to live large.

Missing from western ecology for thousands of years, horses seemed to appear almost magically from the southern end of the Rocky Mountains. The famous revolt of the Pueblos against Spanish settlers in 1680 saw the rebels seize thousands of horses, along with goats, sheep, and cattle. The cattle ended up eaten, the Navajos (Diné) traded for most of the goats and sheep, while the horses attracted many customers. Pueblos and Navajos traded horses (sometimes for children) to the Utes, who traded them to the Shoshones, who dispersed horses throughout the upper West. Within half a century the Nez Perces had horses, and so did the Salish and Black-feet, Crows and Assiniboines and Crees. Some of those groups were in southern Canada, which was about as far north as desert-adapted Spanish Barb horses could survive the winters. By 1730 peoples who had been on foot for 150 centuries were swinging onto horses and preparing to ride them into history.

Horses carried big implications for buffalo, now in a world natural selection hadn't prepared them for. Competing with buffalo for western grass and water, horses were restoring the West to conditions that pre-dated the modern bison, and as new grazers in western ecologies horses began to draw down buffalo numbers. But for tribal bands learning from one another how to ride horses, care for them, breed them, all manner of new possibilities opened. Dozens of groups (a common estimate is three dozen) dropped whatever else they were doing—planting crops and eating vegetables, hunting deer on foot—and rode off to hunt buf-falo, descending the ladder Western philosophers assumed they should be climbing.

Some of these peoples, like the Comanches of the Great Basin Desert and the Siouan speakers of the woodlands around the Great Lakes, had never farmed. Now propelled by horses, they merely switched their focus from jackrabbits or whitetail deer to buffalo. Others, like the Pueblos, the Utes, the Salish, and the Nez Perces, all of whom lived west of the buf-falo range, and the Caddo speakers, Osages, Arikaras, and Mandans on the eastern side of it, remained in their villages but rode off to hunt buf-falo several times a year. The Pueblos were farmers, the Nez Perces deer hunters, but with horses both could now make big journeys to haul bison products home from hundreds of miles away.

But the most surprising of the new buffalo hunters came from villages where their ancestors had farmed for scores of generations. Some of histo-ry's most iconic buffalo hunters—Crows, Cheyennes, and Kiowas—came out of farming backgrounds. Almost all the eastern Indians who went west to hunt were former farmers. Native farming towns often featured sharp class distinctions, and all the evidence indicates that it was mem-

bers of their lower classes who mounted up and rode away. Two groups in
the farming towns tended to resist this move to a pure hunting life. The
elite families had political power to lose and they almost always refused to
leave the towns. And women, who owned the fields and shared political
control in farm villages, evidently realized what they stood to give up—
and take on—as well. Farmers still hunted, so Native women in farming towns hadn't relinquished the ancient tasks of working animal hides,
arduously stripping them of any flesh that could spoil and ruin them, then
devoting days to kneading brain matter into the skins to soften and tan
them. But joining a group that intended to live purely by hunting didn't
just multiply that effort. Hide work was backbreaking enough to push
women to become multiple wives to share the effort. If the hunting band
a woman joined participated in the market, she didn't just become a plural
wife. She joined a labor force.

The groups that had always been hunter-gatherers did very well as
buffalo hunters, though. The Comanches, who migrated southeastward
from the Great Basin to the source of horses in New Mexico and Texas,
established a powerful empire on the Southern Plains. And Siouan speakers who pushed westward out of Great Lakes forests to seize the territories of other tribes did the same on the Northern Plains. One group
that remained in place—the Blackfeet, who'd been plains hunters with
their backs against the Rockies for millennia—also became famous buffalo hunters. The size of the horse herds the men owned and the number
of wives they had defined wealth and status in all these groups. Women
among them seemed rarely to object to the new life on horseback.

To the misfortune of the buffalo, these weren't the only people who
dropped everything to pursue them. From the Red River country of the
north, a mixed ethnic group, the Métis, with bloodlines that were both
French and Native, became buffalo specialists, with cart caravans returning bison products to markets in Canada. In the Southwest, Hispanic New
Mexicans called Ciboleros emulated the Métis, hunting from wooden-
wheeled cart trains and hauling dried meat and hides back to Santa Fe
and Taos. Iroquois and Delawares from the East, like the famous hunters
Ignace La Moose and John Gray, headed west to Montana Territory and
instructed the western tribes in the nuances of the market. As deer herds
played out in the South, bands of Choctaws and Chickasaws and Cherokees moved beyond the Mississippi, looking for more animals. Under

federal Indian Removal policies in the 1820s and 1830s, the United States relocated at least eighty-seven thousand Native people to the Indian Territory (Oklahoma). Many of the young men of these former farmers rode horses into the buffalo country to join the hunt.

The growing multiplication of all these pressures, aimed at buffalo but with collateral damage for many other animals, had been building since 1700. By 1820 the pressure locked into an unsustainable trajectory. Like a vast lake drying from the edges inward, the bison range began to contract. Their destruction in Kentucky and Ohio in the 1780s pressed farther west decade by decade. My natal state was fairly typical. The last time anyone saw a wild buffalo in Louisiana was 1803. That same drying-shrinking effect was happening on the western side of the animal's range, too. As trappers had cast about for ever-scarcer beaver colonies they'd killed off the buffalo herds in Idaho and Utah by the early 1830s. The reservoir of animals even began drying up through its heart. Oxen and horses accompanying the overland migrants to Oregon, California, and Utah ate away the grass along their routes, and heavily armed emigrants shot at every bison—indeed, at every living thing they saw—along the way.

In a final cruel twist, the Civil War unloaded thousands of young men into the country who were familiar with firearms but had few prospects beyond employing their new skill against wildlife. These were the "hide hunters," gearing up wagons and Sharps rifles and skinning knives to impose a gruesome future on the West's remaining animals.

※

IN 1872 the Brooklyn painter John Gast distilled an important assumption about Indians and the country's wild animals into a famous visual image. Gast's *American Progress* portrayed a blonde giant in angelic white garb striding across the West stringing telegraph wires behind her with settlers in her wake. Disappearing off the edges of the canvas were herds of buffalo, packs of gray wolves—and Native people—all stealing away to the margins of the future. Viewers of *American Progress* understood that the country had a policy to acculturate and assimilate the tribes. But what was the future of those iconic wild animals?

Most Americans appeared to assume that in an America modeling itself on the Old World, the fate of animals like this was inevitable. Buffalo stood first in the rank of those incompatible with civilization. Wolves, well, the plan since colonial times had been their total disappearance. Eventually other animals—grizzly bears, cougars, jaguars, wild horses, eagles, and, judging by the reaction to their extinction, passenger pigeons—joined the ranks of the incompatible. Their destruction, Gast's painting implied, was no one's fault. Ancient America simply had to go. Even so committed a preservationist as Sierra Club founder John Muir bought the explanation: "I suppose we need not go mourning the buffaloes," Muir once wrote. "In the nature of things they had to give place to better cattle." Incompatibility was a shame but seemed to comfort us.

As their range contracted to the Great Plains, buffalo as the dominant mammal on the continent confronted a final half-century endgame. As early as the 1840s tribal people could tell something was very wrong with their primary animal. Native historians painting so-called Calendars and Winter Counts (which recorded the most important events in a given year) on bison robes began to document a world in trouble. The Kiowa Calendars from the 1840s referred to shortages of buffalo that made it hard to hold their annual summer Sun Dance. Among Western Siouan groups, their Winter Counts indicated that in 1842, 1843, and 1844 the most significant events were the elaborate buffalo-calling ceremonies their shamans performed.

The tribes realized bison were dwindling. When asked why, they told white interviewers the blame lay with their enemies, or with the Métis, or the whites on the overland trails. One Western Lakota offered that bison were becoming few because they couldn't abide the smell of white people. The Kiowas of the Southern Plains feared that buffalo were once more returning to the earth, their original point of origin. They pointed to Mount Scott in the Wichita Mountains and called it the buffalo's "Hiding Mountain." The Lakota elders whom photographer Edward Sheriff Curtis later interviewed on Pine Ridge in 1905 confessed to him that the whole matter was *wakan*, a mystery. What could make something as fundamental as the stars in the sky disappear? An alteration of the world that profound probably stemmed from divine punishment for some unfathomable transgression.

But the buffalo story was different than either Native people or white citizens thought. Beyond the now-familiar frenzy of turning animals into money, there were other causes, and we know roughly when those causes converged to obliterate an ecology dating back to the end of the Pleistocene.

The competition from growing wild-horse herds was one important new ecological factor. As horses spread and filled in their ancient evolutionary homeland their presence crowded a grazing niche that bison had owned for eight thousand years. On the Southern Plains, where horses initially spread, their numbers reduced the carrying capacity for bison by 25 to 30 percent. Yet another factor was the changing climate of the nineteenth century. During the first half of the 1800s, the Little Ice Age's grass-happy conditions began to dissipate. If the Little Ice Age had been brought on by a collapse of human populations three centuries earlier, it may be that a rebuilding human population cutting forests and again burning wood (and now a fossil fuel, coal) was subtly altering the nineteenth-century climate. Tree-ring studies show that wet conditions continued to about 1820. After that, western weather gradually cycled toward dryer, warmer conditions less favorable for grass, and buffalo. Spotty local droughts settled into a region-wide dry period by the 1850s. An exhaustive study of drought on the Southern Plains argued that the driest decade on record between 1698 and 1980 was not the 1930s Dust Bowl but the years between 1856 and 1864.

And by then the bison's ancient drought refuges east and west of the plains had filled in with people, so there was no longer any place for bison to migrate away from drying, withering grasslands as they'd done during droughts for so many thousands of years.

The overland migrants headed to Oregon, California, and Utah brought another new ecological change. As Old Worlders had done with human diseases, pioneer livestock took west a suite of exotic animal pathogens. Anthrax and bovine tuberculosis, both from the Old World, infected most of the bison that survived into the 1890s. We have no good idea how problematic these diseases were for the wild herds, but in 1867 Texas Ranger Charles Goodnight described a hundred-mile stretch of the Colorado River Valley filled with buffalo dead from some unknown cause. His guess was they had eaten out the grass, refused to leave, and died of starvation. His account sounds more like a disease die-off, and there were similar stories elsewhere across the plains.

But first and always in the historic collapse of America's bison there was the country's market economy for wild animals, still unregulated, still disastrous. The market wasn't just an influence on the thousands of blue-collar hunters who rode onto the plains to shoot down bison to sell tongues and hides after the Civil War. As happened with beavers and deer and black bears, Native peoples became ensnared by the global economy for bison. Resisting the pull of this ubiquitous and potent force was almost impossible, because resistance disadvantaged you compared to your neighbors. But if you embraced it, killed enough animals to create surplus product for the trade, you found your life and culture transformed by metalware, firearms and ammunition, and luxury goods to proclaim your status. Who of us can't relate?

By the nineteenth century, market globalization had been squeezing the planet for three hundred years. It was superb at exchanging goods made in faraway factories for local ecoregional products. Starting in the 1820s the fur companies annually barged (and later steamboated) one hundred thousand Indian-produced buffalo robes to New Orleans, with another eighty-five to one hundred thousand robes going to St. Louis. In Canada the Hudson's Bay trade reached seventy-five thousand robes a year by the 1840s. As with the Beaver Wars among Hurons and Iroquois, intertribal competition for every remaining pocket of buffalo was a feature of the endgame. Expansionist tribes like the Comanches and Lakotas displaced those in prime buffalo pastures. "We stole the hunting grounds of the Crows," one Cheyenne boasted, "because they were the best." As the Nez Perce hunter Yellow Wolf confessed, "I killed yearlings mostly. It was robes we were after more than meat." He could have added that the demand was for softer cow robes, and cows were easier to skin and process into tanned robes anyway. A focus on females didn't do good things for buffalo survival.

While demand for these robes as cold-weather blankets and bedding was limitless in the East and Europe, the supply of buffalo was not. Given drought, horse competition, new bovine diseases, and an insatiable market, buffalo numbers were probably down 40 to 60 percent by midcentury. When the Civil War ended there weren't thirty million bison anymore. Judging by the comparatively slight railroad-shipping figures for bison products after the war, by 1865 more likely only ten to twelve million were left.

At that point all those hardscrabble young American hunters arrived on the buffalo plains. Most were veterans from the war, knew guns and killing, and were armed with firearms technology the Civil War had refined. Others were construction workers the railroads had let go. For tongues cut from the animals' mouths, but mostly for what they called "hides"—the skins of the animals ripped from their bodies and staked out to dry—the hide hunters turned the plains into an open-air slaughterhouse. Just as steamboats had facilitated the demise of beavers, railroads hauled the harvested parts of once-living animals away in boxcars. The commodity parts of the animals went to nearby cities or to the East Coast, where a new chemical process converted buffalo hide into a tough leather used as industrial belting and wagon suspensions. As for their dead, stripped victims, one hunter said that from any eminence their bodies, glistening in the sun like so many glass windows, could stretch to the limits of sight.

Some of the hide hunters may have believed bison were inexhaustible. But most understood what they were doing, and why they did it. J. Wright Mooar, a twenty-something Vermonter who with his brother hunted in Kansas and Texas, claimed the thousand dollars he realized in a month of buffalo hunting was more than he could make in a year in the East. After five years he'd saved enough to buy a ranch and gave up hide hunting two years later. Most of the hunters did understand that the money was temporary, as the majestic creatures they were killing were fast disappearing. But the Mooars, Billy Dixons, Frank Mayers, and John Cooks of the world justified their self-interest in the classic manner. What was good for them was good for America. Mooar, who ended up in Texas, didn't go quite so far as Cook and suggest he deserved a national medal. But he did scoff at conservation and sentiment for the animals: "Any one of the many families killed and homes destroyed by the Indians would have been worth more to Texas and civilization than all of the millions of buffalo that ever roamed from the Pecos River . . . to the Platte." But another hunter, Frank Mayer, saw things differently: "Maybe we were just a greedy lot who wanted to get ours and to hell with posterity, the buffalo, and anyone else. . . . I think maybe that is the way it was."

There are two perfect words for their kind of callous disregard for life, for an attitude that regarded two or three years of returns worth leaving behind a putrid desert of rotting carcasses and blowflies and a deprived posterity.

Fucking pathetic.

THE PUTRID DESERT of the finale fed legions of ravens, magpies, eagles, coyotes, and wolves, but that feast lasted only a few months before only skulls and bones were left, which poor settlers gathered up and sold as fertilizer for a few pennies. Naturally. Birds then made off with the remnant fur still snagged on the sagebrush. The wallows where buffalo rolled and dusted themselves once speckled their country with craters that in low light made it look like the surface of the moon. Now they filled in with sand and the spade-shaped leaves of plants known ever since as "buffalo gourds." The trails over which bison had navigated the West single file for perhaps half a million years just became topography. No one could believe it. In the Dakotas Indian people danced the Ghost Dance to bring the buffalo back and were shot down by the reconstituted Seventh Cavalry for their efforts. In Oklahoma a Southern Cheyenne named Buffalo Coming Out entreated the animals to reemerge from Earth. He finally gave up. As the famed Crow leader Plenty Coups saw it, this was the end of history. Standing out on the empty prairies it must have seemed as if a titanic sound reaching to the heavens had ceased the second before you'd turned to listen.

In Montana Territory in the summer of 1886, a taxidermist named William Temple Hornaday was skinning an unlucky coyote in his campsite when a frontier type who seldom got cast in future westerns appeared in the lantern light. "Doc Zahl," the stranger said his name was. He introduced himself as a buffalo hunter but he seemed to be in need of some serious economic retraining. Zahl and Hornaday had almost nothing in common but pleasure in a warm fire on a frosty Montana night. But Zahl's "profession" fascinated Hornaday, who was in Montana precisely because of Zahl's prey. The National Museum had sent Hornaday out to see if he could find a handful of bison for an exhibit, since stuffed might well be the only way those of us down the timeline would ever see a buffalo. Two years of research and letter writing had convinced Hornaday that, as he and Zahl talked, there were only 1,073 bison left in America.

Zahl would have none of it. In the history of American wildlife, Zahl was the sort of real-life character Cormac McCarthy fictionalized in his great novel of the West, *Blood Meridian*. Zahl had gone all in on the slaughter as the animals had stood perplexed while he shot them dead, and not only did he evince no remorse, he was also in denial, insisting there was no way all the buffalo were gone. Why, just a few years earlier—1883?— he'd watched a herd of fifty thousand cross the Yellowstone River on its way to Canada. Zahl figured they were still up there, soon enough would be back, and the hunt would resume.

Listening to his fireside companion's nonchalance about what he'd done, Hornaday knew (but did not say) that the shipper I. G. Baker at Fort Benton had sent only five thousand buffalo hides down the Missouri in 1883 and none at all in 1884. The days when Zahl or anyone else could experience what a correspondent for the *New York Sun* had described in 1877—standing atop a butte on the Northern Plains for five hours he'd watched a "sea of black, shaggy life rolling like billows at our feet . . . an ocean of buffaloes, surging and swaying like the waves"—was gone forever. Hornaday even knew the fate of the herd Zahl described. One of the bison's instincts as the end approached had been to collect together for safety, but those fifty thousand animals had not even reached the Canadian border before American hide hunters, Métis cart caravanners, and assorted local tribes and homesteaders had murdered every last animal.

The Doc Zahls of the world did have one more ace to play, though. If the buffalo really were gone, what about all those elk, pronghorns, bighorn sheep, mule deer? Couldn't you still get a buck or two for their skins? Or—if you laid in enough strychnine—weren't wolf populations spiking from all those dead, rotting buffalo corpses?

≈

BEYOND HOW THIS STORY truly unfolded in the West, the other question that lingers over America's wild bison endgame is, why did we let it happen?

Western travelers like George Catlin and John James Audubon had publicly expressed their disgust with the callous slaughter of western wildlife before the middle 1800s, but it was Harvard biologist Joel Allen's early-1870s articles "The North American Bison and Its Extermination" and "The Extirpation of the Larger Indigenous Mammals of the United States" that first introduced that unnerving take to the postwar reading public. Another critical new voice emerged that same decade from a start-up national magazine called *Forest and Stream*. New Yorker and Yale graduate George Bird Grinnell, who had dug up the remains of extinct species out West with professor Othniel Marsh, was fresh from an unappealing stint on Wall Street. In 1875 Grinnell launched a celebrated conservation and writing career with an article pleading for U.S. military officers in the West to stop the market hunt for animals. The next year, in a piece he called "Large Game in the Territories," Grinnell blasted the capitalist market as a threat to the survival of *all* the large animals out West.

Although his book on the subject would not appear until 1889, another new activist to join the battle against the market assault was the midwestern taxidermist-conservationist gadfly, William Hornaday. A graduate of Iowa State, Hornaday started his career at Ward's Natural Science Establishment in Rochester, where he became famous for proving there were crocodiles in America by bagging one in Florida. He then moved on to the Smithsonian and became a charter founder, in 1880, of the Society of American Taxidermists. Hornaday was convinced that the country was committing a crime against present and future by allowing its wild fauna to be destroyed. His book, with the no-nonsense title *The Extermination of the American Bison*, began with the hope that it would "cause the public to fully realize the folly of allowing all our most valuable and interesting American mammals to be wantonly destroyed in the same manner." And what manner was that? In the chapter titled "Causes of the Extermination," Hornaday pulled no punches. It was simple. Buffalo were almost extinct because of "man's reckless greed, his wanton destructiveness, and improvidence," played out in a nation where there "was not even one restraining or preserving influence."

Why was there *not* a single restraining influence to halt such profound losses to nature? Where was the federal government while all this was happening?

Simply, it was frozen in inaction. Having revolted against a British Crown that tried on several occasions to restrict the profligate killing of American wildlife, Americans from their origins had been unwilling to suffer federal regulations on the free market for wild animals. States had tried to slow or stop citizens from killing certain valuable species like deer, but always ineffectually. The country's embrace of Adam Smith seemed to demand it look the other way as its citizens pushed one species after another to extinction, near extinction, or regional extirpation. Beavers, sea otters, great auks, and scores of common species had been the initial targets. Following the Civil War the focus shifted to bison, prong-horns, elk, mule deer, passenger pigeons, bighorn sheep, and wolves. Obviously the federal government wasn't secretly targeting all these animals on behalf of the social engineering of Native people. It was defending a larger principle.

That larger principle was what Gilded Age America called "laissez-faire," a sacred belief that governments should never interfere in the "higher laws" of supply and demand. With Reconstruction ending in the South, with Black Southerners losing their right to vote as a result, in the late nineteenth century the Republican Party was especially interested in the support of the new American corporations, a brothers-in-arms bond it has never relinquished. Democrats floundered trying to decide what they stood for, but they, too, regarded the free market in near-religious terms. Confronting a wholesale destruction of wildlife, then, the country's history and beliefs froze the national government into standing aside and letting economics take its course.

The fate of animals in America did have a fellow traveler in the complicated matter of Indian policy, which after the Civil War was in transition from treaty-making to reservation life and acculturation. Native insistence had meant that almost all Indian treaties had included provisions about animals. Washington territorial governor Isaac Stevens's 1850s treaties in the Northwest had built in a widely followed principle of letting tribes assigned reservations continue to hunt and fish in all their "usual and accustomed places" even if those lay outside reservation lands. The United States agreed to these provisions because it hoped the tribes would support themselves with the least possible government assistance. On the other hand, the "ladder of civilization" committed Indian policy to promoting agriculture. In the American tradition that meant private homesteads or "allotments" instead of communal reservations, which intentionally or not struck at the one thing hunting-gathering must have: open space.

Journalistic outrage like Grinnell's eventually did draw the attention of Congress. The killing sprees were taking place in the territories and on unappropriated lands still administered by the feds' General Land Office, and not every member of Congress was willing to let the market have complete sway. So in 1872 Californian Cornelius Cole introduced in the U.S. Senate the first-ever federal bill that attempted to halt the destruction of western animals. Cole's proposed law sought to regulate the "indis-criminate slaughter and extermination" of the whole suite of big western wildlife, including "the buffalo, elk, antelope, and other useful animals." This was the same Congress then taking up the matter of brand-new Yellowstone National Park and its wildlife. But Cole's bill tried for widespread restrictions on economic behavior across all federal lands. It intended to curtail a freedom Americans had taken for granted since Jamestown. The bill died in committee.

In 1874, though, as reports came in of a frenzy of buffalo destruction in western Kansas, a Republican representative from Illinois named Greenbury Fort introduced a new bill in the House to make it unlawful for any non-Indian to kill "any female buffalo, of any age" in the western territories. Fort's bill was hardly a plan to deprive Indians of wildlife, as it proposed allowing Native people to continue to hunt buffalo. Restricting non-Indians from killing female bison meant Fort's bill was aimed at two things: market hunting and the preservation of the species itself. This was a significant step, and after considerable debate, both the House and the Senate passed the bill. But this law—the first federal legislation ever drawn up to protect a single American animal—did not become law. President Ulysses Grant failed to sign it in the allotted time.

Those who later bought John Cook's conspiracy theory of a plot to destroy buffalo cited Grant's inaction as proof he must have been privy to a secret plan. But on the floor of Congress the explanation was that the president supported the bill but he'd been distracted by other business when it had landed on his desk. That led Fort to reintroduce the same bill, without changes, in the 1876 session. Again it passed in the House. But this time the bill to save buffalo never got out of committee in the Senate. As that body was meeting, turmoil over the Little Bighorn battle and the death of George Armstrong Custer and his command deflected Congress's attention. That fall the disputed presidential election of that year, with Reconstruction of the South at stake, dominated the nation. Big political issues seized the stage. No federal bill to curtail the buffalo slaughter ever came up again.

There is good reason to damn 1870s political leaders for not rising to the occasion and protecting the most famous animal in America. But there is a sobering and illuminating Canadian experience from that same decade. In 1877 the Canadian North-West Territories actually did pass a law to protect buffalo, and the outcome of that law suggests that if Fort's bill had passed, it likely would have had little or no effect. Canada's law lasted only a year before it was rescinded. Why? On the vast plains of the West it proved impossible to enforce. In the United States, market hunting of all wild animals would remain entirely unregulated by the U.S. government for another quarter century. That was long enough for the mayhem to extend far beyond buffalo.

In the absence of a policy there *were* plenty of opinions expressed on the buffalo issue. An oft-quoted one came from Grant's Interior secretary, Columbus Delano, who wrote in one of his annual Interior reports that if killing bison sped up Indian acculturation, he "would not seriously regret the total disappearance of the buffalo from our western prairies." Delano's views, while not offered as a policy, found plenty of supporters. GOP representative David Lowe of Kansas, on the other hand, made a speech in the House arguing that letting market hunters kill animals Indians needed for food ran contrary to civilized Christianity. Other members from buffalo country, like Democrat John Hancock of Texas, countered that as far as he was concerned, "saving buffalo" was just misplaced "sentimentality." In the Gilded Age sentimentality was a code word for something only women felt. Men accused of it understood: when someone charged you with sentimentality it meant your manliness had just been insulted.

Most of the military force stationed in the West ignored buffalo. That was especially true of enlisted men. The officers who regarded themselves as hunters, like Custer or Richard Irving Dodge, pursued western animals themselves, wrote books about hunting, and regularly guided European elites on safaris. They wanted to be able to shoot to their heart's content. "There are no game laws," Dodge once wrote. "There can be none; at least none can be executed." Their lack left the heavy, waddling Dodge and the three English sports he took for a hunt on the Cimarron River in 1872 free to shoot down, in twenty days' time, 127 buffalo, 154 turkeys, 11 pronghorns, and an absurd assortment of meadowlarks, robins, bluebirds, hawks, raccoons, rattlesnakes, an obsessively kept total of 1,262 animals. That hunt was "so delightful" that in 1873 they went back for another round and took 1,141 animal lives.

Dodge's "hunts" probably explain his role in an infamous conflict over buffalo on the Southern Plains. As the Fort Laramie Treaty would do for Northern Plains tribes in 1868, the 1867 Treaty of Medicine Lodge

Creek had assigned the tribes of the Southern Plains their reservations. Its Native signatories wanted to keep eastern Indians away from their off-reservation buffalo grounds, and eyeing the hide-hunter destruction going on in Kansas they had the presence of mind to insist on a "dead line" boundary that buffalo hunters weren't supposed to cross. But when Dodge, who was military commander of the region (and working on a hunting book at the time) had J. Wright Mooar inquire whether the market hunters could cross the dead line to get at the wild herds to the south, his response (supposedly) was "Boys, if I was a buffalo hunter I would go where the buffalo are."

In fact, Dodge's own shooting sprees on the Cimarron were south of the dead line. But if he actually said what Mooar claimed, and he likely did, he wasn't reiterating official policy. Indian agents in the region followed the treaty's stipulations and arrested buffalo hunters who crossed the dead line. Claiming they had Dodge's blessing, though, enough buffalo hunters crossed the line and shot down treaty-protected buffalo to precipitate the so-called Red River War of 1874. As further evidence of the lack of any real policy in the military's approach to Indians and animals, though, in the Northwest that same decade military officers were regularly *escorting* Indian groups like the Cayuse and Salish through white settlements so they could "go to buffalo" and continue their treaty-guaranteed traditions of hunting off-reservation out on the plains.

Philip Sheridan, commanding the Missouri Division of the West with headquarters in Chicago, did weigh in on the buffalo slaughter. Sheridan has been vilified as a buffalo arch-villain for most of the past century for the speech Cook claimed for him. But in the real world Sheridan had a naturalist's interest in western animals. In the winter of 1871–72 he, Custer, and Bill Cody were the white guides for Russia's Grand Duke Alexis's safari on the Nebraska and Colorado plains, a hunting party of five hundred, including famous Lakota war leader Spotted Tail and even a regimental band for camp entertainment. But as the buffalo slaughter moved from south to north, in 1878 Sheridan sent a telegram—long hidden away in the National Archives—complaining to the War Department about the market destruction of bison on the Northern Plains. If it continued, he wrote, it would starve the tribes and bring on a level of federal responsibility for Indian welfare he wondered if Washington was willing to shoulder. A man who would spend the last years of his life protecting bison and other surviving animals from poachers in Yellowstone Park went on to say: "I consider it important that this wholesale slaughter of the Buffalo should be stopped."

The most accurate thing to understand about the U.S. government, then, is that it *had* no policy with respect to buffalo, or any of the other ancient American animals dying on the public lands. Washington had taken no steps to preserve the great auk, and it likewise made no effort to stop the frenzy of killing passenger pigeons. Timidly, the federal government stood aside and allowed buffalo and pigeons and dozens of other animals to remain wildlife commons, open to unregulated exploitation. If human pursuit of them led to their entire annihilation, then—however the sentimental might mourn—that was just a by-product of a free-market economy.

As Colonel Dodge observed in the book he ended up writing, *The Plains of North America and Their Inhabitants*, instead of the destruction of the bison leaving Native people no choice but reservation life, the cause-effect actually went in the opposite direction, anyway. Western Indian policy had already located the tribes onto reservations by the 1870s and it determined to keep them there. Once that happened it was safe for parties of market hunters to scatter across the West and shoot down every animal they could find.

None of those who shot down or mourned the last wild buffalo had anything to say about the implications of Charles Darwin's books. Even to Hornaday (who knew Darwin's work) and Dodge (who was acquainted with Harvard's Louis Agassiz Fuertes, who thought Darwin's ideas "a scientific mistake"), the buffalo had invited its own destruction. Hornaday charged buffalo with "phenomenal stupidity" and "indifference to man." To Dodge, bison confronting human predators reacted with "imbecile amazement." Darwin's natural-selection idea could have helped here. That bison might not have had time to adapt to a brand-new world of Sharps rifles, railroads, and stock exchanges didn't seem to register. Instead they were unworthy of life in modern America.

THE WIPEOUT of the charismatic animals west of the Mississippi played out from desert plains to surf-washed coasts in twenty-five horrific years,

1865 to 1890. A few of the ancient western natives—wolves, coyotes, and wild horses—joined passenger pigeons, giant woodpeckers, and native parrots and straggled into the twentieth century. But by 1890 there were five transcontinental railroads from the Midwest to the Pacific, and everywhere a rail line reached (or could be fed by a steamboat or a wagon), animals died by the hundreds of thousands, shot down or poisoned so their skins could be processed into leather or household ornaments.

It happened on the Southern Plains, where market hunters in Kansas tried to keep going by selling pronghorn meat to local butchers for two to three cents a pound. Some thirty-two thousand pronghorn and deer skins went to an Iowa shipper in 1873, but he was able to get only a dollar apiece for them after all the effort of shooting, skinning, and shipping. A Texas cowboy searching for stray cattle mounted the Llano Estacado plateau in West Texas in 1884 to find nothing left but pronghorns and wild horses. No buffalo anymore, no grizzlies, no elk, no deer, not even any wolves. The antelope and horses didn't last, either. Fifteen years later naturalist Vernon Bailey would do a transect by train across the Llano Estacado. He saw no wild horses and barely two dozen pronghorns.

It happened on the Pacific Coast, where in a five-year stretch before the Civil War, hunters killed 423 grizzly bears in the North Cascades. In California a single shooter in a single summer killed 5,000 antelope and deer desperate to get at water during a drought.

It happened in the Southern Rockies, where, exploring the Pecos River headwaters in New Mexico's Rockies in 1882–83, naturalist Lewis Dyche found "that there were no elk in the country except a rare and occasional stragler." He saw no bighorn sheep and he neither saw nor heard wolves, although in an open parkland along the Pecos he did see a "herd" of eleven grizzly bears. A decade before, with buffalo still on the plains, those bears probably would have been out on the grasslands. But by the 1880s grizzlies everywhere, like elk, had fled from the endless peal of gunfire to hide in the mountains.

What initially lured many non-Indians into the West's market hunt was their discovery that it was remarkably easy to collect wolf pelts using poison bait. Wolves were always willing to scavenge carcasses rather than risk injury in a hunt, and dead animals that could serve as bait stations for strychnine poison were everywhere. Distilled from the imported nuts of an East Indian tree, strychnine became available in America when a firm in Philadelphia began offering cheap packages of it in crystal form in 1834. Since there were few predators left in the East by then, most of the poison went west, sold in bulk in every store and trading post. There were no restrictions of any kind on its use, and as a result a whole new frontier occupation sprang into being: "the wolfer." Poisoning animals didn't even require a wolfer's presence. And unlike trapping, poisoning didn't really require any skill. You just baited a carcass or put out chunks of meat laced with poison and then headed to camp to enjoy life while strychnine did its work.

Teams of wolfers driving ox-drawn wagons began laying out strychnine in the Yellowstone country as early as 1864. Approaching a buffalo or horse carcass they'd baited, wolfers would start finding victims, appearing sprayed across the landscape as if by some spinning centrifuge, a quarter mile from their bait animal. The targets were wolves and coyotes but the poison killed everything: eagles, vultures, ravens, magpies, red foxes, gray foxes, swift foxes, kit foxes, skunks. It killed them through asphyxiation, a death preceded by convulsive cramping so violent it twisted an animal to death, with vomiting that sprayed poison across the grass. Poisoned grass could take out collateral victims like horses. Native people hated wolfers for that. Forty dead wolves per baited carcass was common. One party in the Texas Panhandle picked up sixty-four wolves one morning barely a mile from their camp. They made $4,000 in one winter. In Kansas, James Mead once poisoned eighty-two wolves in a single baiting. For more than two decades wolf and coyote pelts traded as money in the West, worth $1 apiece and $2 if you could get them all the way to New York.

There are no figures for this most disgusting of wild-animal economies, and it is little remembered in history. But there is every likelihood that from the 1860s through the 1890s, poisoning wildlife for money killed western animals in numbers that competed with the death tolls of buffalo. In this ability to kill animals en masse, Americans were unmatched.

*

THANKS TO THE EFFORTS of a trio of Park Service historians who assembled accounts of it, the market slaughter of the West's last big animals is documented better than anywhere else in the Yellowstone National Park country. The Yellowstone plateau had ceased to be just another

part of America in 1872, when Congress withdrew from private homesteading a block of two million acres of federal land to create the world's
first national park. Making Yellowstone a wildlife park or game preserve
never came up in the debates. But naturally the question did surface about
whether Congress should allow the pursuit of "game and fish for gain or
profit." Remarkably, the consensus was that there ought to be no hunting
in the park at all, let alone market hunting, for fear of "an entire destruction" of the area's wild animals. This federal ban on market hunting, even
if restricted to a single block of land in the world's first national park, was
an unprecedented move in the United States.

Straddling the Rockies' Continental Divide, the Yellowstone Plateau
was the central place setting of an enormous ecology of island mountains
rising out of the vast grassland deserts of the northern West. It was one
of the most wildlife-rich regions in North America, a kind of Kentucky of
the West, containing all the charismatic animals and birds that had made
John James Audubon swoon. The descriptions of abundance are legion
and they attach to almost every western animal imaginable.

The trapper Osborne Russell, who in the 1830s explored and wrote
about the plateau as a Garden of the World, told of "thousands of mountain sheep" in the surrounding ranges, an account later seconded by a
camper near present Gardiner, Montana, in 1866, who wrote that "you
could see thousands and thousands of mountain sheep, all fat as pigs. We
never went over 100 yards from camp to kill a sheep. . . . You couldn't
look anywhere without seeing sheep." There were grizzlies everywhere,
too. Russell wrote that because of all the wild plums and cherries, grizzlies were more numerous here than anywhere else in the mountains: "I
have frequently seen 7 or 8 standing about the clumps of cherry bushes on
their hand legs . . . merely casting a sidelong glance" at intruding humans.
The Crow Indian name for the Yellowstone River was "Elk River." Millions of pronghorns thronged the nearby plains, with hundreds of thousands migrating in advance of big winter storms, their white rumps
resembling nothing so much as a lake pushed into foamy whitecaps by a
driving wind. Army Corps explorer W. F. Raynolds wrote in 1860 of the
"immense number of wolves." As for bison, when the United States proclaimed the existence of the park there were still probably four million of
them on the Northern Plains. Russell's compatriot in the beaver-trapping
business, Warren Ferris, offered one way to imagine them: "Immense

Stone sheep.   Photograph by Dan Flores.

herds of bison . . . in every direction [were] galloping over the prairie, like vast squadrons of cavalry."

But it seemed that no sooner had Washington set Yellowstone Park aside than market hunters calibrated their compasses for the very spot. The Paradise Valley, carved during glacial times by the Yellowstone River as it exits the park, became the early staging ground for market and safari hunters around both the Yellowstone Plateau and nearby plains. In 1872, the year the park came into being, the Bottler Brothers Ranch there made an initial pelt shipment to New York of 301 elk, 555 wolves, and 250 deer and antelope. By 1870 the nearby town of Bozeman was emerging as one of the important shipping points, the market hunt seeming to strike many early Bozeman residents as "an attractive mode of obtaining a living," as one of them put it. One group in Bozeman reported a wage of $350 a man for a winter's work of killing and skinning.

Then things started getting serious. In 1874 Bozeman market hunters were hip-deep in the big bonanza. That year they shipped out 48 *tons* of elk skins, 42 tons of deerskins, 17 tons of pronghorn skins, and 760 pounds of bighorn skins. In numbers of actual dead animals, that translated to

American elk in the West. Photograph by Dan Flores.

7,700 elk, 22,000 deer, 12,000 pronghorns, and 200 bighorns. The killers supposedly even shot pregnant females and left murdered animals to rot when they couldn't deal with them, in the case of elk wasting 300 to 500 pounds of meat for every animal killed. The 1874 killing orgy had a significant effect on wildlife, as the next year the take in elk skins was down to fewer than 15 tons, and deer and pronghorn skins together dropped to 17 tons. But wolves (1,680 skins), coyotes (520), and bears (225) made up for the falling ungulate numbers. By June of that year Bozeman had shipped out $60,000 in animal parts, a sum that translates to roughly $1.4 million today. At the request of their chiefs, the next year the nearby Crow Agency moved to ban non-Indians "from trapping, hunting, or wolfing within the boundaries of the Crow Reservation."

Midwesterner Philetus Norris, a former trapper himself, became the second superintendent of Yellowstone Park in 1877. Norris quickly conveyed the idea back East that the region around the park possessed a world-class abundance of animals and that they were "fearless of and easily slaughtered by man." Norris estimated that from 1875 to 1877 market poachers killed seven thousand elk within park boundaries, then poisoned

Depiction of the destruction of western animals that appeared widely in the late nineteenth-century press.

the discarded carcasses with strychnine for a second harvest of wolves and coyotes. The "hunter mountaineers," as he called them—a "small but despicable class of prowlers"—told him that so long as the government stood aside they planned to continue doing exactly as they wished. In fact Montana Territory had already proclaimed limited hunting seasons on bison, bighorn sheep, moose, elk, and pronghorns, which couldn't be killed from February to August. But as with similar laws elsewhere, the territory made no effort to enforce these restrictions. The result, as had been the case across America since colonial times, was general sniggering at the futility of government.

So the outrages went on. In 1881 the *Sioux City Journal* described a steamboat called the *Terry* descending the Missouri so draped with products from the market hunt that, save the smokestack and pilot house, the boat appeared as a floating, mounded island of skins. Transferred to trains, the winter's take of hides coming down just this one river filled 350 boxcars. "Nothing like it has ever been known in the history of the fur trade," the paper marveled.

In the plains country not far from Yellowstone, bison were now entirely gone. Pronghorns were down to a few wild bands. There were still some bighorns and elk, but as if to imply how fragile even those survivors were, in the winter of 1884 a group said to consist of "guides and cowboys" shot down 1,500 elk within sight of Mammoth Hot Springs, horrifying the park's tourists. In an effort to protect what was left, Philip Sheridan approached *Forest and Stream* to publicize his proposal to increase the size of the park to include nearby valleys and plains where animals migrated in winter, so "that the noble game of the Rocky Mountains might find a retreat from skin hunters." But the park's size really wasn't going to make any difference so long as no one seemed serious about stopping such horrors. By then tourists visiting Yellowstone rarely saw animals anymore. The few bison were invisible, cougars rare, grizzlies uncommon, and— poisoned and now bountied by the tens of thousands—even wolves and coyotes went unseen and unheard. The destruction of virtually all the large wild animals across the entire region was the most important event in brand-new Yellowstone National Park's brief history, save establishing the park itself.

THE EARL OF DUNRAVEN—his full name was Windham Thomas Wyndham-Quin, 4th Earl of Dunraven—spent much of his life in elite English circles. Born to privilege, Dunraven had fellow aristocrats and politicians as friends but he also consorted with painters and actors, even scientists. His circle made up the audience at the famous evolution debate at Oxford in 1860 when Robert FitzRoy, formerly the captain of the HMS *Beagle*, stood up and, waving his Bible, exclaimed of Darwin, "Had I known then what I know now, I would not have taken him aboard."

While Darwin and evolution were the talk of the scientific world, the American West and its animals provided a classic elitist escape for Dunraven. Hunting first in Colorado and Nebraska, Dunraven was soon drawn north by stories of wildlife in primeval abundance in Montana. Eventually he found the Bottler brothers in the Paradise Valley and hired them as guides. Colonials in East Africa claimed one couldn't go native without taking on a lion, which in America Dunraven translated into a grizzly bear hunt. Word was that if left alone a grizzly rarely engaged humans, but if attacked the bears tended to respond in kind, and with unparalleled vigor. So Dunraven's guides dutifully put him on to a grizzly. But when he fired and the enraged animal whirled to locate its tormentor, the nobleman's bravado collapsed. "I never heard any beast roar like it before, and hope I never may again," he shakily wrote in his journal. "It was the most awful noise you can imagine." As the wounded bear searched for him and bellowed its final, blood-drenched breaths, "I lay on the ground as flat, by G-d, as a flapjack," he admitted.

A pathetic finality was in the air as animals vanished on all sides. And Dunraven was sensitive enough to sit beside a fire one night and pen a premonition of the consequences. On this particular day he and his guides had run their horses into a huge herd of elk. For a few moments it had been exhilarating: "This elk running is perfectly magnificent. . . . We gallop after them like maniacs, cutting them off till we get in the midst of them, when we shoot all that we can," he exulted. But as the adrenaline ebbed, Dunraven experienced an odd ennui. At the end of the shooting he got a glimpse of the future. He tried to fight it off but he knew it was real.

"In a second it was all gone. . . . There was not a living creature to be seen, and the oppressive silence was unbroken by the faintest sound. I looked all around the horizon; not a sign of life; everything seemed dull, dead, quiet, unutterably sad and melancholy."

That was the silence and emptiness that had come for wild America.

CHAPTER 8

LAST RIVERS
ACROSS THE SKY

When Galileo offered his telescope to Catholic leaders so they might see Jupiter's moons, the church elders refused to look, declining to witness with their own eyes that not everything in the universe orbited around the Earth. Charles Darwin also gave humanity a revolutionary new way to understand nature and how we fit into it. Evolution made it possible for the Western world to emerge from a long fable wherein we stood separate and exceptional from the rest of Earth's life. Science had now confirmed forgotten intuitions about ourselves, that with respect to animals, we were them and they were us. But offering was one thing, accepting entirely another.

The American tragedy in the decades either side of the opening of the twentieth century, when the United States was inflicting a telescoped destruction of its wild creatures, is lodged in that Darwinian irony. Despite the epic breakthrough in discovering evolution, the philosophical and economic—and biological—momentum of the American story had become unstoppable. Initially accepted only among a small circle of scientific elites, evolution's insights had little chance of blunting that trajectory. We humans have often struggled to show empathy even for others of our own species. Seeing other creatures as kin and with compassion was going to be a difficult step.

But Americans needed such a step, because from the 1880s to the outbreak of the Great War, iconic birds and animals from the continent's deep
past were disappearing from view like grains of sand in a howling wind-
storm. There was, at least, a small but powerful cadre of activists who took
up a defense of the incomparable natural bounty Old Worlders had inherited in America. Yet vast numbers of people appear to have been uncom-
prehending, or engaged in spurious denials, or just oddly unmoved. Maybe
humanity's self-inflicted crises have always played out this way.

≈

IN THE SUMMER OF 1889, two career naturalists were developing a grand
new theory about American wildlife. The senior of the two was a native
of New York City and a graduate of Yale and the medical school at Columbia. Back in the early 1870s, as a precocious sixteen-year-old, Clinton Hart
Merriam had accompanied government explorer Ferdinand Hayden onto
the Yellowstone Plateau and written a report on the birds of that region.
Now he was rotund and sported the wild sideburns of a Civil War general, but he was still studying birds in the classic nineteenth-century manner: shooting them and preserving them as "specimens." In 1889 Merriam
was heading up the U.S. Department of Agriculture's new Division of
Economic Ornithology and Mammalogy. The division was a first glim-
mering of what would one day become the U.S. Fish and Wildlife Service,
and Merriam was hoping to make a bold contribution to its new publication series, North American Fauna.

Merriam's intent was to spend the summer investigating the bird
and mammal life of a fascinating part of Arizona Territory, namely
from the depths of the Grand Canyon and the stark, mounded badlands
of the Painted Desert, up the slopes of the nearby San Francisco Peaks
to their snowy summits. When Merriam wrote about that transect in
the resulting publication, he claimed the trip's unanticipated surprise
was the "discovery that there are but two primary life areas in North
America, a northern (boreal) and a southern (subtropical), both extending completely across the continent and sending off long interpenetrating
arms." Merriam believed that these life "areas" weren't based on eastern-
versus-western conditions, as scientists had long assumed, but instead
had formed entirely around altitude, humidity, and sun exposure. As the

party climbed from the hot, arid deserts to the cold, humid mountain summits, the New York scientist decided they had passed through seven distinct life "zones" of the southern area. Remarkably, it seemed that the highest elevations on mountains even in Arizona harbored life similar to species found in Alaska! At the same time, tropical species from far south in Mexico reached to the lowlands of the Painted Desert and the bottom of the Grand Canyon. As Andrea Wulf, author of a definitive biography of Alexander von Humboldt, reminded me over dinner in Santa Fe in 2019, Merriam's idea wasn't exactly original. Humboldt had famously proposed a life-zones model from Mount Chimborazo in Ecuador. But that was plant based. Merriam's was different.

The other naturalist who accompanied Merriam that summer was almost the New Yorker's polar opposite. Vernon Bailey was a slight, owl-eyed, chinless farm boy from the tiny settlement of Elk River, Minnesota, with a total of seven years of grammar- and middle-school education. While Merriam was urban, educated, and cosmopolitan, Bailey sprang from the kind of rural population whose young men often earned cash killing animals for the market. He was also a churchy puritan who made it a point to find a chapel to attend wherever he was every Sunday. Bailey never relinquished those attributes. He would spend his entire life as a teetotaler who no other American scientist (so they said) ever heard let fly even a mild curse. That summer of 1889 he was in his twenties, likely had little idea (yet) who Charles Darwin was. But he was very good at catching animals and preserving them as specimens.

Merriam and Bailey were forming a union to establish the direction of a new federal agency, which by 1905 would bear a famous name: the Bureau of Biological Survey. They more than anyone else would also dictate how the federal employees in America's brand-new national forests and national parks would promote and "manage" the wild animals in their care.

❧

THREE YEARS BEFORE Merriam and Bailey camped in the San Francisco Peaks, an amateur ornithologist named Frank Chapman went girl watching on the Upper East Side in Manhattan. Over two wintry days in February Chapman looked over more than a thousand women on New York's city streets and noticed that 700 of them were wearing hats. Of those 700, he saw 542 women's hats adorned with bird feathers and in several instances with entire birds, making it appear that female humans were strolling the streets of the nation's largest city with living creatures in their hair. Chapman scribbled down the bird species he saw and came up with 160. Bluebirds and robins and Baltimore orioles were among them, but the women who drew the most envious sideways glances wore hats bearing the dazzling-white plume feathers of tropical wading birds. Feathers from those birds—snowy egrets, great blue herons, roseate spoonbills—were almost exclusively from semitropical Florida and Louisiana. Chapman publicized what he'd seen in his Manhattan walks and as a result became a major player in a recently formed scientific organization called the American Ornithological Union.

As the 1800s drew toward a close, the business of hat-making that had once extirpated beavers had transitioned to a new target. Having abandoned fur, hat fashion now sought ornamentation, in particular the colorful, recurving feathers of a variety of large birds and the stuffed bodies of small ones. The so-called millinery trade, with global headquarters in Paris and New York City, evinced a cultivated myopia, a conscious nearsightedness on behalf of fashion. The men and women who were its customers displayed a complacent cluelessness, or perhaps it was a mere indifference about an industry that threatened creatures with outright extinction so that humans might impress one another on city streets.

It was America's heron family of wading birds that suffered most from the turn-of-the-century millinery market. The heron family includes several species of egrets that hunt fish and frogs in shallow waters. These are birds that nest together in large colonies called rookeries, which these species had learned to site on islands where patrolling alligators kept raccoons and other predators at bay. Herons and egrets are what is known as altricial nesters. Their chicks hatch naked, remain helpless for weeks, and their parents feed them by regurgitating food. The chicks require intensive care, and in Florida and elsewhere along the Gulf Coast in the 1880s and after, the care adult birds lavished on their young turned out to be a fatal weakness.

What drew human attention to heron rookeries in the first place was bird sex, or at least the display accoutrements that accompanied avian sexuality. During the breeding season, herons of all species attract mates by rapidly growing lacy, plume-like feathers on their backs, heads, and necks, the so-called *aigrette* feathers. Aigrette plumes can be twelve to sixteen inches long, recurved, filamentous, and gorgeous. On the North American continent, the aigrettes of the snowy egret, in particular, are one of the most beautiful things in nature. In the nineteenth century they rivaled the glossy-soft pelts of sea otters in their ability to transfix admiring human attention.

Thus was another animal market born in America, stretching from marsh to city, starting with rural plume hunters and local shippers and including fashion designers, unskilled workers in millinery factories, newspaper advertisers, and shoppers visiting retail stores on Saint-Germain in Paris or Fifth Avenue in Manhattan. Outlawing a market like this was sure to have economic consequences. A plume that brought a Florida hunter anywhere from 10¢ to $10 graced a hat that in Boston cost $130. By 1900, making those hats employed eighty-three thousand workers, nearly all of them women. Plume birds in the Deep South weren't the only targets, either. Dismayed ornithologists reported that in one four-month stretch a single town on Long Island supplied 70,000 shorebirds for the millinery trade in New York. In 1886 the skins of 400,000 American hummingbirds sold in London in a single week, and in the early 1890s a hatmaker in Paris commissioned 40,000 birds from a shipper in Virginia. London's Commercial Sales Rooms in 1902 offered packages of plume feathers from 192,960 Florida and Louisiana herons killed at their nests.

As early as the mid-1880s some of the same activists who had protested the slaughter of bison, elk, and antelope in the West started writing exposés about this shocking new destruction. As a young student growing up on Manhattan Island, future *Forest and Stream* editor George Bird Grinnell had learned about ornithology from Lucy Audubon. Now he used his position to take on America's market hunt for birds. In 1886—the same year Chapman went girl watching around Central Park—Grinnell created the first iteration of a new activist organization he called the Audubon Society. At that point, by his calculations, the number of birds dying for fashion had reached five million a year.

A more professional scientific organization, the American Ornithological Union, also emerged from outrage over Grinnell's exposés. In a special issue of *Science* the next year the AOU decried the avian market hunt, noting the similarity of the South's "jungle river trips" to shoot

tropical birds and alligators to train trips in the West where passengers had murdered bison and pronghorns for fun. The AOU went so
far as to draw up a model law for state adoption banning market hunting for all birds. Through its new publication, *Bird-Lore*, edited by New
Englander Mabel Osgood Wright, it recommended limited open seasons
and daily kill limits for "game birds": ducks, geese, cranes, quail, grouse,
and pheasants. More problematic was the AOU's designation of several
tiers of bird categories. It recommended a different kind of treatment for
"song birds," that is, no pursuit at all, particularly not for the market. On
the other hand, the AOU's "birds of prey," which included eagles, condors, hawks, falcons, kestrels, and owls, fell into an ancient Old World
category of "harmful" species. These birds rated no protection at all. In
fact, they should be killed by every means possible. Wading birds like
herons and egrets, trumpeter swans, and a great many shorebirds fell
outside the three groups.

The horrors of the plume hunt among these latter species increasingly
left early activists dumbfounded. An ornithologist named William Scott,
writing in *The Auk*, the new journal of the AOU, related that after visiting
a massive, two-hundred-acre nesting island in Florida known as Maximo
Rookery he'd discovered that one notorious plume hunter was actually
settling nearby. Over the next five breeding seasons Maximo Rookery
ceased to exist. Scott believed the town of Fort Myers was the primary
connection between northern plume buyers and rural southern hunters.
Some buyers outfitted forty to sixty hunters a season, guaranteeing prices
for plumes and bird skins, which sold to museums and taxidermists. The
best-paying birds were always snowy egrets, but the millinery agents
would buy almost anything, even hawk and owl feathers. Viewing a rookery destroyed by this economy, the ornithologist wrote: "I do not know
of a more horrible and brutal exhibition of wanton destruction than that
which I witnessed here."

In Bartram's and Audubon's America, herons and egrets had nested all
across Florida. But by the 1890s market hunters had shot out their rookeries in the northern part of the state, driving the birds southward into
the Everglades and to the islands and mangrove swamps of the Florida
Gulf Coast. Here unfolded a national scandal. After an extensive search,
a hunter named George Cuthbert ferreted out the largest rookery left in
Florida. Droppings and discarded food rotted in the sun as herons, egrets,

Snowy egrets, Florida's Cuthbert's Rookery, postcard. Digital Commonwealth.

spoonbills, and ibises swooped in and out, feeding their chicks. Cuthbert smelled the rookery first. It was on an island so dense with birds that from a mile away it looked like a roiling cumulus cloud resting on the water's surface. In their first two shoots, Cuthbert and a companion took 1,165 plumes. "Cuthbert's Rookery" became famous in part because it was one of the last big ones. Even as the real version was shot up, in New York Frank Chapman attempted a kind of salvage simulacrum of it as an exhibit in the American Museum of Natural History. In the 1890s that seemed the best America could do.

As for the law, it was the same old story. From colonial times onward various governments had put laws on the books about killing wild animals. The citizenry had just as often ignored all such attempts to curb what they regarded as their natural right. In rural America killing animals to sell was one of the few ways men with no training for much else could bring home some cash. George Cuthbert provided a model. Following two years of shooting up "his" rookery, he was able to retire, purchase a schooner, half an island, and a houseful of fine furniture. His "success" was the 1880s moral equivalent of a successful Florida cocaine deal a century later.

In 1901 the AOU finally persuaded Florida to pass a law outlawing killing birds for the market. But in keeping with tradition, the state made no effort to provide any enforcement. So the next year the AOU itself hired a young man named Guy Bradley, a former plume hunter, to enforce Florida's bird law. Guy Bradley was thirty-two years old and a family man from the tiny coastal town of Flamingo, a burg named because it was the final stopover for the imperiled flamingos that plume hunters were soon to chase entirely out of Florida. The AOU asked Bradley to supply the names of any New York plume buyers he could discover and to confront any tour-boat shooters he could find. Bradley knew his primary task was to bring plume hunters, many of whom he knew personally, to justice. He also knew that a warden "must see them first, for his own sake." It was Guy Bradley's destiny to show America what happened when at long last someone actually tried to enforce wildlife restrictions.

The cocaine analogy isn't a stretch. By the first decade of the twentieth century, with populations of the birds plummeting, the price for egret plumes had reached thirty-two dollars an ounce. That made heron feathers more precious than gold. So it didn't surprise the bird activists to learn from Bradley that other plume hunters had located Cuthbert's Rookery and the last great heron nesting site in Florida was now in rotting ruin. What came next was predictable. On the morning of July 8, 1905, Bradley attempted to arrest a hunter he knew well. That morning the man had arrogantly killed plume birds right in front of him. When Bradley intervened, the hunter's father shot and killed Florida's first game warden.

Bradley's death became a national story. It also had a coda: a Florida grand jury set his murderer free. That same year, up in Pennsylvania, poachers fired on fourteen game wardens employed by the state, wounding seven and killing four. Three years after that, another AOU/Audubon Society warden in Florida, Columbus McLeod, disappeared. Searchers never found his body. A few months later someone ambushed and killed yet another Audubon warden, Pressly Reeves, in South Carolina. Enforcing wildlife laws was dangerous as hell. Animals didn't shoot back, but those who stalked animals for profit definitely did.

Growing consternation over the plume-bird market produced results the slaughter of western animals hadn't. Large numbers of women worked within the new activist organizations and they began to refocus the blame, raising consumer awareness at the *demand* end of the equation. The bird activist Minna Hall took this appeal directly to upper-class women in the Northeast, imploring them to turn away from a fashion statement that destroyed millions of birds.

The United States was now on the cusp of the modern age. There would soon be automobiles parked outside houses, roads connecting cities and towns, electric lighting, motion pictures. Watching as so many legendary American birds and animals were going under, aware citizens began doing a modern thing. They began to question whether the market was really justified in thinking of living animals as little more than commodities awaiting exploitation. For the first time there was widespread anxiety about what Americans were doing to the natural world. Some historians have called this "frontier anxiety," a growing fear about the loss of an abundance Americans had exploited for three hundred years. To many who were alive then, it seemed as if the country was confronting a loss of its innocence.

AT LEAST FIFTY THOUSAND years ago a creature Old Worlders would one day call the passenger pigeon emerged as the single most numerous bird species on Earth. How numerous was a moving target, since the coming and going of ice ages and forest types in prehuman America had expanded and contracted their range for millions of years. By five hundred years ago they were confined to the forests east of the Great Plains, perhaps by Indian-set fires that burned back encroaching woodlands. At that time America probably held between three billion and ten billion passenger pigeons. A billion is a thousand millions and at those numbers passenger pigeons made up between 25 percent and 40 percent of all American birds. Today house sparrows, European starlings, ring-billed gulls, and barn swallows are the only birds whose global populations top a single billion.

None of us alive in the twenty-first century has ever experienced anything like the full-body impact of a large passenger pigeon flight. It was a multisensory overload, and it often left people shocked, strung out. Millions of beating pigeon wings created a roar like a tornado shaking down a forest, or a hurricane hitting shore. The air they moved was a wind on the hair and skin. The flocks emitted a peculiar scent witnesses struggled to assign, something like the smell of a very large chicken farm, though gamier. With hearing, touch, and smell all engaged to their limits, the visual impression added the unbelievable, and beautiful. The sight (so Alexander Wilson wrote) often resembled "the windings of a vast and majestic river." It was one of the continent's most profound natural spectacles. John James Audubon said of these feathered rivers that when a hawk or falcon swooped into one, the whole body of birds performed a curvilinear swerve. Then, like a snowmelt stream routing around a newly deposited boulder, all the succeeding flocks would reenact the same movement all day long. Audubon saw one bird "river" that ran for three days.

From the generations that got to hear, smell, and see passenger pigeons, there are hundreds of open-mouthed accounts, even if few were written with the observational skills of a Wilson or an Audubon. But basic math could demand literary flourish. Digest this: the largest nesting site ever reported, near Sparta, Wisconsin, in 1871, spread across *850 square miles, and many of the trees in that nesting contained 400 nests per tree.* Here's another to absorb. The midwestern passenger pigeon historian, Arlie William Schorger, reprised an Alexander Wilson technique and translated an account of a vast, migrating flock in 1860 based on flock width, bird density, flight duration, and a guessed-at speed of sixty miles per hour. He came up with a figure of *3.7 billion* pigeons. In one flock.

Stories by eyewitnesses invariably focused on the effect of pigeon numbers, how their droppings fell like snowflakes and covered the ground beneath roosts like a blizzard blowing out, how the birds descended on a forest in the manner of swarming bees. And, in the end, how the pigeon "harvest" sent staggering numbers of bodies by boat and train to supply New York and Boston and Chicago with cheap protein. The individual birds got less attention, but they were beautiful creatures, large and athletic. People compared them to mourning doves, but they resembled doves about the way a wolf resembles a coyote. Male pigeons could be a foot and a half from beak to tail, with a two-foot wingspan, a tail pointed in a dramatic V, and a mass approaching a pound. The consistent coloring of females was a bluish-gray, but the males added an iridescent breast and a throat and tail of flashing maroon and copper, along with orange eyes and coral-red feet. Viewed in bright sunlight when a flock wheeled and landed, then fed in a series of "waves"—the birds in the rear continuously leapfrogging those in the front in a "rolling cylinder" that could present a quarter-mile front—the enameled flashing and glinting of color was spellbinding.

Passenger pigeons were so numerous because they had evolved a survival strategy similar to the one evolution had selected in bison, one that, in a true irony, made both species unlikely candidates for extinction. Both presented themselves in such massive aggregates that their sheer numbers overwhelmed predators. At pigeon nestings and roosts, fast, short-winged accipiter hawks like peregrine falcons and goshawks and Cooper's hawks had forever preyed on them. So had eagles and foxes and wolves and raccoons. But more regularly available prey regulated predator numbers, and passenger pigeons were ephemeral, appearing only when forage was available to support them. When they did appear, their adaptation produced what biologists call "predator satiation." Native people, the first human predators in the passenger pigeon's world, exploited them for five hundred human generations without materially reducing them.

Biologists have differed about how pigeons could keep their numbers so high when they seemed to lay only one egg. But reading the rapidity and synchronicity of pigeon nesting behavior, recent ecologists have speculated that it points at two nestings a year. Passenger pigeons were the poster birds for what ornithologists call precocial rearing behavior. Covered in down, with eyes open and mouths gaping to be fed, their hatchlings were the opposite of heron young, coddled and fed for months. The parent pigeons seemed almost frantic to get the whole process done with. Nests were flimsy, thrown-together affairs, the eggs hatched in a remarkable thirteen days, and fourteen days after that the adults abandoned their squab to flop onto the forest floor and figure out how to forage and fly on their own. Wham, bam, hope you make it.

Again similar to bison herds, pigeon populations rested on the landscape's varying ability to support them. Indians knew, and Old Worlders came to understand, that some years pigeons arrived in numbers so dense they destroyed miles of forests. The weight of thousands of pigeons toppled grown trees and broke limbs with a sound like gunshots, and their foot-deep droppings could bury seedlings. The next year, in the same country, there might be no pigeons at all. Ecologists now argue that millions of years of passenger pigeons' feeding habits and droppings "engineered" the species structure of eastern forests. Their specialty was hardwood mast, especially acorns but also beechnuts and chestnuts. Regional hardwoods didn't produce mast every year, so the pigeons had to shop across the landscape to support themselves. Pigeon country reached as far south as Texas and Louisiana, but tended to center around Kentucky and Ohio and northward to the Great Lakes, then eastward to Pennsylvania and New York. Just as wild bison populations ultimately evaporated down to eastern Montana, the passenger pigeon endgame played out in the upper Midwest.

Of all the grim human crimes against American animals, the passenger pigeon story occupies a special place on the shelf of historical horrors. We now realize these were birds that had been living and dying in America for fifteen million years, before our species was even a gleam in some arboreal primate's eye. Yet somehow, despite their mind-blowing numbers, passenger pigeons could not survive a mere three hundred years of the American market. Buffalo and passenger pigeons stand today as the premier case study of a core economic principle. Without regulation, free-market forces inevitably drive the species they target to extinction. As a pair of economists wrote in a widely cited international article in 1998: "The rational behavior of economic agents in a free market economy . . . is likely to lead to the loss of biological resources." America's bison and pigeons, they argued, demonstrate all too well "the inability of economic markets, and market-based public policies, to preserve biodiversity." This was a little like claiming as an epiphany that the full moon rises at sunset.

Almost from the moment of their arrival, the new Americans from the Old World suffered a form of mass psychosis when confronted with passenger pigeon abundance. Pigeon flights over the continent's new towns sent residents rushing outside to shoot them from porches and balconies, filling streets with a cannonade of urban gunfire. In 1860 residents of Cleveland shot not just guns but fireworks into overhead flocks. Pigeon flights panicked farmers—in Montreal a Catholic bishop actually "excommunicated" the birds—but the widespread replacement of broadcast sowing with seed drills in the mid-1800s ended pigeon depredations on seed crops. Pigeons did compete for acorns with the thousands of razorback hogs southerners released into their woods, but since millions of pigeons ended up as hog feed, swine won this matchup.

Colonial efforts were nothing compared to the assault on pigeons from 1850 to 1900, when enterprising Americans brought their full range of invention and technology to bear. Ordinary rural folk like Vernon Bailey's neighbors in Minnesota still made some cash killing pigeons, but professional pigeon hunters dominated the game along newly extended railroad lines across the Northeast and Midwest. Among the earliest uses of the telegraph was tracking and locating nesting pigeons. Market pros later pushed for refrigerated boxcars, allowing them to ship the birds to city game stalls and restaurants in Chicago, Boston, and New York. This is often told as an admirable story of American economic ingenuity. Except it precipitated the disappearance of a fifteen-million-year-old species.

The innovations continued. Shooting or clubbing or whacking birds out of the air with poles was amateur hour. Instead the professionals (of whom there may have been as many as five thousand by the 1880s) expanded and perfected a technique Indian hunters had used. They netted the birds, and from waterfowl hunters they borrowed the idea of bait and decoys to lure flocks down from the skies. Market pros wired the feet of live pigeons to one end of a seesaw, whose movements could induce the decoy birds to flutter as if they were landing. That drew the attention of wild pigeons to a potential food source. In pigeon-hunter jargon, the seesaw was a "stool." "Stool pigeon" lives on in American English to this day.

Once a wild flock was on the ground, the new technology was the net itself, which evolved into a spring-loaded linen mesh measuring as much as forty feet across. Released to arc over unsuspecting feeding pigeons, nets like these could capture astonishing numbers of birds. During the final years of pigeon nestings in Wisconsin, one throw netted 7,200 birds. A team of netters at a legendary 1878 nesting near Petoskey, Michigan, captured 50,000. Two years later a netter hauled in $650 worth of pigeons with six throws. Success like that produced a patented American invention, an execution pincer designed to crush the skulls of netted pigeons, saving market hunters the annoyance of what modern computer users call "repetitive strain injury."

Half a dozen American companies—the Allen Brothers of Michigan and W. W. Judy of St. Louis were two prominent ones—formed to provide urban dwellers with wild game. Once netted and pincered, pigeons traveled efficiently to distant markets. Hunters hauled the gutted, dressed carcasses by wagon to the nearest railhead. There rail employees would pack the birds in iced wooden barrels, load the barrels onto boxcars, which whisked pigeon parts off to Chicago, Boston, Philadelphia, or (the biggest market of all) New York City. In the latter case, horse-drawn wagons would fetch the barrels from Grand Central Station and deliver them to streetside game dealers such as Fulton's, which provided pigeons (and many other wild creatures) to restaurants like Delmonico's on Manhattan's Fifth Avenue.

The companies also took up the challenge of providing live passenger pigeons for the emergent shooting contests known as trapshooting. A British import, trapshooting emerged as a significant market for living passenger pigeons after the Civil War, including in southern states like Texas, which held shoots in Dallas, Austin, and Houston in the early 1880s where as many as five thousand passenger pigeons were shot during tournaments. The Texas versions paled against trapshooting contests in New York, though, where the New York State Sportsmen's Association went through forty-five thousand passenger pigeons in 1874. By the late 1880s passenger pigeons as convenient live targets was proving too cruel even for nineteenth-century American sentiments. New England states took the lead in outlawing the contests, with clay discs replacing the birds. Trapshooters still call these "clay pigeons."

After millions of years of blotting out the continental sun, the passenger pigeon's last wild flurry took the form of remarkable nestings in the Midwest from the 1870s to the 1890s. As bison had done in their final stand, when pigeon numbers dwindled the birds sought protection by congregating into single flocks of staggering size. That 850-square-mile nesting in Wisconsin in 1871 may have included 136 million birds, 75 percent of the passenger pigeons left in America. Human numbers were staggering, too. A hundred thousand people descended on this Wisconsin nesting, and as an observer put it, the "slaughter was terrible beyond any description. . . . The scene was truly pitiable."

Americans' response to these last huge nestings was purely economic. Nearby small towns boomed. In 1874 Shelby, Michigan, saw pigeons inject $50,000 into its local economy. Seven hundred thousand birds lost their lives for that particular boost, enough that in Chicago the price for a full barrel of pigeons dropped to less than 50¢. With prices that low, netters and shippers dumped tons of undervalued birds into local rivers. The huge 1878 nesting in Petoskey, Michigan, between Lake Michigan and Lake Huron, produced a similar effect, an infusion of money locally but a collapse of prices on the national market. Petoskey also drew the first real resistance to the slaughter. Michigan was one of the few states that tried to regulate the killing of pigeons, and sheriffs arrested the editor of the

Grand Rapids newspaper for setting up his nets too close to the nesting. At trial he was acquitted of all charges.

Passenger pigeons tried to nest in Wisconsin as late as 1887 but hunters drove them away before they could lay their eggs. By then passenger pigeon billions were reduced to dozens. The last accounts are pitiful, pathetic. In 1895 a hunter shot three birds near Houston. Three passenger pigeons died in Michigan in 1898, the year after Michigan became the only state to ban killing them. Indiana saw its last pigeon die that year, too. That bird's need for social interaction had led it to join a flock of mourning doves. The last three wild birds known to have been killed in America were all shot, one by a small boy near Sargents, Ohio, in 1900, another near Oakford, Illinois, in 1901, and one final pigeon near Laurel, Indiana, in 1902. By that point nature writer John Burroughs was offering a reward for anyone who could provide a reliable report of a passenger pigeon nesting colony anywhere in America. He gave up the effort in 1912.

Somehow nobody's heart—or at least nobody's heart that mattered— seemed that into pulling passenger pigeons back from the brink. The country's response to the last of the birds was almost bizarre in its lack of sentiment. The famous female named Martha, who was destined to be the last of the billions, was born into one of three captive flocks that survived into the twentieth century. Martha joined a score of other pigeons at the Cincinnati Zoo, but one by one the birds died until only Martha and a male named George (as in the country's original First Couple) remained. George made it to 1910. For the next four years Martha was the only passenger pigeon left in the world. She died of old age, at either seventeen or twenty-nine—no one knew, exactly—on September 1, 1914. The outbreak of the Great War a month prior was stealing the headlines, but in America the end of the most numerous bird on Earth produced little anguish. Only a few newspapers published Martha's obituary. The New York *Evening World* did so, but scoffed: "While we may lament in sentiment the passing of the last Passenger Pigeon, we have good cause to rejoice that nature did not fit him to adapt to civilization and stay with us like the grasshopper." That was it, of course. Like the buffalo, the passenger pigeon was not fit for an America making itself into a clone of Europe.

Why *had* passenger pigeons disappeared? Theories at the time centered on two ideas. Either passenger pigeons weren't *actually* extinct or, if they were, we had nothing to do with it. When year after year went by with no pigeons, the weight fell on the second theory. The birds had all drowned in the Great Lakes, some insisted. Automaker Henry Ford's pet theory was that they'd fled for Asia, and the reason no reports of their arrival surfaced was that a storm in the Pacific had destroyed them all. In the 1920s an ornithologist asked one of the market companies for figures on the birds it had shipped. The company issued the terse reply that the pigeon economy was over because the birds had abandoned America for Australia. The Cherokees claimed that the pigeons disappeared when the deity, seeing that white people were destroying them, whipped up a great storm that whirled the birds to sea and drowned them. The one thing everyone seemed to agree on was that these were birds it would take an awfully big body of water to absorb.

Even today theories persist to absolve us from direct blame. A 1980s article in *Human Ecology* asserted that passenger pigeons were never numerous until "released" by the impact European diseases had on Native people. Subsequent genomic population modeling has disproved that theory. British ecologist Mark Avery believes it was the loss of habitat that killed off the species. By 1872, he argues, settlement had cleared forest cover in eastern America to 48 percent of what it had been in 1620. That might indeed have been why the bird's range shifted westward, from New England and New York to the Midwest. But a 2014 article in *Biological Conservation* tested this idea and concluded that because of the wide range of foods pigeons consumed in the eastern forests, a 50 percent forest destruction could not have wiped them out. Still another effort to absolve market hunting rested on what biologists call "Allee effects," when species that are density dependent experience a decline in numbers that produces a feedback loop leading to further decline. But a paper in *Conservation Biology* in 2017 concluded from the DNA evidence in twenty-seven pigeon specimens that Allee effects had not influenced pigeon decline. One last attempt to let humans off the hook is based on the predator-satiation idea. Once pigeon numbers fell, the theory goes, predators were finally able to take a toll, wiping them out. A 2020 article in *Geography*, though, this time looking at the genetic evidence in birds from archaeological sites, implies that a speculation like that is valid only if you include human predators, who descended savagely on the birds during all their final nestings.

In the nineteenth century John Stuart Mill expressed the perfect American equation for the market in wild animals. Mill's classic definition of freedom was "doing as we like, subject to such consequences as may follow, without impediment from our fellow creatures, as long as what we do does not harm them." For Mill, fellow creatures obviously meant fellow humans only, since in America the freedom to "do as we like" was wiping other species off the face of the Earth. And that, as Thoreau articulated so well, also harmed other humans down the generations.

What happened to passenger pigeons? *We* happened to them.

ON BEHALF OF the federal government's wildlife division in the Department of Agriculture, Vernon Bailey had already seen an enormous amount of the country. In the years since they had been in Arizona Territory, his boss, C. Hart Merriam, had developed the life-zones theory into a detailed series of maps he believed would lay out the larger biological patterns of the country.

Bailey's job was to ground-proof the idea, and it sent him everywhere, including across the East and South. Somehow Merriam's altitude-based life zones didn't seem as applicable there as in the West. But as Bailey traveled, observed, trapped and preserved specimens, and wrote reports and monographs for the bureau's North American Fauna series, his boss kept directing him to forgotten corners of the country, neglected pockets of surviving nature like the Colorado Plateau canyons, the badlands of the Northern Plains, and island mountains stretching from West Texas across the Southwest to California. These were all places that nineteenth-century natural history had missed. Essentially Bailey was getting to be the Meriwether Lewis of turn-of-the-century America, with Merriam as his Jefferson. As he naturalized around the country, Bailey's family worried that while his letters had things (often not flattering) to say about French Creoles, southern Blacks, and "Mexicans," Vernon almost never mentioned girls. It would surprise everyone when he announced his intention to marry.

Unlike Jefferson, who stayed in Monticello and tried to upload a huge country through the eyes of others, Merriam himself was off to see a little-known part of North America. In 1867 an inquiry to Russia about fishing rights for residents of Washington Territory had led to an unexpected

C. Hart Merriam, Third Provisional Bio-Geographic Map of North America, Showing the Principal Life Areas. U. S. Department of Agriculture, 1893.

offer. Strapped for cash and trying to modernize his vast country, Czar Nicholas I decided to have his ambassador to the United States propose a counter: Would the Americans be interested not just in fishing rights but maybe in *possessing* Alaska? We would, although, beyond considering the Far North a propitious spot for a penal colony, we struggled to know what to do with the place. Alaska answered that question itself with a series of gold strikes—Juneau, Nome, and the Klondike in the adjoining Canadian Yukon—from 1880 through 1897.

With primeval Alaska and its rumored grand abundance of wild creatures now invaded by rapacious miners who literally lived off local animals—a boreal reprise of what had befallen the mining camps of the Sierra Nevada and Rockies—men of Merriam's kind thought to send a grand expedition of scientists and writers to Alaska who could at least observe, collect, and preserve a memory of pristine nature in the Far

North. Railroad magnate E. H. Harriman underwrote and lent his name to this "Harriman Expedition." His true intent was to assess Alaska's resources for potential exploitation, ultimately generating money his wife would later use to fund eugenics research. It was that kind of age.

This expedition got underway in the summer of 1899, with the promise of a new century in the air, and the party was a rarified group indeed. There were luminaries like writer-naturalists John Muir and John Burroughs and outdoor editor George Bird Grinnell. The young, aspiring photographer Edward Sheriff Curtis was among them, as were geologist Grove Karl Gilbert, bird painter Louis Agassiz Fuertes, ornithologist Albert Fisher, and Alaska veteran William H. Dall (who had already described the snowy-white, thin-horned mountain sheep ever since linked to his name). Merriam, who had selected them, was the primary mammalogist. He would spend the next dozen years shepherding into print a planned thirteen volumes of baseline Alaskan science, with volumes six and seven assigned to Alaska's mammals under his own pen.

Except Merriam never wrote his volumes. While many of Alaska's hundred-plus summer breeding birds—long-tailed jaegers that wafted buoyantly over the Arctic tundra, colonies of murres and waddling horned puffins in the ocean cliffs, harlequin ducks performing histrionics along the mountain rapids—appear in the Harriman Reports, mammals receive only token mentions. Harriman himself was on the trip primarily to shoot a bear, and shot a sow with a cub on Kodiak Island. The party looked for but never saw polar bears, which the industrialist also badly wanted to bag. Muir pointed out a valley where in 1890 he had heard hundreds of wolves howling. None howled there nine years later. In Alaska as everywhere else, no one paid any attention to existing wildlife regulations. The group visited a federally regulated fur seal rookery on the Pribilof Islands where, it turned out, 75 percent of the animals had disappeared over the previous decade. Grinnell noted with disgust that Alaska's fur trade was done, the "fur seals are practically gone . . . no sea otter . . . the deer are disappearing." Fur entrepreneurs were now farming foxes. So much for a pristine wilderness in Alaska in 1899.

While Merriam was in the Far North, his protégé Vernon Bailey was camping and trapping and writing reports about the landscapes and animals of the Lower 48, collecting stories like the one by a self-described "spiritual" group of West Texas ranchers, who kept a camp where "never

Vernon Bailey working on Merriam's life zones on the rim of the Grand Canyon.
Courtesy the American Heritage Center, University of Wyoming.

an oath was heard nor a bottle of whiskey" present, whose "jolliest" account was of shooting dead the last grizzly bear in Texas. The nation itself was now entering both a new century and a new stage. In 1890 the U.S. Census had announced that three hundred years of settling the American "frontier" was over. Insofar as homesteading went, that wasn't entirely true—taking up government land would go on for another forty years, longer in Alaska. The country's relationship with its frontier was psychological, though, often based on the idea that its ending meant a loss of anarchistic freedom of action. That anguish may have had deep roots. Those anxious that wildlands with few people were disappearing may have been mourning a loss of space that signified our predatory past. Open space and its animals had drawn us around the world. Now space was dwindling even here. And many of the animals were long gone.

Merriam and Bailey knew there were still some remaining pieces of

wild country out there. But not many. In 1800, when all the wild continent and its creatures had seemed to stretch out forever, the country had
a population of only six million. That was roughly equivalent to Native
America in those final five centuries before Columbus. By 1900 there
were seventy-six million people in the United States and there was no
going back.

⁖⁖⁖

BORN TO A New York family of the patrician class three years before the
outbreak of the Civil War, with a Southern mother for whom the whole
War/Reconstruction thing was conversationally off-limits, Theodore
Roosevelt early settled on nature as a suitable dinner topic. Wild animals
became the focus of long discussions among him, his younger brother
Elliott, his father, Theodore Sr., and a naturalist uncle, Robert Barnwell
Roosevelt. The family indulgence of such topics was almost reflexive. On
a Nile River descent with his father when he was thirteen, Teddy and a
new shotgun collected more than a hundred Egyptian birds for his personal "museum." Teddy's uncle, meanwhile, was restoring shad to the
Hudson River, while his father—a forceful promoter of Darwin and of
the anti-cruelty animal-rights movement—became a founder and trustee
of the American Museum of Natural History in 1877, which eventually
opened on Manhattan's Upper West Side. In a situation of many options,
Teddy surprised no one in his family when he announced he wanted his
college education to train him as a naturalist.

What Roosevelt really wanted was to roam the wild places of the world
and hunt and collect like Humboldt and Meriwether Lewis. You suspect
there were times when he would gladly have traded places with Vernon
Bailey. But attending Harvard, Roosevelt's wilderness romance collided
head-on with the reality of academic science, designed by then to produce
lab-bound Spencer Bairds, not Audubons. By his junior year Roosevelt
concluded that what he desired as a profession lay outside the natural
sciences.

He had long hoped to be a writer and by the end of his time at Harvard
he had discovered history, which turned out to be a way to explore and
hunt and collect in the imagination. This led him into a multiyear writing
project he would call *The Winning of the West*, in which he conceived fron-

tier hunters as the true heroes of the whole colonial process. Roosevelt's book came out at almost exactly the same time that the Wisconsin historian Frederick Jackson Turner used Darwinian ideas to argue in a famous essay that the American wilderness had turned Europeans into a sort of Homo americanus, a new and exceptional people forged by their relationship with nature. Roosevelt's and Turner's ideas shaped and prefigured many of the developments around American animals and wilderness for the next six decades.

But first the future president had to figure out how to save the animals that had made the country the frontier adventureland he believed it to be. This was a battle his friend George Bird Grinnell had been fighting without much success for a decade. So in December of 1887 Roosevelt invited Grinnell and ten other friends, virtually all wealthy northeastern capitalists, to launch an organization Roosevelt named after his two favorite frontier hunters.

Distinct from hunting for the market, so-called sport hunting had first emerged in America in the 1840s with the writings of an English immigrant named William Herbert, who insisted that what had made Brits the great imperialists they were was a hunting tradition among upper-class men. In Herbert's view, hunting was a noble recreation that kept men from succumbing to a threat from within, namely the creeping feminization of modern life. For men in cities, Herbert recommended a return to the virtues of colonial hunters but with some of the aspirations of the upper class. In his view modern rural hunters were "pot hunters" at best, meaning they hunted for food, and market hunters at worst. They knew nothing about the natural history of the animals they shot. But middle-class hunters could emulate the elite European model and endeavor to understand animal science. As "sport" hunters they could look on animals benevolently, with love, not as a means to easy cash. They could engage in morally superior behavior, practicing "fair chase," refraining from shooting nursing females or their young. They could even observe government laws regulating the hunt.

Restrictions like these were a hard sell to many Americans, especially in rural parts of the country. In Maine, Pennsylvania, and Minnesota, rural and immigrant hunters who wanted to keep killing animals for food and money clashed almost yearly with the new "sport" hunters, most from cities, who increasingly saw hunting as weekend recreation.

Roosevelt's Boone and Crockett Club was washed in the blood of William Herbert's philosophy. Its initial roster of a hundred full-time members and fifty associates came from the right class. All were prominent, all were easterners, all urban. They were also all men. There were celebrity painters and writers like Albert Bierstadt and Owen Wister, military heroes such as Philip Sheridan and William Tecumseh Sherman, prominent scientists like Clarence King. Animals did figure into the club's constitution, as its primary purpose was to "work for the preservation of the large game of this country" and a second was "to assist in enforcing the existing laws." Promoting natural history science was there, too.

But the core of the club was hunting. To be a member one had to have pursued and shot three different species of that "large game" with a rifle. Indeed, firearms—especially the rifle—figured as prominently in the constitution as the animals themselves. The very first of the Boone and Crockett's bylaws made clear: the club's purpose was "manly sport with the rifle."

AS ROOSEVELT AND HIS elite friends set about trying to save some elk and bighorn sheep to shoot, yet another of the country's most dramatic birds—albeit not one anyone considered a "game" bird—was in serious jeopardy. One of the many ironies about Carolina parakeets is that they were our most colorful native bird yet we hardly know them. Somehow, across three centuries Americans never bothered to learn their natural history. The birds took knowledge about their numbers, courtship, nesting behavior, and life span in the wild to oblivion with them. We do know from sightings where they ranged, and we know something about the ecologies they preferred and what they ate. We can infer from historical records that they could be tamed fairly easily. Recent genome sequencing of their DNA reveals that they were closely related to South America's sun parakeet, from which they split three million years ago. A few million years of adaptation to North America made them tolerant enough of cold climates that the eastern subspecies did not migrate. Early Dutch settlers in New Amsterdam saw the birds along the Hudson River in winter and reacted with alarm at their exotic appearance in the midst of snowy landscapes.

Carolina parakeets *by John James Audubon.*   From *The Birds of America.*

John Lawson first associated them with the Carolinas when he wrote
from there in 1714 that the "parrakeetos" were more brilliantly colored
than any birds in Europe. Mark Catesby cemented the association by cap-
tioning his painting of the bird "The Parrot of Carolina." But the birds
occurred in two subspecies (as the only members of the genus *Conuropsis*)
and actually ranged from the Eastern Seaboard to the Great Plains. A 2017
re-creation of their ranges in *Ecology and Evolution* indicates the eastern

and western versions occupied different climatic niches in geographically separated populations. Eastern Carolina parakeets hugged the Atlantic coastline and spread throughout Florida, while the western subspecies occupied an interior range from Louisiana to Nebraska. The western birds may have even migrated seasonally. Those differences gave the two versions of the birds different endgames, too.

Carolina parakeets were sizeable, roughly a foot long from beak to the ends of their forked tails. Seeing an individual bird was rare. Virtually all the descriptions we have are of flocks, which flew in compact bodies that undulated and pirouetted through the trees at high velocity making "loud and outrageous screaming," said Alexander Wilson. The flocks dazzled everyone who saw them, for Carolina parakeets wore a wild mélange of tropical-fruit-colored feathers, the shades of lemon, lime, tangerine, and turquoise blues merging and overlapping as they swirled through somber riverine forests. And in sunlight? "It is a pleasant sight to see a flock of them suddenly wheel in the atmosphere, and light upon a tree; their gaudy colors are reflected in the sun with the brilliance of a rainbow." Those were the words of midwestern ethnologist Henry Rowe Schoolcraft, in 1819.

In leafy forests America's parakeets interacted with one another in classic parrot fashion, grasping limbs with both feet and beaks, hanging upside down, twirling through the canopies like feathered gymnasts. On the ground they seemed awkward and sore-footed. They nested in large colonies in cypress swamps, and unlike passenger pigeons they raised young that appear to have been relatively altricial—the hatchlings were born blind, featherless, and required adult feeding and care for a long period. The best evidence for Carolina parakeet roosting behavior is accounts of them packing into woodpecker holes and lightning-strike cavities in old-growth trees. That adaptation turned into tragedy when Europeans introduced honeybees to America and the bees appropriated exactly those locations for their hives.

So what took out our only native parrot? Why have modern generations not been able to enjoy them in our forests and be dazzled? Various researchers have blamed the honeybee swarms or a range of other suspected threats across the seventeenth to nineteenth centuries. Others have targeted the logging of old-growth bald cypress swamps, favored roosting and nesting sites for eastern Carolina parakeets. A southern representative of redwoods, old-growth bald cypress grew 130 feet high, lived for 1,200 years, and belonged in "the first order of North American trees," said William Bartram. A skeletal bald cypress giant, a forlorn sentinel from the ancient forest, still towers over the twentieth-century woodlands near my boyhood home in Louisiana, and I've no doubt parrots once swarmed through it. European settlers targeted groves of cypress for lumber early on and razed them throughout the nineteenth century.

Invasive species and potential habitat loss appeal to us as causes because they're indirect human impacts. Unfortunately, as with passenger pigeons, humans also directly attacked parrots. A core feature of the birds' tropical origins was a thick, powerful beak, a key adaptation to eating tough fare like cocklebur seeds and cypress cones. American farmers decided that those beaks made parrots a threat to agriculture, especially to fruit crops like apples and peaches, which the birds insulted by discarding the fruit mass so they could crack the heavy seeds. Then, in the decades following the Civil War, the millinery trade discovered their brilliant feathers. Egg collectors noticed them, too, and since Carolina parakeets were reputed to be tamable, so did traffickers in exotic pets. Scientific naturalists also went after them with a vengeance, a phenomenon of the age hardly limited to this species.

Modern genetic research in the form of a trio of recent (2019–21) scientific articles in journals like *Current Biology* has finally provided the smoking gun for the loss of the Carolina parakeet. Analyzing the extinction probabilities of several eastern North American birds, among them Carolina parakeets, passenger pigeons, heath hens, and ivory-billed woodpeckers, one group of biologists found no evidence any of them had a fundamental susceptibility to extinction, no hints for any of these birds that "their time had come." Focusing specifically on extinction modeling for the two subspecies of Carolina parakeets, other authors concluded that the two types went extinct thirty years apart, western parakeets around 1914 and the eastern subspecies in the late 1930s to mid-1940s.

Science got particularly lucky with the discovery of extant DNA from a parakeet collected by a Catalan naturalist, Marià Masferrer. The researchers who sequenced the genome of that parrot went straight for the heart of the extinction mystery. They uncovered no genetic signals of the inbreeding typical of endangered species as they decline from habitat loss, and

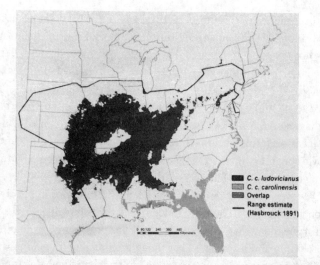

Original range of the two subspecies of Carolina parakeets. Courtesy *Ecology and Evolution.*

no evidence that the birds were in a state of slow decline in the past. "As such," they wrote, "our results suggest its extinction was an abrupt process and thus likely solely attributable to human causes."

If we humans killed off our only native parrot, tragic elements of both human and parrot nature may have played roles in dooming them. As a highly social species, the birds possessed a powerful emotional empathy for their own kind. Whenever farmer or millinery-trader or sport-hunter gunfire brought down a single parakeet from a flock, such was the birds' attachment to one another that entire flocks surrendered their caution to aid the fallen. As Alexander Wilson told it: "At each successive discharge, though showers of them fell, yet the affection of the survivors seemed to increase. . . . They again alighted near me, looking down on their slaughtered companions with such manifest symptoms of sympathy and concern as entirely disarmed me."

However disarmed he was, Wilson didn't stop shooting. Nor did anyone else, until there were no parrots left to shoot. The birds caught the eye and were easy to slaughter. One example of the common human response was a shooter in Florida in 1889 who killed every single parrot in a flock of two hundred. Frank Chapman, that counter of Manhattan's plumebird hats, may have shot some of Florida's last parrots on collecting trips for the American Museum of Natural History. Near Lake Okeechobee in April of 1904 Chapman saw thirteen and shot four of them. A remnant flock possibly frequented that area for two more decades. But there were no more sightings after 1926. There were claimed glimpses of Carolina parakeets as late as 1938 in a part of South Carolina's Santee Swamp slated to be inundated by Lake Marion Reservoir, but no confirmation. The last native American parrots in captivity—a pair named Lady Jane and Incas—left this world as thirtysomethings in the years 1917 and 1918. They'd spent their lives in the same zoo in Cincinnati where Martha the last pigeon had died a few years earlier.

They were representatives of a bird family famous for impressive cognition and memory. There are modern bird specialists, like David Sibley, who argue today that "crows and parrots perform as well as dogs in tests of reasoning and learning." That thought chills me. It could mean that Carolina parakeets were aware enough to understand what was happening to them.

⁂

IN ITS FIRST TWO decades the Boone and Crockett Club moved heaven and earth to save some of the country's wild animals. Its efforts began when the club became a force supporting the U.S. Army's thirty-two-year role in ending poaching and market hunting in Yellowstone and other early national parks. It continued when the club advocated on behalf of a wholly new idea in America. Rather than seeing the General Land Office continue to privatize every square inch of federal land through the homestead acts and giveaways to encourage rail lines and other infrastructure, the club supported the federal government retaining important parts of its public domain. And urged on by George Bird Grinnell, the Boone and Crockett Club promoted yet another unprecedented step for the federal government, the passage of a bill designed to curb the market hunting that had made American animals commodity resources for three centuries. That definitely moved the heavens.

These were revolutionary steps for a country steeped in Adam Smith economics, John Stuart Mill freedoms, and the anarchy of the frontier. The movement to end the market slaughter was truly revolutionary. Shooting, trapping, and netting animals and birds for profit had been central to Americanism since colonial times. At the turn of the twentieth century its role in wrecking the natural world had grown impossible to ignore. One New England state where the market had destroyed beavers and heath hens and episodically wiped out wild turkeys and whitetail deer became the first to try to shut down the killing economy. In 1896 Connecticut passed a measure that prohibited transporting its game birds to markets across state boundaries, in effect to New York's game marts. A challenge to that law reached the Supreme Court, and the case produced a landmark ruling. In *Geer v. Connecticut* the high court determined that the common practice since 1607 wherein colonies had assumed proprietary rights over local wild animals was legally correct. *Geer* ruled that states owned the wildlife within their borders, which meant they had the legal right to protect wildlife on behalf of their residents.

Inspired by Connecticut, in 1900 an Iowa congressman named John Lacey stepped forward with an attempt to engage the federal government in halting the abomination of market hunting. Lacey was well known among activists. Elected as a Republican from the little burg of Oskaloosa, Iowa, he had risen to chair the House Committee on Public Lands, which brought him to Grinnell's attention and into the Boone and Crockett Club. Lacey earned the members' loyalty by helping design a true reversal, the 1891 Forest Reserve Act, which had removed from homesteading and private sale fifteen mountain reserves in the Rockies, the Sierra Nevadas, and the Cascade range. That initial thirteen million acres was the start of something huge. Congress set aside the forest reserves because settlement had now reached the arid West, and the reserves protected the snow fountains that were the region's source of water. But Grinnell and Roosevelt knew mountain forest reserves preserved perfect habitat for the big animals they wanted to protect. Elk and grizzly bears, in particular, had already fled there from the fusillade on the Great Plains.

Lacey's political status with the wildlife activists improved further following a series of celebrated incidents in Yellowstone in the early 1890s. U.S. Army "scout" (i.e., warden) Ed Wilson had arrested a notorious park poacher whom a local judge promptly released, citing a lack of any law to prosecute him. A few weeks later Wilson vanished without a trace. Other scouts found his hidden remains a year later. That story brought a *Forest and Stream* reporter to Yellowstone, who soon elicited from another poacher that his own arrest and release had involved the massacre of

eighty buffalo, most of the bison in the park. At that point Congressman
Lacey introduced and secured passage of the 1894 Yellowstone National
Park Protection Act, with $1,000 fines for killing wildlife.

Six years later, with the Court's *Geer* ruling in place, Lacey decided
to take on the biggest monster of them all, a practice that since colonial
days could have made fair progress filling the Grand Canyon with the
bodies of all the American birds and animals killed for money. White-
haired, bearded, slender, Lacey was about to turn sixty. In history and in
his career, the time had come. Calling on the federal government's con-
stitutional power to regulate interstate commerce, Lacey drew up an act
he called the Wild Birds and Game Preservation Act that at last landed a
body blow on the wild-animal economy. This *federal* law stopped the shipping of slain birds, animals, and their body parts between states where
market hunting was still legal and states where it wasn't.

It was already too late for passenger pigeons and wild bison. It verged
on too late for snowy egrets, Carolina parakeets, and a lengthening list of
others. But the Lacey Act finally put a long-hesitant U.S. federal government in play. In his speech describing the bill before Congress, Lacey referenced both passenger pigeons and bison, but he proclaimed that all was
not lost, that "there still remains much to preserve." In truth, the Lacey
Act put the burden for enlightened laws on the states, which as a result
of the Supreme Court now began moving to create "game" departments
and enacting policies on behalf of some of their wild animals. For guidance in that endeavor they looked to the Bureau of Biological Survey and
to people like Merriam and Bailey.

In 1901, when an assassin's bullet cut down President McKinley, his
vice president was a forty-two-year-old former New York congressman
and city mayor whom Republican Party operatives referred to as "that
damned cowboy." Teddy Roosevelt was the youngest man ever to serve
as president, and his energy was prodigious. Here was a stunning opportunity to take on the issues that had dominated his life since childhood.
If celebrated individual freedom was the very cause of the destruction
of America's animals, not to mention its forests, its grasslands, its fisheries, then expanding government power to regulate such freedom was
the only course that could preserve the country. What Roosevelt and his
policy-makers eventually called "conservation" became his personal campaign to save the animals and wildlands he believed had created Homo

americanus in the first place. "The whole standpoint of the people toward the proper aim of government," he later wrote, simply, "were brought out first by this conservation work."

Histories of the American presidency tend to label Roosevelt's conservation programs as one of the great American triumphs of the twentieth century, and they're not exaggerating. Conservation's successes were breathtaking. To begin, Roosevelt oversaw the creation of six new national parks, a form of preserved wildlands America invented. Starting with Pelican Island in Florida, which he designated by executive order in 1903 "as a reserve and breeding ground for native birds," continuing with a pair of national bison ranges in Oklahoma and Montana, then right on through his lame-duck period, when new bird reservations (a total of fifty-one) seemed to precipitate from the heavens, Roosevelt created a National Wildlife Refuge system that preserved millions more acres of prize bird and mammal habitat. The forest reserves he and Grinnell celebrated in the 1890s were at 42 million acres when Roosevelt assumed the presidency. Renamed "national forests," they had swelled to 172 million acres (including Tongass, a 17-million-acre monster in Alaska) by the time he left it. Using the new Antiquities Act of 1906, he proclaimed another extensive body of preserves from the public lands in the form of eighteen national monuments. The Grand Canyon was one of the early ones.

In just eight years Roosevelt removed more than 230 million acres of the country's public lands from private entry, retaining them in federal ownership and transforming the future of the country in the process. Providing refuges for America's big mammals and birds was a primary reason he did almost all of it. *The Wilderness Warrior*, the title of what is probably the best recent biography of him, lays that triumphant story and sense of TR before the reading public.

But Roosevelt was human and, as are we all, a product of time and place. While he could explain the tragedy of wildlife loss with lovely analogies—"The destruction of the wild pigeon and the Carolina paraquet has meant a loss as severe as if the Catskills or the Palisades were taken away. When I hear of the destruction of a species I feel just as if all the works of some great writer had perished"—he resisted drawing the conclusions from Darwin that many disciples thought the great naturalist intended. Roosevelt had grown up versed in Darwin, read Darwin himself when he was a young teenager floating the Nile, believed from that point onward that humans were indeed primates that had evolved from earlier forms. Nonetheless, the animals he hunted did not—perhaps for him simply could not—become evolutionary near kin of humans.

While Roosevelt was president, his writer friend John Burroughs precipitated a public controversy about the accuracy of a popular new genre, a form of literary natural history that attempted to provide insight into the lives of wild animals. Most of the new authors of animal stories were not scientists. But many of them, notably John Muir, Jack London, and Canadian writer Ernest Thompson Seton, had extensive outdoor experience and employed scientific methodology. Seton, ironically enough, won the John Burroughs Medal for his comprehensive *Lives of Game Animals*. Both Frank Chapman and William T. Hornaday praised his work.

Looking back on this "Nature Faker" controversy (labeled so by Roosevelt) from today's vantage, what seems apparent is that its practitioners were trying to take on the big implications of Darwinism in the world around them. Theirs was a response—and it strongly appealed to the public—to the bleak "Nature, red in tooth and claw" conclusions Thomas Huxley, Herbert Spencer, and Alfred Lord Tennyson drew from evolution. Instead of finding humans condemned because we'd turned out to be mere animals, nature writers like Seton looked for examples among birds and mammals of traits humans admired: compassion, cooperation, loyalty, an ability to reason and to transfer cultural learning across generations. And most of all, individuality. In one of Seton's stories in *Scribner's* in 1900, a captured female coyote named Tito sits chained in a ranch yard observing the techniques stockmen are using to wipe out "the coyote kind." When she escapes and has litters of her own, she passes on her knowledge about how to avoid traps and poison, which (in Seton's telling) was why coyotes were surviving when so many other creatures were not. In his famous wolf story, "Lobo, the King of Currumpaw," Seton wrote about a canny New Mexico wolf with one fatal flaw, his fidelity to his mate. As he put the matter in *Wild Animals I Have Known*, his theme was: "We and the beasts are kin."

Given the perspective of contemporary, twenty-first-century bird and mammal research, with its emphasis on intelligence, tool use, theory of mind, perceptions of fairness, cultural transmission, and *especially* individuality, all that seemed to be lacking in the best writings of the Seton/London/Muir genre were double-blind, replicable experiments. But at the time the well-respected Burroughs, furious at what he saw as chicanery, went on the attack. Nonhuman animals, he thundered, were purely creatures of *instinct*, nothing more.

That drew Roosevelt into the fray. Simply enough, the "Nature Fakers" were misleading their readers, he told the public. In a private communication to the editor of *Collier's* magazine, Roosevelt went further. He echoed the lines in Darwin's *The Descent of Man* when he wrote, "I believe that the higher mammals and birds have reasoning powers, which differ in degree rather than kind" from those of humans. But, he continued, he meant different in degree "from the lower reasoning powers of, for instance, the lower savages." In other words (if we can take this astonishing comment literally), Roosevelt believed the resonances and parallels between animals and humanity did not really apply to modern, First World humans. Jack London, at least, would eventually get revenge, writing in *Collier's* that "President Roosevelt does not understand evolution, and he does not seem to have made much of an attempt to understand evolution."

John Muir got in a twist of the knife, too. Urged on by C. Hart Merriam, the president had the exquisite experience of camping for three nights in Yosemite with Muir. It gave the legendary writer-activist (no shrinking violet himself) a chance to confront Roosevelt around their campfire. His question was something many had wondered about but only the Scottish poetico-trampo was willing to broach: "Mr. Roosevelt, when are you going to get beyond the boyishness of killing things?" Author of numerous books and magazine articles on hunting, probably already planning the African safari he would take as soon as he was out of office, Roosevelt responded with an evasive, "Muir, I guess you are right." Then, as ex-president, in an eight-month expedition to Africa in 1909, he proceeded to shoot 5,013 mammals, 4,453 birds, and 2,322 reptiles and amphibians. All for science and America's museum cabinets, of course.

⁓

ONE TRIUMPH turn-of-the-century wildlife activists did effect was to rescue from the brink the most iconic American animal of all. From a population of millions, buffalo numbers had plummeted to barely a thousand panicked individuals. Could so few rescue the species?

Given the intimate bonds buffalo and tribal people had enjoyed for so long, it is appropriate and satisfying that so many of those who acted to save the animal from extinction were Natives. One was a Montana Salish named Latati, son of Mary Sabrine and a Pend d'Oreille stepfather named Samuel Walking Coyote, who coaxed six orphaned bison calves to follow them from eastern Montana home to the mountain valleys of the Flathead Reservation. Those six, three bull calves and three cows, became the nucleus of a famous salvaged group, the Pablo-Allard herd, managed and bred by another part Indian, Michel Pablo. The mixed-blood Lakota Fred Duprees was another Native rescuer. He caught five calves on the Yellowstone and sold them in Canada. A pair of Canadians named James McKay and Charles Alloway were also important in the early recovery of bison, and James McKay was a Métis.

Several others who loom large in this story had a different relationship with buffalo. The former Texas ranger Charles Goodnight (the inspiration for Tommy Lee Jones's "Captain Call" character in Larry McMurtry's *Lonesome Dove*) was one. Goodnight didn't start out to save buffalo. Founding a ranch in Texas's Palo Duro Canyon, in 1878 he invited hunters from Kansas to come shoot the fifteen thousand bison in the canyon to save grass for his cows. But his wife, Molly, pleaded to let four or five calves live and Goodnight relented. Those calves became the nucleus of an important purebred herd of the smaller Southern Plains bison. The former hide hunter Charles "Buffalo" Jones of Kansas was another central figure in this story. Stricken with remorse over the buffalo he had killed, Jones concluded that as wild as the last animals had become, the hide hunters could not possibly have found every last one in the tangle of Southern Plains canyons. Between 1886 and 1889 he searched that country, at one point scoring fifty-six bison calves in one roundup. Explaining his bison rescue, he said, "I am positive it was the wickedness committed in killing so many that impelled me to take measures for perpetuating the race which I had helped almost destroy." To characterize Jones as colorful barely does him justice. Appointed supervisor of wildlife in Yellowstone National Park, among other stunts he once roped and spanked an adult grizzly to teach it to stay away from campers.

It was William T. Hornaday, having left the Smithsonian to help launch the New York Zoological Park, who supervised exchanging animals among these herds in pursuit of genetic diversity. With Roosevelt's blessing, in 1907 Hornaday launched an organization called the American Bison Society, whose initial plan was to create a series of national bison ranges to save the animals. Eventually the Texas, Kansas, and Montana buffalo who were left—some biologists argue a frighteningly small group of only 88—became the nucleus of animals released onto the new national ranges. There easily might have been a larger genetic pool, but Hornaday refused to deal with the "half-breed Mexican Flathead" Michel Pablo, who responded to that treatment by selling his Montana herd—now grown to 672 animals—to Canada. So America's national bison ranges came with some baggage. They weren't large, ended up fenced to keep predators out, and only the Oklahoma range was actually in classic bison habitat. The Montana range, set in the Rocky Mountains, was not only west of the bison country, it draped across the very Mission Valley where Michel Pablo's herd had roamed before Canadian authorities hauled them away!

Disbanded in 1936 when it believed its project finished, the American Bison Society performed heroic work. But the society, with a membership overwhelmingly male and eastern (of its 723 members in 1908, 717 were from the Northeast and Midwest and there was not a single Indian), did not intend that buffalo be restored to the West as wild animals. No one in early twentieth-century America saved bison to return them to the open grasslands. The western lands the country was preserving as wildlife habitat were mountain ranges and deep canyons, poor bison country. Members of the society accepted without question that the former bison plains were now a privatized empire of farms and ranches. Grinnell thought that "necessary." Ernest Seton concluded bison were "incompatible" with the developing country. Even Hornaday, who lamented that "with the loss of liberty [the buffalo] becomes a tame-looking animal," invoked Darwin and argued, astonishingly, that natural selection favored cattle.

Jack London was right. Blinkered by their biases, *none* of these people actually understood evolution. Now, where enormous congregations of bison had migrated with the seasons, cows milled inside fenced pastures and required supplementary feed to get them through winters. The riparian zones buffalo had maintained for thousands of years fell apart under the pressures of water-needy domestic stock, and junipers, oaks, aspens, and mesquite soon invaded the grasslands that had coevolved with the vast wild herds. But these were all steps in making America in the image of the Old World.

What the Bison Society meant when it said it was "rescuing the buffalo" was that buffalo would now exist in parks and refuges so twentieth-
century tourists in sunglasses could gawk for a few minutes before
looking at their watches or phones and turning away in boredom. Thanks
to its rescuers, the wild creature whose massive herds once symbolized
America to the world was on its way to becoming another consumer
trope of Old West nostalgia, no different from roadside rattlesnake pits,
Indian curios, and Wall Drug billboards shouting at you about the amazing sights you just missed.

※

THE DECADES THAT brought an end to the centuries-long destruction
of America's birds and mammals by the market straddled World War I.
At that juncture it was people like C. Hart Merriam, Vernon Bailey, and
William T. Hornaday who had laid out the template for the country's
twentieth-century relationship with its wild animals. All three were enigmatic products of an age we struggle now to understand. But we and
most of America's animals and birds have been living for more than a
century in the world they helped create.

Hornaday is simultaneously the most transparent of the three and
the most difficult to decipher. Like Bailey's, his background—a midwesterner, graduate of a state college—was modest. The elite Boone and
Crockett Club refused membership to him. But Hornaday's successes
fashioned a formidable personality. He was a moralistic crusader, and like
many who've sought the high ground, he was arrogant, belligerent, quick
to take offense at minor slights, and slow to forgive. And he was given to
public gloating over those he defeated, an annoying tic. In short, Hornaday was a successful, self-absorbed narcissist.

It gets worse. Like several other conservationists of the period, Hornaday was also a racist, and he was public about it. He became director of
the Zoological Park because fellow conservationist Madison Grant hired
him. A blue-blood trust-funder with a variety of conservation successes,
including promoting an Alaska game law that in 1902 undermined market hunting there, with a background that ought to have produced a
genteel activist in the mold of a George Bird Grinnell, Grant somehow
morphed into an advocate of eugenics, whose major book, *The Passing of*

the Great Race, favorably impressed Adolf Hitler. (In America it appeared in time for F. Scott Fitzgerald to lampoon it in *The Great Gatsby*.) As for Hornaday, his early books brimmed with praise for upper-class, gentleman sportsmen, "the very bone and sinew of wild life protection," while devoting whole chapters to describing how immigrants from southern Europe, Blacks in the South, and working-class whites were all "human mongooses" and primary villains in the destruction of America's animals. He singled out Italians as "born pot-hunters" who were wiping out songbirds everywhere they settled and were "pouring into America in a steady stream. . . . The Italians are spreading, spreading, spreading."

This was a deep hole Grant and Hornaday were digging for themselves, but they managed to bottom it out. In 1906 they placed an African left in their care, Ota Benga, who was a member of a Pygmy tribe, on display in a cage in their zoo. The layered implications of that abhorrent move boggle the mind.

For all the prodigious weight of such baggage, Hornaday fought to save America's wild animals like no one else of the age. "To allow a great and valuable wild fauna to be destroyed and wasted is a crime against both the present and the future," he argued. Beyond his role in saving bison, he was the most vocal proponent in the country for ending the damage the market was inflicting on wild animals. As he proclaimed in bold italics in his 1913 book, *Our Vanishing Wild Life*, "Here is an inexorable law of Nature, to which there are no exceptions: *No wild species of bird, mammal, reptile, or fish can withstand exploitation for commercial purposes.*"

Hornaday's campaigns led to mop-up successes against centuries of market hunting. In 1911 he drew up a new law for New York that outlawed all sale of game meat in the state unless it came from "game farms." States like Massachusetts and California followed suit within the year. That same year the United States, Britain, Japan, and Russia signed a Fur Seal Treaty that ended pelagic sealing (killing seals in the water, which led to the wholesale deaths of females and cubs). Hornaday managed to add a five-year ban on fur seal hunting on land, to let the population rebuild. He then worked on a rider attached to the Underwood Tariff of 1913 that amounted to a federal ban on commerce in bird feathers. And in the feds' Weeks-McLean Act of the same year on behalf of migratory birds, Hornaday managed to protect *all* migrating birds, not just the game birds hunters were interested in.

But more than anything else, Hornaday was a one-man actor for endangered species half a century before any law existed. In 1889 the Smithsonian had listed just three American species it thought were extinct: the Labrador duck, the Steller's sea cow, and the elephant seal. By 1913, when *Our Vanishing Wild Life* appeared, that list was both longer and more dramatic. Of course no one, including Merriam at the Biological Survey, had any real method for doing censuses of animal populations, but that didn't stop Hornaday. His bold list of American birds and animals "certain to become extinct" included the Carolina parakeet, the Eskimo curlew, the whooping crane, the trumpeter swan, the California condor, the heath hen, the sage and sharp-tailed grouse, the pronghorn, the bighorn sheep, the mule deer, and a pair of subspecies, the Merriam's elk and the California grizzly. Hornaday's prediction was that "the big game of the northwest region, in which I include the interior of Alaska, *will go*. It is only a question of time."

All this was horrific, an epic tragedy for any country, and the pessimistic zookeeper was convinced no one could stop it, including the feds' own gamekeepers at the Biological Survey. But while Hornaday was focused on rural and ethnic "game hog" hunters as the threat, Merriam and Bailey had come up with more ancient villains in America's wildlife story. And they knew exactly how to deal with them.

CHAPTER 9

GOLDEN-EYED LIGHTNING ROD

In the 1920s, as flappers and jazz and Hollywood captivate American cities, a man named Bill Caywood is engaged in a different cultural project. At fifty Caywood is a stocky stump of a man with a face like a granite cliff. He is a professional assassin of wolves but he says he loves the animals he watches die—"He's a real fellow, the big gray is. Lots of brains. I feel sorry every time I see one of those big fellows thrashin' around in a trap bellowin' bloody murder." Caywood is the sort of American that writer D. H. Lawrence, getting his first extended exposure to this country, will describe as "stoic, a killer," and what he is doing is mop-up work. Where the continent only three centuries before had easily held several hundred thousand wolf packs, by the 1920s few packs remain anywhere in the United States outside Alaska, the Great Lakes country, and the Lower South. Caywood is after the last survivors in the West, few enough animals that ranchers and government hunters hired on their behalf have started giving the animals individual names. They call two of these last gray wolves Caywood is tracking down "Rags" and "Greenhorn."

Animals that had once lived in packs, once had mates and pups, Rags and Greenhorn are enduring lives of lonely desperation. Like a significant percentage of gray wolves who turn to livestock, they are too old and frail to bring down elk without a pack's help. Younger wolves who ended up stock-killers often had suffered crippling injuries, frequently by losing multiple toes or an entire foot escaping the serrated jaws of the Newhouse number 4½ steel trap. Rags had seen one of the two mates he'd had during his lifetime panicked and helpless in a trap. He learned from that and is himself unmaimed. Rags is an old wolf—the ranchers say seventeen but he's probably closer to ten or eleven—and now travels either alone or with two younger wolves who are far less crafty. As for Greenhorn, this female wolf named for a local mountain near Caywood's Front Range home has teeth so worn she's been reduced to strangling her prey. In her past she's escaped traps and spit out a strychnine bait before it could kill her. When Caywood goes after her in 1923 the ranchers claim she's eighteen years old. Whatever her real age, she is slowly starving to death.

These are wolves the federal agency Caywood works for should leave to die natural deaths. But Rags and Greenhorn live in a nation that cannot brook a single wolf remaining alive, anywhere. It's Rags's turn first. Across weeks of time Caywood sets his traps and Rags digs them up. With a wolf that smart, the former bounty hunter rigs a trap-set designed to snare a wolf by a back leg as it digs up other traps with its front paws. It works. With a trap biting into a rear leg and a second trap sprung on the dragline of the first bouncing after him on a three-foot chain, the old wolf spends a final day in tortured flight. In the end, hemmed into a box canyon, he confronts a fate he's escaped for a decade. Purposefully he limps straight toward Caywood, yellow eyes fixed and staring as the metal clanks over the rocks behind him. Caywood stoically shoots the equivalent of an octogenarian wolf in the head.

Next, Greenhorn. It's December, cold and snowing on the Front Range, and with her teeth mostly gone the elderly wolf can't down a deer, let alone a cow. She's desperately hungry. She knows the scent of strychnine but Caywood has attracted her with a horse's head wired to a juniper, around which he's placed chunks of fat suet soaked in poison. Greenhorn shies away from the smell again and again. She knows from her own experience and from wolf culture that this scent means tragic danger. She has witnessed the thrashing, vomiting endgame more than once. But she is starving to death. She circles back, picks up a chunk of suet, swallows it. Then another and one more.

It's the day after Christmas, 1923. Caywood believes she is the last wild wolf born in the state of Colorado.

IN THE 1920s America started looking like the country we recognize today. By that decade more people were living in towns and cities than in the countryside. Americans owned and drove automobiles. Airplanes were in the skies. Electrification powered home appliances like washing machines and radios. We had telephones, although we didn't walk around staring at them. Sports had become a national obsession and advertising was converting us into status-crazed product influencers. Meanwhile, immigration, discrimination, and the school curriculum turned politics toxic. We were recovering from a global pandemic, a spillover flu virus we'd caught from birds that in two years killed almost three-quarters of a million people in the United States.

By 1920 there were 106.5 million of us, a figure that would balloon to 152 million by 1950, thirty times the number of Native people who were living in America in 1500. The remaining wildlands Vernon Bailey was studying were shrinking year by year. The space required for a hunter-gatherer lifestyle still existed in parts of Alaska but virtually nowhere else. The western canyons, and mountain ranges in East and West, where Roosevelt's conservation programs set aside public lands, were bright spots for the wild. But elsewhere intact landscapes that for millions of years had been habitat for America's birds and animals were dwindling to remnants, ecological versions of Rags and Greenhorn.

For scientists and the conservation minded, the demise of wildlands was less immediate than the troubling disappearance of so many of the continent's historic animals, which spoke to an ongoing environmental crisis. The public largely tried to ignore that as peripheral and inconvenient. There was also the brand-new idea, unsettling to those who imagined the continent with a past that began with Columbus, that America was actually a very old place. Somewhere in the Southwest scientists on an archaeological dig had found evidence of a history going back thousands of years. People also began talking about a new science. It was called ecology and most Americans knew little about it except that it was vaguely connected to evolution, itself much in the news because of the "Monkey Trial" over teaching Darwin in schools. Whatever ecology was, it looked suspiciously subversive of the normal order.

Out of sight and mind for most except when a federal agency wanted

Gray wolf. Courtesy Shutterstock.

to crow about its eradication program, there was also an animal, a sleek, golden-eyed lightning rod that somehow distilled all of it—the loss of so many species, the United States as a clone of Europe or an ancient place with its own historical arc, the insights of a new science versus the old and accepted order. That animal was the wolf.

BILL CAYWOOD was from the class of men Teddy Roosevelt once thought heroes of the continental conquest. They'd shot whitetails for the leather trade and buffalo for money, sprung nets over flocks of passenger pigeons, and in Florida driven snowy egrets nearly to extinction for their feathers. As wolfers they'd come to live off bounties paid for dead predators by livestock associations and the territorial and state legislatures that served them. Bounties were the Old Worlders' first wildlife laws in America and by the twentieth century every state but one had them. In New York between 1871 and 1897 bounty hunters had turned in ninety-eight "wolf scalps." It was one of the few kinds of government expenditures men like Caywood could get behind.

Governments at all levels paid money for the heads or ears or "scalps" of a suite of animals—wolves, coyotes, mountain lions, grizzly and black bears, jaguars, bobcats, lynx—for the single purpose of promoting agricultural economies. Sometimes both governments and stockmen's associations paid, and if you could make a few dollars selling pelts, a rural kid could carve out a kind of blood-soaked market niche. If ranchers decided a particular wolf or wolf pack had "gone rogue," prices went up. Montana's government was so committed to shielding cattle and sheep raisers from predators that the state supported a whole class of wolfers. Between 1883 and 1928 Montana shelled out payments on 111,545 wolves and 886,367 coyotes, a subsidy for ranching that when Montana was still a territory had eaten up two-thirds of its annual budget. Bounty hunters in Wyoming and Montana told one another the story of a Colorado hunter who killed 140 wolves in one year, earning him $7,000. He was able to retire and buy a ranch and stock. That hunter was Bill Caywood.

But in the early twentieth century the bounty era began to wind down. A new institutional player was emerging to confront wolves and other predators in the United States. You would have thought in 1905 that C. Hart Merriam's Bureau of Biological Survey was sitting pretty. Teddy Roosevelt was president and the bureau was dear to his heart. But Congress was growing less interested in funding an agency that seemed mostly interested in pure science and mapping species. At livestock association meetings and among their many representatives in Congress, western ranchers were arguing that Roosevelt's vast public lands had become refuges for predators that attacked their stock. Since the feds had created this mess, they believed, the feds ought to fix it. If Merriam wanted the bureau funded, he and Bailey were going to have to turn their attention from life zones and do something practical.

So in an act of self-preservation, the Bureau of Biological Survey remade itself into the solution to the country's "predator problem." Between 1907 and 1909 it issued four reports on the "predator–big game–livestock relationship" in and around the new national forests. Merriam had Vernon Bailey author the most important of them. To prepare the stage, the bureau first had to undercut the existing model of state bounties for carnivorous animals. That inefficient system, it claimed, was shot through with corruption, graft, and fraud. A federal agency like the bureau, on the other hand, could bring "orderly and scientific control" by employing "trained hunters and trappers." Like Bill Caywood, who was one of the first hunters the bureau hired. To demonstrate bureau expertise, Merriam dispatched Vernon Bailey to teach national forest managers how to solve "the wolf problem." In the wake of Bailey's seminars on finding wolf dens and best strategies for destroying pups and packs, Forest Service rangers killed 1,800 wolves and 23,000 coyotes in the national forests within a year.

Once a naive country boy from Minnesota, Bailey was now in the act of transforming himself. President Teddy Roosevelt would hereafter refer to him as "Wolf Bailey."

At the time there was virtually no opposition to a federal campaign against carnivores. America's most beloved nature writer, John Burroughs, opined that predators "certainly needed killing," since the "fewer of these there are, the better for the useful and beautiful game." Wildlife agitator William Hornaday insisted that when it came to predators, "firearms, dogs, traps, and strychnine [are] thoroughly legitimate weapons of destruction. For such animals, no half-way measures suffice." Not even John Muir spoke out against the feds' gathering war against America's carnivores, although he did worry that slaughtering them before studying the issue might induce a "penalty for interfering with the balance of Nature." The hero of conservation, Teddy Roosevelt famously referred to wolves as "beasts of waste and desolation" and went on hunts for them himself. And coyotes? *Scientific American* branded the coyote "the original Bolshevik," an animal that ought to be killed on sight if for no other than patriotic reasons.

There was one additional constituency for the war on predators and they went all in. Public relations experts within the bureau mounted a campaign to spread the idea to sport hunters that destroying wolves, lions, and coyotes—which state after state was classifying as unprotected "non-game" animals, or "varmints"—would produce bumper populations of deer, elk, and other game to shoot. No science existed on this question, but promulgating the idea that human sport hunters could replace predators in managing deer and elk populations was a stroke of genius. It brought all manner of sportsmen's groups, firearms manufacturers, and state game and fish agencies to the cause of wiping out every wolf and mountain lion on the continent.

Claiming that America suffered from (the bureau's words) "wolf-infested national forests and the federal public domain," the Biological

John Grabill, Roping Gray Wolf. John Grabill Collection, Library of Congress.

Survey had engineered its own public support. In June 1914, Congress approved an initial appropriation of $125,000 for a new bureau program, Predator and Rodent Control, universally referred to as "PARC," to destroy wolves, coyotes, mountain lions—and prairie dogs and ground squirrels—on both public and private lands. Within two years three hundred federal hunters working for PARC destroyed more than thirty thousand predators. Among its other endeavors, like managing the national wildlife and bird refuges, the bureau under Merriam's leadership had become a crucial federal agency. It even became an indispensable arm of the United States for the world. The one field of twentieth-century wildlife science in which Americans became acknowledged global leaders was in the destruction of "undesirable species."

THE WOLF STORY, however looming and hotly contested it grew, was part of a much larger maturation in America's twentieth-century relationship to its wild creatures. The country by midcentury was still a long way from shedding assumptions and habits the Old World had bequeathed us, yet nonetheless we proudly announced our wildlife policies were "a triumph." In truth, remaking America on the European template was an unconscious design choice that deemphasized the wild and assumed that many of the country's most charismatic species were incompatible with modern life. It was as if America's biological past and unique evolutionary gifts, exotic enough to dazzle colonial naturalists, had somehow disappointed and alarmed us. So we released European hogs to run wild in our forests, imported Old World birds like starlings and English sparrows because Shakespeare wrote about them, and initiated all manner of other introductions, like Chinese pheasants and Hungarian partridges, to give sport hunters more things to shoot. Hoping to manage nature and create an America in Europe's image became an impulse with no room for animals that seemed *too* wild. Europe after all had no grizzly bears, no jaguars, no free-roaming bison, and close to no wolves. The British Isles had none of any of those.

The triumphant narrative about "saving America's wildlife" was actually a movement to save huntable game. We chose to protect songbirds and certain shorebirds (like snowy egrets), the former because they were innocuous backyard ornaments, and the latter because the horrors of the plume hunt had shamed us into action. Following the *Geer v. Connecticut* Supreme Court case, every state created game departments empowered to impose seasons and bag limits on ducks, geese, and other migrating waterfowl. Once the targets of "punt guns" that killed hundreds with a single shot, even of explosive charges set off beneath swimming flocks, which rained down birds to the applause of admiring onlookers, waterfowl now came in for special protection. As migrants across state and national borders, ducks and geese merited additional efforts by the federal government under Interstate Commerce provisions. The Weeks-McLean Act of 1913 established that principle and the Migratory Bird Treaty with Canada in 1918 cemented it.

One of the winter delights in my part of the Southwest is an annual trip to a federal migratory bird refuge, the Bosque del Apache, set up in 1939. A decade before Congress designated this New Mexico bird wonder it passed a bill that created eighty-five such refuges, totaling three-quarters of a million acres. This initial round of federal waterfowl refuges banned hunting, a major win for a growing population of Americans who wanted to see wildlife but not over gunsights. The hunting ban roused a famous organization, Ducks Unlimited (originally called "More Game Birds in America Foundation"), to advocate for one-dollar "duck stamps" to finance waterfowl refuges for hunters. Ducks Unlimited was a darling of firearms manufacturers, whose chief organization—the innocuously titled American Game Protective Association—poured big money into sport hunting.

That iconic forest game animal of America, the whitetail deer, had long seen its populations almost wiped out in many states. By the late 1800s their numbers had collapsed to fewer than three hundred thousand. But whitetails were prolific. Their restoration successes for game commissions in states like Pennsylvania became an example for other northeastern states. Trying to rebuild mule and black-tailed deer from remnant populations in the West was harder. Big ungulates in the West migrated from high peaks in the summer to lowland river valleys in the winter, and the lowland valleys were now broken up by ranches and towns, where humans had taken over wild-animal winter ranges. Another big ungulate, caribou (except for their most southerly populations) fortunately inhabited remote country that survived into the twentieth century. The same trajectory prevailed for moose in the Northeast and Midwest. Minnesota still had two thousand moose by 1920, which became a nucleus for expansion throughout the upper Great Lakes country.

America's grandest deer, elk, were in a far more precarious situation. By 1900 an animal that two centuries before roamed across much of America had dwindled to only 150,000 animals, all in the West. Saving them became a special sport-hunter project. In 1912 the Taft administration established the National Elk Refuge in Jackson Hole, Wyoming, and the Biological Survey started feeding the animals to get them through winters. This was a practice the Wyoming game commission would eventually expand dramatically. No non-game creature of any sort would come in for the kind of assistance elk got until the twentieth century was nearly over.

There were other successes where we managed to collect our wits enough to pull beautiful, iconic animals back from the brink. Another casualty of the market hunt, the wild sheep species called California, desert, and Rocky Mountain bighorns were all nearly extinct by the early twentieth century. Market hunters had in fact exterminated a Rocky Mountain subspecies, the Audubon bighorn, that the great bird painter had hunted in the Dakota badlands in the 1840s. Along with their contraction of various diseases from domestic sheep flocks, wild sheep had long served as an expendable trophy commodity. But scattered bands survived in remote mountains in Wyoming and Montana, and starting in the 1930s state wildlife commissions began live-trapping bighorns to restock them in Southern Rockies locales where they'd been gone for decades. Similar projects restored desert bighorns to canyon country in the Southwest. The Dakotas would eventually get wild sheep again, if not precisely the animal that had evolved there.

The modern United States also drew very close to losing entirely one of the most uniquely American animals of all. The pronghorn had evolved on the western savannas and became a Pleistocene relict outliving its predators. But it proved no match for gunsport capitalism, nor for another bit of technology new to the America of the nineteenth century. Like other western ungulates, pronghorns had evolved to migrate, in their case big flights to outrun winter blizzards driven by Arctic air surging down the Rocky Mountain Front. The barbed-wire fences demarcating new swaths of private property uncovered the pronghorns' major evolutionary weakness. Across their twenty-five-million-year family history, there had never been natural-selection pressures for pronghorns to learn how to jump obstacles. With the new wire fences blocking their migrations, bad winters devastated them. Following one fierce winter blizzard in 1882, western homesteaders discovered 1,500 freezing and dying pronghorns, like a deck of cards blown against a curb, stacked behind a fence across the Texas Panhandle. Even hard-bitten homesteaders were horrified.

In 1800 there had been many millions of pronghorns, probably more of them than buffalo. In 1909 the New York Zoological Society estimated five thousand were left. That was precariously close to the number of buffalo left in the 1880s. But two lucky breaks helped pronghorns. One was the growing number of wildlife refuges, which organizations like the Boone and Crockett Club and American Bison Society stocked with remnant animals, including two refuges in Nevada designated specifically for America's striped antelope. The other break was a true stroke of luck in a nation where economics trumped all. As forb rather than grass eaters, pronghorns didn't compete with cattle and only marginally with sheep. So the ranching community didn't hate pronghorns the way they did other wild animals. That meant an expanded habitat beyond wildlife refuges. By 1925 a three-year study by the Biological Survey indicated a resurgent population of nearly three hundred different herds with a total population of some thirty thousand animals. By 1950 Wyoming had become the pronghorn capital of America with almost eighty thousand of them in the state. Naturally, hunting seasons followed.

The country was successful with other wildlife restorations that attempted to correct our more egregious historical mistakes. Some of those corrections focused on smaller mammals and birds, including remnant species that had once been targets of the fur-market frenzy. Sea otters, fur seals, even beavers made limited comebacks (fairly considerable comebacks in the case of beavers) from the mayhem we'd once inflicted on them. But most recovery and restoration efforts were driven by hunting priorities. Wild turkeys, one of the first creatures Old Worlders had hunted to local extirpation in their colonies, took a lot of effort to bring back. But there was restoration money available from the New Deal. In the 1930s the Pittman-Robertson Act insured that taxes on firearms and ammunition sold for sport hunting would go to state wildlife agencies to help them restore huntable species, and the wild turkey was eminently huntable. You could even go after turkeys in the spring, an otherwise fallow season in the new world of managed hunting. Not long after midcentury, state restoration projects brought eastern wild turkeys back in the Northeast and Midwest, where hunters had killed them off a century earlier. With their reintroduction, western Rio Grande wild turkey numbers exploded in Texas, and many western state wildlife departments stocked a wild turkey named for C. Hart Merriam—a bird originally found only in the deep Southwest—into all sorts of locales wild turkeys never inhabited before. Naturally, hunting seasons followed.

*

BUT WHAT ABOUT American animals the hook-and-bullet magazines rarely wrote about? What about all those horses that had run wild into the landscapes of their evolutionary past? No one went after trophy horses with a rifle, so what would be their fate in the twentieth century? What of all the birds no one trained dogs to point, especially the predatory raptors, the eagles, owls, hawks? Or that breathtaking icon of colonial America, the ivory-billed woodpecker? What of the big carnivores that were as synonymous with America as lions were to Africa or tigers to India? In Alaska you could still hunt grizzly and brown and polar bears, which walked a boundary line between game and vermin, depending on locale.

By 1900 black bears were virtually eradicated everywhere outside Alaska and the dense forests of the western mountains and Great Lakes. Black bears did benefit from the decline of grizzly bears, and like wild turkeys and whitetails they would eventually reemerge as game animals even across the East and South.

In the contiguous states, though, America's most imposing big predators—grizzlies, cougars, jaguars, wolves—experienced the twentieth century far differently. These wildest of our carnivores struck many as too dangerous to exist in an America modeled on Europe.

The grizzly bear was a special case. Everyone tended to agree that the huge bears needed to go from all settled country. Naturalist Elliott Coues, who collected a grizzly in the San Francisco Mountains of Arizona and also had extensive experience farther north, believed grizzlies were originally most numerous of all in the Southwest: "The southern Rocky Mountains, and the ranges of California, seem to be particularly the home of the huge Grizzly . . . which becomes less numerous farther north," he wrote. There were good reasons for that. Grizzlies in California, where the bears fed on foods from a profusion of habitats, including carrion washed ashore on the Pacific Coast, did not need to hibernate. And as elsewhere in the Southwest, their numbers shot up dramatically with the carrion possibilities from Spanish-introduced cattle and horses. An early American pioneer in today's Napa/Sonoma wine country, George Yount, wrote of grizzlies there that "it was not unusual to see fifty or sixty within twenty-four hours."

The sheer size of grizzly bears terrified some people. In California they grew as large as Kodiak bears in Alaska. Northern West or Southwest, ranchers, cowboys, and hunters shot every grizzly they saw from the 1850s onward, for many as a rite of masculinity (a story in *Harper's* in 1861 claimed that "the ladies" much admired a "'chawed up' man"). Like other carrion eaters, the big bears died by the thousands from eating poison baits. By the twentieth century settlers had driven grizzlies off the Great Plains and from the open country in California. The last grizzly to die in Texas, in 1890, was the one Vernon Bailey wrote about in the Davis Mountains. From a vast population of grizzlies that only seventy-five years before had numbered more than ten thousand, the final one of California's totem animals died near Sequoia National Park in 1922, its kill certified by Merriam himself. The last grizzly in Utah fell in 1923, and the last bears in Oregon and New Mexico in 1931. Arizona's last grizzly died in 1935. A hunter shot the last of Washington's original grizzlies in the North Cascades in 1967. As for Colorado, a state that supposedly produced a killing machine of a bear (ranchers claimed a grizzly called "Old Mose" killed eight hundred cattle and five humans) hosted bears in its San Juan Mountains well past midcentury.

In 1950 a national census estimated that from a population of 56,000 bears at the time of Lewis and Clark, only 750 grizzly bears remained alive in the contiguous states. Those were in two distinct places that, unfortunately for grizzly genetics, were hundreds of miles apart: in and around Glacier and Yellowstone National Parks in Montana and Wyoming.

The retreat of the big cats from America rivaled the disappearance of the giant bears. Cougars had once hunted every part of America, from the swamps of Florida and Louisiana to New England's rocky hills, from the West's deserts and canyons to its mountain glaciers. By the twentieth century cougars were gone from much of the country and pursued with hounds everywhere they remained. Cats prefer fresh kills to carrion, so government hunters lacing carcasses with poison were ineffective in killing them. Along with wolves, big cats had once been the predatory control for wild horses. No hound man gave a thought to ecological connections like that, of course, and we had plans for wild horses, anyway. But the cougar wasn't alone. America's great cat, the jaguar, whose numbers had seemed healthy when naturalists had probed the Southwest in the 1850s, by 1900 also seemed to be dwindling away to nothing.

When America's giant steppe lions and saber-toothed cats had died out in the Pleistocene extinctions, jaguars had assumed the mantle of North America's most imposing big cat. With a range that stretched deep into South America, jaguars were at the northernmost limits of their range in the southern United States. Leopard-like in appearance but heavier and more muscular—male jaguars can weigh more than three hundred pounds—the great cats were still establishing territories and breeding in the United States into the twentieth century. This northern range was not merely a place male jaguars roamed into. Among more than sixty historical records of jaguars in the United States from 1880 to 1995, females and kittens are well represented. American jaguars denned and hunted in deserts, oak foothills, and piñon-juniper and ponderosa pine forests, country strikingly unlike their jungle habitats to the south. Such open terrain made them vulnerable to human eyes, though, especially the eyes of stockraisers and the bounty and government hunters employed to protect cows in the twentieth century.

Aware that jaguars were declining, Vernon Bailey collected as many accounts of jaguars as he could find. From them he concluded that the Black Range in the center of New Mexico had long been jaguar territory. A bounty hunter had killed jaguars in that range of choppy, vertical ridges in 1900 and 1902, and Bailey collected several other accounts from there. Other New Mexico ranges—the Sangre de Cristos, the Sacramentos, and the San Andres—also held jaguars, and into the 1920s ranchers were still shooting them. There were accounts of jaguars nearly to Colorado, as well as out on the Great Plains. Arizona, where settlers reported jaguars from Zane Grey's Mogollon Rim country to the Grand Canyon, had a similar jaguar record. The Biological Survey's first recorded jaguar kill was there: a federal hunter shot a jaguar in Arizona's Santa Rita Mountains in 1918. The bureau's operatives in Arizona had few doubts about the sources of the jaguar "threat." Its position was that "all Lobo wolves and jaguars will be taken as fast as they enter this State from Mexico and New Mexico, as one hundred per cent of them live on livestock and game." The bureau recorded five such renegades crossing into pristine Arizona from 1924 to 1927. Two were females. All ended up killed. Jaguars were still dying at human hands even on the Gulf Coast of Texas as late as the 1940s.

That long history of America's grand spotted cat, venerated by groups like the Comanches, with rumors of "tygers" reaching back to the naturalists of the colonial South, seemed to have reached an emphatic conclusion by the mid-twentieth century.

Finally there were the canids, the gray wolves, red wolves, and coyotes, the latter two found nowhere else in all the world. The ideologies we inherited from the Old World and the Great Chain of Being drew a bright line between animals that were "beneficial" and those, like wolves and coyotes, that were "destructive." However passé the Great Chain was by the twentieth century, the Bureau of Biological Survey struggled to evolve. You would have thought that C. Hart Merriam's fine education would have enlightened him. But saving the bureau meant he had to make it an arm of ranchers and the livestock industry, and Merriam was committed to that course. Bears, lions, jaguars, prairie dogs, eagles, magpies, and especially wolves and coyotes—Merriam was willing to erase them all from modern America.

~

THE STORIES OF THE "non-game" wild animals had distinctive plot elements even if the planned ending was the same. But the wild-horse story had dimensions all its own. American evolution had produced the horse. The horse had disappeared from the continent for several thousand years. It had returned in domesticated form as a result of human agency, then exploded into its evolutionary homeland as a Pleistocene reprise. In the mid-nineteenth century a few camels got a similar shot at life on their old grounds, but that recovery didn't take. The horse recovery took, and then some. We can only guess at how many horses ran wild from the 1600s through the 1800s, but as the bands steadily spread into their ancient habitat northward and westward the guesses range up to two million by 1800 and as many as seven million by the 1880s.

Wild horses were an ancient human prey, but the value we attached to them ultimately rested on their domestication. In the West Native people and Americans like Philip Nolan had fashioned a lively trade economy around wild horses from 1680 to 1880. But as the automotive age loomed, the new horse question became, what to do with *millions of wild horses?* Americans disdained eating horse meat and there was little market for their pelts or hooves. There was also a bigger problem. U.S. policies had earmarked the Great Plains and the mountain valleys of the West for cattle. Just as wild bands of horses had previously cut into bison numbers in that country, horses ate the same grasses and drank at the same water holes as cattle. So the horse as wild animal crossed the line. It was beneficial only as a domesticate. Wild, it became "destructive."

The solution seemed to be to shoot wild horses on sight. In the 1880s some fifty thousand wild horses still formed stallion bands, engaged in battles for the right to breed, and ran free in the canyonated breaks of the Texas Panhandle. But by then the JA and XIT Ranches had made that cattle country, so the ranchers paid cowboys to go out with rifles and wipe out wild horses. Shooting an animal identical to the one you were riding must have been odd, something like stories of dogcatchers of the time taking their own dogs along as they rounded up strays to dump into drowning tanks. But some of the big ranches pushed the envelope even further. They had their cowboys shoot down wild horses, then bait their carcasses with strychnine to kill wolves, coyotes, and eagles. Cold-blooded but efficient if what you dreamed at night was a world populated just with humans and cows.

There were parts of the West where the old capture economy persisted. Most Americans no longer needed riding stock, but militaries did. During the Boer War in Africa the British military bought thousands of horses from American mustangers. They did so once again during the Great War, when remnant wild herds provided Allied buyers some thirty-two thousand wild horses that sold, barely broke, for $145 to $185. (This was a history that gave filmmaker Steven Spielberg the chance for a wildly sentimental movie, *War Horse*, a century later.) In the 1920s the postwar economy again came up with a way to turn wild horses into commodities. One of the markers of modernity in the Roaring Twenties was the remarkable growth in companion animals. So when the first of the national pet-food companies, Kennel Ration, began to build its plants in the Midwest, wild horses confronted a new fate.

Some of the mustangers building capture corrals in the badlands of eastern Montana and Wyoming's Red Desert evidently were unaware of the fate of the horses they were selling to buyers in fancy suits at the railroad stations. One mustanger, a young man named Frank Litts, learned the truth around a Montana campfire one night. When he realized the animals he was driving to Miles City were going to pet-food slaughterhouses he bought 150 sticks of dynamite and a train ticket to Illinois with the intent of performing eco-terrorism on a dog-food plant there. Guards caught him before he could set off his charges, but it's tempting to think that in the Frank Litts story Spielberg could have made a movie truer to the moment horses were in.

Our myopic disregard for the wild new world we seized from its Native caretakers finally began to change in the twentieth century. Sometimes, as with biologist Vernon Bailey, the changes took place in individual lives and lifetimes. The scientific way of knowing the world—the growing acceptance of Darwinian evolution, the emergence of ecology, and the professionalization of natural history—finally liberated modern Americans from old, anthropocentric ways of understanding nature and ourselves. But the lag stretched out decades. Our most engaging traits of national character, our optimism and self-confidence, sometimes morph into a surety that spills into arrogance. That seems the only explanation for some of the policy decisions we made about wild animals in the early twentieth century.

LIKE HER BROTHER who was eight years older, Florence Merriam grew up to love the natural world around the family's opulent home in upstate New York. Their mother was a graduate of Rutgers, so there was no doubt that Florence would join her brother and become college educated. Since she could not follow C. Hart to Yale, Florence enrolled at brand-new Smith College in Massachusetts. Photos show an appealing young woman with an open countenance who wore her brown hair up and spent leisure time hiking with her dog. Like her brother, she was not traditionally religious. In fact she planned to write her senior thesis at Smith on evolution. But the events of the time—it was the early 1880s—angled her toward the preservation of birds. After seeing Smith classmates return from weekends in New York exclaiming over all those gorgeous feather-decorated hats, she decided to found an Audubon Society group at her college. It ended up enrolling a third of the campus and brought John Burroughs in to speak about the destruction of birds for feathers. It was the start of her career.

Graduating college at twenty-four, Florence went west for a health affliction she never identified. But her mother died from tuberculosis, and what Florence called her own "great shadow" was most likely the same lung disease. Seeing California for the first time led to her initial book, *Birds through an Opera Glass*. While writing her second, *My Summer in a Mormon Village*, about Ogden, Utah, she got in a brief visit with her brother's field naturalist, Vernon Bailey, who passed through in 1893. The two had met in New York the year before. Bailey hadn't exactly been smitten, finding her "pleasant" but "not striking in appearance." Meeting her in fact sent him on a search for the "ideal girl" a friend had suggested he meet. The ideal girl turned out to be taken.

Florence continued to write, and the next book—*A-Birding on a Bronco*, illustrated by the fine avian painter Louis Agassiz Fuertes—became her first hit. It was an account of a California summer, a favorite horse, and the

Florence Merriam Bailey. Courtesy Smith College Archives.

bird life of the San Diego region. The search for a drier climate had now brought her to both Arizona and New Mexico, where the "climate is wonderful," and her great shadow began to lift. It was now the late 1890s. She had been out of Smith for a decade, had authored three books, and now her fourth, *Birds of Village and Field*, came in 1898. Fellow ornithologists were saying that American bird populations had declined 50 percent in fifteen years. She was on a mission to save those that were left.

In 1899, though, accompanying her brother on a scientific trip to Mount Shasta, Florence found herself in Bailey's company again. Neither sister nor naturalist received an invite to go on the Harriman Expedition to Alaska that year, but maybe this was calculation by C. Hart, since his sister and his assistant spent the ensuing autumn together in Washington, DC. Vernon Bailey and Florence Merriam married that December, raising eyebrows across the world of American natural history at what seemed an unlikely pairing. Their honeymoon the following spring was a camping trip across Texas to New Mexico. Still a territory in 1900, New Mexico was a crucial destination for both their careers. While Bailey worked on faunal publications about the region, Florence immersed herself in the birds, identifying ninety-four new species in the Southern Rockies. It was a part of America that left an altogether different impression than the East: "We felt everywhere in New Mexico, [that] while to us the country was new, in very fact this land of *poco tiempo* is an old, old land."

Florence may well have gone on to write the two books she is now remembered for—the first field guide to the birds of the West, *Handbook of Birds of the Western United States*, and the first close study of the birds of an interior state, *Birds of New Mexico*—had she never married Vernon Bailey. Vernon Bailey, on the other hand, who was about to become the federal government's point man on wolves and predators, quite possibly would not have evolved in the direction he finally did without Florence in his life.

~

THE EARLY TWENTIETH CENTURY was the age of the bureaucratic professional, and professionalism prevailed at the Biological Survey. If your assignment was to mass-kill wolves and other predators, the way to go was not shooting or trapping individual animals but poisoning entire populations, and you did that by strewing poisoned baits by the thousands across the American landscape. At one point Bailey inquired about the strychnine dosing. He wanted to know how much to conceal in a bait so a poisoned wolf might die more quickly. He thought within three minutes would be a humane span. Knowing that any expression of mercy or tenderness toward wolves could become a political liability, Merriam shot back: "You had better go at once to the hospital in Albuquerque. . . . Inasmuch as no sane man could possibly make such an absurd and utterly preposterous statement as this you are obviously in need of mental treatment." Merriam went on: "We want the cattle man behind us. Sabe?"

With its new funding the bureau proceeded to build a plant in Albuquerque to produce strychnine baits in volume. Chillingly, unsentimentally, they called it the Eradication Methods Lab. In 1921 Washington moved this federal killing facility to Denver, in which location it eventually perfected an amazing witches' brew of ever-deadlier predacides. But for the next two decades federal poisoners relied on strychnine, whose fatty bait-cubes also turned tens of thousands of nontarget animals into collateral damage. Bait stations surrounded by strychnine victims—

dead wolves, coyotes, eagles, ravens, magpies, even bears, in a mile-wide arc—was soon a common visual image in the countryside. Victims often required considerably longer than three minutes of violent vomiting and convulsions to die. The poison infamously twisted wolves and coyotes into a "strychnine signature," their bodies wrenched and their tails shocked straight out as if they'd been struck by a bolt of lightning.

Federal hunters quickly grasped the wolf's fatal flaw. All of us who have become close to a dog understand one fundamental thing about canines. Like us, they are social animals, so much so that they exemplify a trait we clearly prize: they are loyal to their social groups. Our ability and theirs to attach as loyal companions is how we two species joined forces twenty-three thousand years ago. The smaller American canids, coyotes, had evolved an old adaptation to threats, one human evolution bequeathed to us as well. Like humans, coyotes are fission-fusion animals, living in social groups when possible (fusion), but capable of scattering when ecological pressures call for it (fission). But wolves are strongly pack based. Wolf killers realized that such was the wolf's affection for mates, pups, and pack members that killing one meant (as a bureau hunter put it) you could quickly kill "all of the members of whole families of wolves," with "unmistakable evidence that the remaining members of the wolf family have been seeking the lost member."

At the moment when the federal government first launched its campaign of eradication against these ancient American species, Congress created the National Park Service (1916) to manage its parks and monuments and to allow the public to experience wild American nature in its pristine state. Yellowstone, the country's and the world's first national park, and Glacier, created along the Continental Divide in Montana in 1912, emerged as symbols of just how far the predator war intended to go. In Yellowstone the army rangers who had kept poachers at bay began poisoning lions, wolves, and coyotes inside park boundaries as early as 1898. Then in 1914 the park's managers invited Vernon Bailey to come and show Yellowstone personnel the proper extermination techniques. With Bailey's help, by 1916 Yellowstone rangers had destroyed 83 coyotes and 12 wolves in the park. Stephen Mather, the wealthy New Englander who became the Park Service's first director in 1916, went so far as to invite in PARC hunters. Yellowstone's tally 'til the death of the last gray wolf in the park in 1926 was 136 wolves, 80 of which were puppies. Between 1918 and 1935 Yellowstone National Park, the world's first grand wildlife park, issued a death sentence to 2,968 coyotes.

Glacier Park's early superintendent, James Galen, initially tried a different approach. In 1906 the state of Montana had experimented with biological warfare against predators. It had passed a law requiring veterinarians to infect any live wolf or coyote that came their way with sarcoptic mange. So Galen wrote the state veterinarian with a request: "I am desirous of inoculating, with mange, some coyotes to turn loose here in the park, with the idea that I may eventually kill off all the coyotes in the park in this manner." Montana obligingly sent a pair of mange-infected coyotes to Glacier, and Galen turned them loose with best wishes for biological mayhem. But his Washington superiors decided poison was the better choice for ridding the park of its wild canids. By 1920 Glacier had taken out fourteen gray wolves using Bailey's strategies, although its proximity to a healthy Canadian wolf population would eventually make the park a wolf-colonization destination later in the century.

With the Eradication Methods Lab cranking out the strychnine, only a few years after Glacier killed its last gray wolf PARC hunters had distributed an unimaginable 3,567,000 poison baits across the country. In the mid-1920s the bureau turned its poisoning focus to coyotes, whose numbers somehow seemed as high as ever. But there were still wolves to kill, and as their numbers dwindled, evidence of Darwinian natural selection amid a new wolf culture began to emerge. Stanley Young, an early bureau hunter who would rise through the ranks, believed of these last animals that "it is probable that never did more intelligent wolves exist." They represented the accumulated learning of scores of generations that, like eighteenth-century sperm whales famously figuring out how to avoid whaling boats and harpooners, passed down information to their pups about rifles, traps, and poisons. No wonder the last ones were so smart.

Ranchers and federal hunters named many of them. Along with Rags and Greenhorn in 1920s Colorado, there was also Sycan (Oregon), King Lobo (New Mexico), Snowdrift and the Pryor Creek Wolf (Montana), Aguila (Arizona), the Custer Wolf (South Dakota), Old Lefty, Old Whitey, Unaweep, Big Foot, Phantom Wolf, and Three-Toes (all from Colorado). Names like "White Wolf" and "Three-Toes," the latter referencing all the animals crippled by traps, were applied widely to remnant local gray wolves. There were red wolf "renegades," too, including Traveler (Arkansas), Black Devil (Okla-

homa), and Crip (Texas), another maimed animal. Given the accounts federal hunters left of their protracted efforts to kill these wolves, there is little
doubt these were, indeed, remarkably intelligent animals. And why not? As
they were experiencing on all sides, their kind was being wiped out.

*

AT ALMOST THE SAME TIME Congress authorized the Biological Survey
to eradicate wolves, scientific naturalists in the United States founded
the Ecological Society of America, which met for the first time in 1915,
in Philadelphia. America's founding ecologists—Frederic Clements, his
wife Edith Clements, Charles C. Adams, Victor Shelford—agreed at that
gathering on several basic strategies for their field. Among these were
the study of adaptation that had been so critical to Darwin's insights, an
investigation of the flow of energy through nature, an analysis of "climax
conditions" (the inclination for disturbed nature to recover via a series
of stages leading to optimal life-forms—a climax—which fascinated the
Clements couple), and an attempt at better understanding the consequences when humans disturbed the natural world. Shelford, who had
just published his landmark *Animal Communities in Temperate America* and
became the Ecological Society's first president, pushed his fellows to recognize and work on "biotic communities," an approach quite different
from the Humboldt/Merriam life-zones models. By the end of 1915 the
society counted 307 members.

Ecology's most old-fashioned topic—an idea Western culture had
known since the time of Herodotus and Plato as "the balance of nature"
and Linnaeus had called "the economy of nature"—first pushed science toward rethinking predators. The Biological Survey's policies had
assumed the European folk position: predators were entirely disposable,
and with wolves and cougars and coyotes banished from America, a civilized Elysian Fields for deer and elk and ranchers and sheepmen would follow. The ecologists who met in Philadelphia believed there might instead
be dynamic equilibriums at work in the natural world. That assumption
would become the crux of a raging battle in American and Western science for the next half century.

Ironically, the organization that first challenged the bureau over the
feds' predator policy looked in the beginning like a natural ally. The

American Society of Mammalogists, which met for the first time at the Smithsonian in 1919, brought zoologists, ecologists, and bureau personnel together for annual meetings around the country. Its members elected C. Hart Merriam as the society's first president, with Vernon and Florence Bailey as charter members. Major E. A. Goldman, an outspoken bureau operative who headed up Biological Investigations and was a good friend of Bailey's, was also there and would be a major presence in the predator story (Goldman's ancestors had migrated from Pennsylvania to California in the nineteenth century and along the way changed the family name from Goltman to Goldman).

Like Bailey, Goldman had field experience but lacked a college education. Nonetheless, the bureau tasked these field-trained naturalists to explain and defend the predator campaign against university-trained and PhD scientists. Among that group were Charles Adams and a young forester interested in wildlife, Aldo Leopold. Another was Joseph Grinnell, a cousin of the legendary George Bird Grinnell and first director of the Museum of Vertebrate Zoology in Berkeley, California. Grinnell was an original thinker whose ideas quickly challenged the very premises on which the bureau based its predator war.

As the bureau sent its hunters after those remnant, superintelligent wolves, and Goldman finalized plans to expand the predator war to coyotes next, the ecologists attending the conferences prepared their attack. The first big blowup occurred at the 1924 meeting. That year Joseph Grinnell published his argument for ecological niches, a fundamental insight into nature and a critical idea in appreciating the "job description" of predators in a functioning ecosystem. By this point the bureau was slaughtering predators at a rate that forced scientists to consider what might happen when major niches suddenly went empty. Grinnell and other ecologists, one of whom was his graduate student E. Raymond Hall, insisted that the bureau was wiping out animals that occupied a critical role in the continental balance of nature. While the ecologists were absorbing that idea, both they and representatives from the bureau were unaware that American nature was about to offer powerful examples of the implications of emptying America's ancient predator niches.

Goldman and Bailey decided at this point in the debate that they had heard about the balance of nature a few too many times. Preparing a rebuttal, Goldman solicited Bailey's opinion. Bailey's intriguing response began by channeling an idea current in the 1920s, that Native peoples had merely been another kind of natural predatory fauna rather than human managers of pre-European America. Bailey told Goldman that perhaps in a "pure state of nature" such as had existed before Columbus there might have been a balance of nature. But once modern European humans were on the scene, any such balance vanished. If the United States let predators live, the feds' wolf expert went on, there would soon be nothing but carnivores left. All America's other creatures would be gone. The bureau's predator war was "sound and unassailable and if better understood would be more popular," he insisted.

Goldman folded Bailey's ideas and his own into a throw-down essay he wrote for the *Journal of Mammalogy* that he called "The Predatory Mammal Problem and the Balance of Nature." It began by acknowledging that predators' role in America was a question about which there was "some difference of opinion." But there shouldn't be. The coming of "civilized man" meant that "the balance of nature has been violently overturned, never to be re-established." Ergo, "large predatory mammals, destructive to livestock and to game, no longer have a place in our advancing civilization." Any scientist who believed otherwise was going to alienate both ranchers and hunters. The timing of this essay couldn't have been more ironic. As scientists opened their journals and read it they were also hearing about the new Folsom archaeological discovery that left no doubts humans had coexisted with the American bestiary, predators included, for untold thousands of years. In Bureau of Biological Survey thinking, though, the only continental history that mattered had begun with Pilgrims and Puritans. Those folks had warred on predators from the day they stepped ashore, and properly so.

Meanwhile, Joseph Grinnell had yet another idea, and it vexed the bureau as much as niches and the balance of nature. There was now a National Park Service, with an organic act that charged it with "preserving nature" for all future generations. But how would the parks do that? Grinnell and zoologist Tracy Storer had a suggestion, which they laid out in an article, "Animal Life as an Asset of the National Parks," they co-wrote for the journal *Science*. What they suggested was a far cry from what Yellowstone and Glacier, at the bureau's urging, had done. In their opinion, preserving nature meant that "predaceous animals should be left unmolested and allowed to retain their primitive relation to the rest of the fauna." As naturalists they were convinced that there was a long-standing, organic balance between predators and their prey. No park superintendent should worry about sacrificing game to predators. Besides, as tourism had shown for thirty years, wolves, coyotes, bears, and lions were themselves "exceedingly interesting" to park visitors. If the parks became refuges for game animals *and* predators, the ecologists said, the grand cycles of ancient America could remain forever intact.

The idea of offering wolves, mountain lions, bears, and coyotes permanent refuge in America's national parks dumbfounded bureau personnel. Charles Adams, one of the founders of the Ecological Society, stepped up to second Grinnell: "Without question our National Parks should be one of our main sanctuaries for predacious mammals," he told the scientists. In 1927 the predators-in-parks idea produced an official statement from Mammalogy's annual meeting that predators *ought to be* preserved in national parks. Furthermore, the statement read, the bureau and livestock interests were exterminating America's carnivorous animals without having done the science to support that policy. Bailey and Goldman refused to sign the statement. The bureau then issued a public release that sounded as if it had first consulted a copy of the Great Chain of Being. It had every intention of suppressing animals that were "economically or biologically injurious," it read, and no "sentimental grounds" would cause it to hesitate.

New director Paul Redington was incredulous at the ecologists' position: "We face the opposition," he told his employees, "of those who want to see the mountain lion, the wolf, the coyote, and the bobcat perpetuated as part of the wildlife of the country."

〰️

ONE OF THE SCIENTISTS keenly interested in these meetings was a young Yale-trained forester named Aldo Leopold. When he was barely out of college Leopold had published an essay he called "The Varmint Question," which heaped abundant praise on Bailey and the bureau for their "excellent work." Posted to the Carson National Forest in New Mexico and becoming its superintendent in 1912, Leopold met Estella Bergere, the woman who would become his wife, in Santa Fe. Her family was a prominent one and a new brother-in-law was the president of the New

Mexico Woolgrowers Association. But Leopold was someone who soaked up the world with an open mind, and one strand of thinking he was particularly open to was William Hornaday's. He read *Our Vanishing Wild Life* while recovering from an illness and it rearranged the furniture in his head. The book impressed him with a brand-new idea: that all wild creatures were worthy of respect and consideration, not just the ones Americans called "game." A couple of years later, at the age of twenty-eight, Leopold heard Hornaday speak at the University of New Mexico. From that point on Hornaday became a friend who shaped Leopold's views across the next two decades.

As Leopold's interests shifted from trees to animals, he left the Southwest in 1924 and the Forest Service in 1928. Attending the tense mammalogy conferences in the 1920s he was not a vocal critic of the bureau, but as conference chairman at the firearms manufacturers' American Game Protective Association's meeting at the end of that decade, Leopold set down his own evolving position with respect to predators. "No public agency" (guess which one?) should control predators without substantial scientific research first, he argued. Poisons should be used only in an emergency. And "no predatory species should be exterminated over large areas."

Hired in the early 1930s as a professor of game management at the University of Wisconsin, Leopold dropped more deeply into ecology, the field worried critics were already calling "the subversive science." One of the issues he particularly studied was the response of game animals to predator removal, and he did so for the reason that the ecologists very much seemed to be right about it.

The 1920s witnessed the rise of a new word to describe a process in nature. That word—"irruption"—had everyone from national park superintendents to Hollywood actors up in arms. An irruption referred to a sudden population explosion, primarily among animals like deer and elk, that caused local herds to become so huge they ate themselves out of forage. And then, shockingly, they died of starvation, with tens of thousands of rotting carcasses littering whole landscapes. It was the Kaibab Plateau episode among mule deer on the North Rim of Grand Canyon National Park that introduced the word to the American lexicon and a new mental image to the mind. But once on peoples' lips and in their minds, it kept cropping up because it kept happening. Why? The cause-effect seemed pretty clear. The Kaibab irruption was preceded by earnest bureau hunters erasing 30 wolves, 781 mountain lions, and 5,000 coyotes from the North Rim. Over a handful of years the resident mule deer then exploded from roughly 4,000 animals to 100,000, leading to an initial catastrophe of a 60 percent die-off. In a much-ridiculed move, Arizona novelist Zane Grey organized nature-loving friends, including Hollywood celebrities, in a failed attempt to drive the survivors (whose numbers finally plunged to ten thousand) to a new range. That made Kaibab a national story.

Irruptions became a moral that wouldn't go away. With predators gone, deer and elk began experiencing new, crazy population oscillations somewhere in the country almost every year. Something in nature obviously was amiss. Leopold would go on to do a study of irruptions, finding reports of only two before 1900 but a whopping forty-two between 1900 and 1945, the number rising sharply after 1920. They took place in Utah, Oregon, Texas, California, and Yellowstone Park, and not just among deer and elk. Moose, particularly moose that had recently colonized Isle Royale in Lake Superior, where no wolves were present, also underwent irruptions and die-offs. One of the most famous cases was in Pennsylvania, where by 1905 hunters had killed whitetail deer down to fewer than one thousand animals. Pennsylvania had also wiped out its predators, and with the deer herd under human management and no wolves or lions to cull the excess, the herd exploded to a million by 1927. Then it collapsed spectacularly for lack of food.

Vernon Bailey visited both Kaibab and Pennsylvania. In Arizona he posited, weakly, that reducing the number of cattle on the range might help. In Pennsylvania he advised shooting seasons on doe deer. The irruption lessons seemed so obvious to ecologists that they served as evidence for the so-called Lotka-Volterra equations, ecological models of how prey and their predators follow an oscillating algorithm of rising and falling populations. But no one at the bureau, certainly not Bailey, advocated for restoring wolves or lions, or calling off the war against them. The inertia of that war had become an unstoppable undertow in American policy.

THE BUREAU'S PROMOTION, and Congress's passage, of the Animal Damage Control Act in 1931, which promised the agency a million dollars a year for ten years so it could "control" (i.e., wipe out) coyotes and finish off wolves, dismayed the scientific community, which mounted a petition against the bill E. Raymond Hall called "the largest organized attempt at destruction of wildlife undertaken anywhere." Goldman told the stunned ecologists that while it was true that the policy was preceding the science, good studies were on their way and would "confirm the conclusion that the coyote is the archpredator of our time." With PARC now a full division within the bureau, its operatives weren't backing down. They took their message to the American public with a series of canned articles akin to those lauding the G-men hunting crime celebrities like Al Capone and John Dillinger. "U.S. Agents Stalk 'Desperadoes' of Animal World," the headlines read.

The scientific work Goldman referred to had gone to a pair of mammal ecologists destined to be legendary figures in American conservation. To begin their careers as government biologists the Minnesota brothers Olaus and Adolph Murie had taken advanced degrees in the study of wildlife at the University of Michigan. The Biological Survey had hired the older of the two, Olaus, in 1920, and he had quickly impressed his superiors with landmark studies on elk in Jackson Hole and on the Porcupine caribou herd in Alaska. When Goldman, Olaus's superior in Biological Investigations, appointed him to study the relationship between coyotes and elk in the Jackson Hole Elk Preserve, Goldman's assumption was that Murie would finally validate the coyote's death sentence. Meanwhile, the Wildlife Division of the National Park Service had hired younger brother Adolph and assigned him a similar investigation of coyotes and game animals, but in Yellowstone Park. The science Goldman had such high hopes for, Olaus Murie's *Food Habits of the Coyote in Jackson Hole, Wyo*, appeared in print in 1935. Adolph Murie's *Ecology of the Coyote in the Yellowstone* saw print in 1940.

The Murie brothers would become famous as ethical scientists who let their evidence lead them where it would. Unfortunately for E. A. Goldman, that evidence did not lead to the conclusions about predators he hoped for. After four years of collecting data, the bureau's own, Olaus Murie, concluded that, far from being archpredators of game, coyotes in fact were omnivorous generalists. As far as elk predation went, coyotes were "unimportant" even in taking elk calves. Murie's data indicated that by eating rodents, rabbits, and insects, 70.3 percent of a coyote's actions produced a net *benefit* for human economies, and another 18 percent had a neutral effect. It was far from what the bureau wanted to hear. But surely Adolph Murie's Yellowstone study would provide ammunition. He would be studying far more game species, and the Park Service had after all sent him out with the idea of refuting the ecologists' argument for making the national parks refuges for predators.

Like a modern climate scientist, Adolph Murie knew he was conducting critical research in a highly politicized atmosphere. He responded similarly by being exceedingly careful. *Ecology of the Coyote in the Yellowstone* was a more thorough and more sophisticated study than his brother's, and it took on more issues. Murie began in the park's archives, reconstructing the history of coyotes on the plateau. The first thing that leaped out was that fears of predators swamping the park if control was ended, as it had been in 1935, were unfounded. The coyote population in 1940 mirrored that of 1935. Nor did the park seem to be acting as a predator breeding ground. Coyote populations in the surrounding countryside appeared unchanged. Murie even weighed in on Goldman's insistence that Old Worlders had made moot any balance of nature that once existed in America. So far as Murie could tell, the "relationships of the coyote to the rest of the fauna is today similar to what it was formerly."

Yellowstone was America's version of a game park and its superintendents had always insisted it should be a showcase for game animals. Like Goldman, many in the Park Service thought allowing predators to remain would turn out an absolute disaster for elk, deer, antelope, and bighorn sheep. But after two years of study Murie had to conclude that it was impossible for coyotes to wipe out game animals. Did coyotes destroy elk? "All available data indicate that the coyote is a minor factor in the status of elk," he wrote. What about mule deer, an animal the bureau said could not survive without coyote control? Sometimes, in crusted snow, coyotes ran down the weakest fawns, but those fawns probably would have died of winter weather. "There was no evidence that the coyotes molested any deer except fawns." Pronghorns, those beautiful, striped survivors from ancient America? The bureau insisted that coyotes were a threat to their very existence. But while coyotes caught some fawns,

Murie believed it was in no way problematic: "The coyote is not at the present time adversely affecting the antelope." Bighorn sheep? "Coyote predation is at most an unimportant mortality factor."

When scientific evidence, supposedly a holy grail of policy decisions, conflicts with long-standing received wisdom, there's often a common reaction, which goes something like "OK, let's do another study." That coyotes had been bountied and trapped and poisoned for decades before anyone attempted a scientific investigation of their relationship to prey was shocking enough. That we had almost *erased* wolves, then were baffled as deer and elk populations irrupted, spoke to our arrogance about received wisdom. But the powers that be did reach for a new study, and again it was Adolph Murie, dispatched by the Park Service with support from the bureau, who shouldered the burden of proving "truths" that had driven wildlife policies since the days of the Massachusetts Bay Colony. But the evidence would do no such thing.

If coyotes in Yellowstone were still interacting with prey the way they always had, what about gray wolves? By that time Mount McKinley in Alaska was the only park that still had wolves, and given the controversy over predators, park managers needed guidance. Maybe it was time for a scientist to figure out something about wolf ecology. So in 1939 Murie went to Alaska and spent three years engaged in the unthinkable: actually observing and studying wolves (and other predators) interacting among themselves and with their prey. It was arduous. Murie said he walked 1,700 miles that first year. But everyone who reads his classic, *The Wolves of Mount McKinley*, thrills to his excitement at studying the mythic American animal Europeans reflexively had tried to exterminate for three centuries.

Murie again began his work with history. Wolves had been abundant in Alaska in the 1880s, but as the Harriman Expedition had discovered, some unknown cause had drastically reduced their numbers early in the twentieth century. Murie speculated this was likely an epizootic disease, maybe canine distemper introduced by sled dogs. Wolf numbers were recovering by 1925, then two severe winters crashed the Dall sheep population. By the time Murie arrived in Alaska, "it appears that the sheep and wolves may be in equilibrium," he wrote. After three seasons studying the two species, he was prepared to go much, much further than that.

Though bureau policy had been too arrogant, or myopic, to see it, the relationships Murie was witnessing between the wolf and its prey were

Wolf personalities. From Adolph Murie, *The Wolves of Mount McKinley.*

ancient, pre-dating Europeans by thousands or even millions of years. Those ancient relationships were there for an ecologist to observe in the 1940s because natural selection had long since sorted out the details. Yes, coyotes ate pronghorn fawns. So pronghorns had evolved a solution: they gave birth to twins. Yes, wolves did prey on Dall sheep and caribou, but they primarily caught the very young and the very old. Wolves held sheep numbers in check and kept them from overgrazing their mountains. Rather than destroying their prey, "wolf predation probably has a salutary effect on the sheep as a species," Murie wrote. Similarly, wolves did kill caribou calves, but as he studied the relationship Murie realized "the caribou herds are no doubt adjusted to the presence and pressure of the wolf." Grizzlies robbed wolves of their kills but themselves posed little threat to sheep or caribou. Golden eagles took caribou calves but were hardly relentless attackers of Dall lambs, a predation Murie found "negligible." All these interactions were ancient. They pre-dated Europeans in America. They pre-dated humans *coming* to America.

Murie did one other thing readers of his book never forget. He brought the wolves he studied to life as individuals. "Dandy," "Robber Mask," and "Grandpa" all had unique personalities. Pack members seemed affectionate and caring of one another. A pup he captured before her eyes were open, "Wags," grew tame, friendly to both people and dogs and "always gentle." These were not the wolves Old Worlders feared from the animus they brought ashore with them. They weren't the wolves the bureau's public relations articles implied were gangsters of the animal world. They were the wolves that had been in North America all along, the wolves the Native peoples had known and folded so richly into thousands of years of stories.

The Wolves of Mount McKinley was the country's first entrée into a modern sensibility about wolves and predators generally. And once the blindfold was off, it was hard for the scientifically literate ever to put it on again.

In 1942, two years before Murie's wolf book appeared in the Fauna of the National Parks series, C. Hart Merriam and Vernon Bailey passed away within a month of each other. The predator campaign had deflected them both from further refinement of Merriam's life zones, which was altogether too bad. In that same decade a forester named Leslie Holdridge would outline a more nuanced life-zones proposal, based on vegetation instead of animal life, that—amplified by computing power—has become a staple of modern ecology.

Vernon Bailey died just shy of seventy-eight. He had experienced an interesting and revealing final decade. In 1933 the members of the Society of Mammalogists had elected him president of the organization. But less than a month later, the bureau where he had worked for forty-six years asked him to retire. The timing intrigued many at the time. For years there had been gossip that Florence had changed him. Others thought that years of witnessing fear and suffering in the eyes of wolves and coyotes had gotten to him, that he could no longer support his superiors' predator policies. Indeed, a rumor has persisted across the years that Florence used her connections with Eleanor Roosevelt to try to persuade FDR to bring a halt to the government's predator war. What is more certain is that Bailey eventually invented and patented an injury-free trap, the "Verbail." And that among his papers researchers would later find a photo of a trapped gray wolf with this caption in Bailey's hand: "A Big Gray Wolf, in the snow. Caught and Held in two no. 2 Steel

Traps. Feet Frozen but no less Painful. Yes, he killed Cattle, to Eat. But, Did he Deserve This?"

Despite the continuing controversy over predators, and the role "Wolf Bailey" had played in it, well into their seventies the Baileys continued to hold sumptuous dinner parties at their Washington, DC, home where they discussed everything related to America's wild animals. But there was no cocktail hour, no wine served with dinner, not even coffee served with dessert. Florence, who passed away in 1948, may have changed Bailey's views about predators, but his small-town puritanism prevailed at dinner.

〰

VERNON BAILEY may finally have seen predators in a new light, but his federal compatriots remained loyal to the predator war. In 1944 Stanley Young and E. A. Goldman finally published their two-volume collaboration, *The Wolves of North America*. By this point the Bureau of Biological Survey had become the U.S. Fish and Wildlife Service. But PARC staffers Young and Goldman proudly pointed out that by this time, from New England to Virginia, wolves were entirely gone. That's right, of North America's apex predators, not one was left in the East. The Rocky Mountain states with all their public lands barely held 100 wolves (there were only 6 in Wyoming). The Pacific Northwest still held maybe 140, largely because of nearby Canadian populations. The authors thought California still held 50, but more likely a hunter had killed California's last wolf in 1924. In the Southwest there were still 60 Mexican wolves, a gray wolf subspecies named after Vernon Bailey. There were only 7,000 wolves left in Alaska in 1944. The largest number of wolves remaining in the Lower 48 in the 1940s were in the upper Great Lakes country, with more than 1,400. And red wolves were still hanging on in the Mid-South, where there were some 450 left. The authors did not realize it but those red wolves were on the verge of high-drama transformation.

Young and Goldman were unrepentant wolf killers. They laid out their book, especially the second volume, like a military campaign, a grand adventure promulgated against a worthy opponent. It was full of accounts of wolf depredations on livestock, photos of stock killed by wolves, claims that game animals were disappearing because of wolves. They expressed

dismay that stock-raising Hispanic settlers in the Southwest and California had never attempted predator control and had coexisted with wolves
for hundreds of years by relying on shepherds and guard dogs to protect
their herds. Meanwhile, they claimed that Anglo-American settlement of
the continent had been drastically slowed by the presence of wolves! As
to the "opinion held by many present day game conservationists" that
wolves made ungulate species healthier, they weren't buying it. Today
their naivete seems breathtaking.

To their credit, though, the one charge of atrocity they did not level
against America's wolves were attacks on humans. Not that they didn't
look hard. But except for the occasional rabies story, or a plainsman's
account of falling off his horse while chasing a wolf and having to wave
his hat in the curious canid's face, wolf culture in America just didn't
include a prey template for humans.

In a case of future, meet past, Aldo Leopold reviewed *The Wolves of
North America* in the *Journal of Forestry* in 1945. For starters, the most
famous ecologist in American history wondered how it happened that
Young and Goldman hadn't acknowledged the deep history of their subject. If wolves were as destructive a force as they implied, why had the
continent's wolf population failed "to wipe out its own mammalian food
supply" millennia before Europeans arrived? Somehow the authors obsti-
nately refused to confront such a troublesome question. Leopold had
visited several European countries and studied their wildlife policies,
and he knew the bureau men never questioned the Old World model.
But European countries had nothing comparable to the vast, wild public
lands Americans had set aside. And their model was based on folk tradition established long before ecology was born. It was "not scientific." The
new wolf book hardly acknowledged those differences. It reflected "the
naturalist of the past, rather than the wildlife ecologist of today," Leopold
wrote. Ouch.

Yet the United States emerged from World War II with the anti-predator
ideology within PARC in full song and progressing to a new, Dr. Strange-
love territory. With coyotes having developed a culture of avoiding the
strychnine baits that sent their pack members into vomiting convulsions,
the lab in Denver had designed three new wildlife poisons. Poisoners had
used thallium sulfate on rodents since the late 1920s, and thallium proved
immensely effective on smart animals like coyotes because it cruelly took

Coyote. Photograph by Dan Flores.

days to kill them. In the process of dying from thallium, coyotes lost most of their hair. The pads fell off their feet. But the interval between ingestion and collapse obscured cause-effect. Another new poison was the lab's "humane varmint killer," a .38-caliber cartridge case loaded with sodium cyanide that sprayed a cyanide mist into an animal's face when it tugged on an attached baited cloth. But the researchers' pièce de résistance was something they called 1080 because it had taken a thousand and eighty tries to perfect. Animals that ingested 1080 baits tended to break into a panicked run, blind to their surroundings, uttering strange cries. Until they dropped.

With this new trio of poisons and chemical companies like Monsanto enlisted to mass-produce them, not just coyotes and the few remnant wolves, but eagles, hawks, vultures, condors, magpies, ground squirrels, badgers, prairie dogs (and the predators in prairie dog towns like black-footed ferrets), even dogs and cats, all became victims. Whole food chains were at risk. Millions of animals died. In the 1950s writer Arthur Carhart published an article he called "Poisons—The Creeping Killer" in *Sports Afield* magazine. "An area equal to one sixth of all the crop land in this nation is now being treated with deadly new poisons whose total effects are dangerously and shockingly unexplored," Carhart wrote. He told readers that since the 1940s a federal program had spent $500 million to spread these new poisons across one hundred million acres in the United States. The result was an unimaginable parade of death.

THE ANNIHILATION of wolves and coyotes and eagles was one thing: in their cases "control" actually aimed at *deliberate* extinction. Losing passenger pigeons and Carolina parakeets was not so intentional. But unfortunately, "accidental" extinctions had kept happening. On Martha's Vineyard back in the 1920s ornithologists had watched, helpless, as the eastern prairie chickens called heath hens shrank to this single island and, finally, to an all-male population. The end was quick. In 1923 there were forty-six of the birds. That dropped to thirty-one in 1925, thirteen in 1927, and by 1929 to one last bird. The final confirmed sighting of "Booming Ben," the only heath hen in the world, was on March 11, 1932. Later that year the American Ornithological Union proclaimed that both the heath hen and the ivory-billed woodpecker were extinct. Were prominent species simply going to wink out like this, one American original after another?

The announcement seemed to set a new trajectory. In 1935, when biologists discovered that the largest waterbird in North America, the trumpeter swan, was down to its last hundred birds, the bureau moved to create a swan refuge in the valley of Red Rocks Lake, seven thousand feet high in the Montana Rockies. Two years later there was another shock: the continental population of snow-white whooping cranes had dropped to a mere sixteen individuals! No one had yet figured out where whoopers bred and nested (it turned out to be in Canada's giant Wood Buffalo National Park), but in 1937 the bureau designed Aransas Wildlife Refuge on the Texas coast around a whooper wintering ground. Then in 1940, alarmed that bald eagles as well might be "threatened with extinction," Congress passed the Bald Eagle Protection Act, which prohibited killing, capturing, and selling the national birds. Despite that federal law, the active warfare livestock interests had long waged on bald and golden eagles never let up. It just moved a little more into the shadows. The numbers of both the big raptors continued a steady decline.

The country's cultural direction was ever so slightly shifting, and in

Aldo Leopold, his wife, Estella Bergere, and their dog Flick. Courtesy of the Aldo Leopold Foundation and the University of Wisconsin Archives.

1949 America's star biologist finally published the book that would set a great many Americans onto the new path. Aldo Leopold's *A Sand County Almanac* became both a best seller and a crucial philosophical foundation for the ecology movement that would soon sweep America as part of the sixties' cultural revolution. We were damned lucky Leopold got it out—he died of a heart attack battling a grass fire that same year—because *A Sand County Almanac* was an absolute game changer. In vivid, poetic passages, it introduced the world to the insights of an observant mind that had tracked every breakthrough in ecology. By the time he wrote *Sand County*, like many ecologists Leopold had concluded that the balance-of-nature concept actually lacked the flexibility to account for a natural world that was endlessly changing. He was now thinking of natural settings as interlinked communities of species, with predators at the top. We know those communities today as ecosystems.

Leopold's ideas were near epiphanies for many. To start, he laid out an ecological philosophy for living. He called it "the Land Ethic" and it included his Golden Rule of Ecology: "A thing is right when it tends to preserve the integrity, stability, and beauty of the biotic community. It is wrong when it tends otherwise." The genius that built the United States had always been self-interest. Adam Smith and John Stuart Mill had hit on a trait fundamental to our evolution. But Leopold did *not* say that an act was right when it preserves *humanity*, or *economics*, the easy position of a self-absorbed species. Instead he called on readers to think of the innate rights—among them the simple right to exist—of other species in an ecological community that *included* us. We'd long done things for anthropocentric reasons. Leopold's admirers called his new idea *biocentrism*.

A *Sand County Almanac*'s most unforgettable scene was Leopold's own myth of personal change and redemption. "Thinking Like a Mountain" was not merely a poetic rendering of his view that the United States had an opportunity to create a distinctively *American* policy toward predators and big, wild creatures as opposed to continuing to follow the Old World's template. Over the next quarter century it became far more than that. For a population of readers soon to be immersed in painful soul-searching about so many unexamined assumptions in American life, Leopold's story of shooting a wolf, watching the "green fire die in its eyes," and coming to realize what a miscalculation he had made about the ancient centrality of predators in the biotic community offered America a whole new destination. We were wrong—I was wrong—his story said. But it's not too late.

Aldo Leopold didn't get to see the country begin to think like a mountain. But he had pointed us toward a new destination in history where we would try to save species for all eternity.

CHAPTER 10

A SPECIES OF
ETERNITY

From its daggerlike bill to the stiff, forked tail that propped it upright on tree trunks, the ivory-bill was a nearly two-foot-tall woodpecker. In flight its wingspan extended two and a half feet. The ivory-billed woodpecker was the second-largest woodpecker in the world, exceeded only by its genetic kin, the imperial woodpecker of Mexico. Native people admired its disposition and courage. Audubon said that every time he saw one fly it reminded him of an Anthony Van Dyck painting.

Attired as if in a tuxedo, the ivory-bill's black body was artfully set off by a pair of white stripes extending from its yellow eyes to a large patch of white feathers on its back. Matching white swaths on both sides of the trailing wing edges made it easy to identify in flight. Both sexes had topknot crests but the male's was livid scarlet, giving the bird an air of formal, self-aware magnificence. Beneath the crest was a skull that was like a compressible sponge, built to absorb shocks. When not in use the eight-inch tongue recessed into storage around the back and top of its head. There were three eyelids, one of which was transparent and remained over the eye to protect it from flying debris. Its flight was direct and fast with slight up-and-down undulations, wing beats then glide, wing beats then glide. All who wrote about seeing one mentioned the elegance of

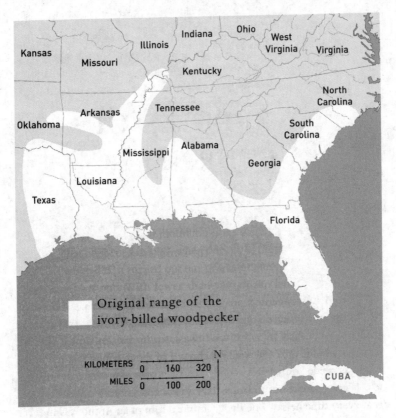

Original range of the ivory-billed woodpecker.

an ivory-bill's passage through the forest, and its strange, primal, toy-trumpet cries.

The ivory-bill's range—from the Carolinas down and around the Gulf Coast plain as far west as the Brazos River in Texas, and then in a band up the sandy swipe of the Mississippi to Illinois—was as southern as NASCAR. As tropical migrants, they'd spent two million years adapting to the region's old-growth riverine forests, which once covered fifty-two million acres. Among sweet gums and oaks, where storms, fires, and floods often killed thousands of trees, ivory-bills became specialists, focusing on the beetle larvae that infested freshly dead trees. The birds were transient, endlessly searching out the South's latest natural disaster.

Blessedly, neither of the ivory-bill's colonial biographers, Wilson and Audubon, lived long enough to see ivory-bills follow so many of the country's primeval creatures toward oblivion. By 1885 their calls and unique double-drumming had vanished from a third of their original range. Then they retreated from three or four states every decade until, in 1932, ornithologists reluctantly announced they were extinct. It was not that they were market targets or game birds. People did sometimes shoot and eat them, and naturalists collected far too many. But what pushed ivory-bills toward the brink was not the boom of guns but the chop of axes and the whine of logging trucks. With the Civil War resolved and loggers exhausting trees in New England and the Midwest, the industry had turned to the South, first wrecking its longleaf pine forests, then the towering bald cypresses of the coastal plain, finally the oak and sweet gum giants of the river bottoms. The vanishing of ivory-bills after 1875 directly tracked the felling of thousand-year-old trees across a "modernizing" South. Did the industry provide jobs for the rural class that once shot birds for their plume feathers? It did that. Did it care if it was wiping out one of the South's most impressive and iconic birds? It's hard to find evidence it did.

As creatures of the deep wilderness, ivory-bills became bona fide unicorns and they raged against the dying of the light. They hid from us in the green hells of the South's swamplands until, as emblems of Americana, they became targets of a biological treasure hunt. No one claims to see a passenger pigeon anymore. But repeatedly, yearly, Americans still catch glimpses of ivory-bills, invariably quick screen-grabs at the periphery of sight, or a replay of some collective memory. They always turn out to be from the periphery of our imaginations.

But there was a template for losing ivory-bills, then finding them again. In the late 1930s, a few years after science proclaimed them extinct, a young graduate student with an Audubon Fellowship from Cornell University found himself actually watching live ivory-bills in a part of Louisiana not far from where I would grow up. His name was James Tanner and he had come to Louisiana in 1937 because two years earlier a Cornell–American Museum of Natural History expedition had miraculously found a remnant population of seven pairs, a discovery confirmed for the world by the only uncontested movie footage ever shot of ivory-bills. The birds were in a place called the "Singer Tract," 120 square miles of old-

Rare photo—an ivory-bill, still from video shot in Louisiana in the 1930s. Courtesy of
Macaulay Library, Cornell Lab of Ornithology.

growth loggers had missed near the Mississippi River that were owned by
the Singer Sewing Machine Company.

Americans thought all they'd ever know about our premier woodpecker
had come from Wilson and Audubon. Through Tanner, the birds had a lot
more to say. From 1937 to 1939 he didn't just study the Singer Tract birds,
he examined forty-eight other possible sites, most in South Carolina and
Florida and elsewhere in Louisiana. His assessment everywhere was the
same stark conclusion: "Now the swamp is completely logged over and I
found no signs of the birds." While Tanner did believe there might still be
twenty-four of the giant woodpeckers across the South in the late 1930s,
the only living ivory-bills he actually saw were those seven pairs in Louisiana. It was a fantasy act of salvage research, along the lines of some talented journalist having gotten to Stratford-upon-Avon in time to manage
an interview with Shakespeare with the ink still wet on the last plays.

Tanner believed there were never great numbers of ivory-bills. His
estimate of their original density in their habitat was a single pair per six

square miles, compared to as many as seventy-two of the slightly smaller pileated woodpeckers that often resided in the same stretch of woods. He theorized that ivory-bills were as few and as large as they were because they evolved to such a narrow niche. Freshly killed standing timber was widely scattered, and attacking the beetle larvae they preferred required real strength because the birds had to chisel away bark that was still tight. They weren't gregarious, but roamed the South as paired companions mated for life. Tanner found active nests anytime from January to May, with an average of three eggs. Both males and females incubated the eggs, brooded the young, and fed them. Nestlings were altricial and took nearly six weeks to venture into the world. Occasionally, Tanner found, a hawk or owl might nab a young bird. But as adults ivory-billed woodpeckers had no predators. They were noisy and visible, he said, and not especially wary of humans.

Tanner's remarkable experiences, which the Audubon Society published in 1942 as *The Ivory-Billed Woodpecker,* had a tragic coda. It turned out the sewing machine folks had plans for the Singer Tract, which Louisiana was managing for the company as a wildlife sanctuary (it's now the Tensas River National Wildlife Refuge). Yet even when scientists had found the only certain ivory-bills left in the world on its land, amid a cacophony of desperate pleas the sewing machine company refused to set aside plans to log the place. In 1937, as Tanner was heading into the dawns to spend his days studying those last ivory-bills, the company sold the logging rights to the Chicago Mill and Lumber Company. In 1943, the year after Tanner's book came out, its loggers arrived to destroy the last old-growth bottomland forest in northern Louisiana. One final ivory-bill, a female bird, the reports say, lingered in the area until the spring of 1943. Then the loggers felled a tree containing a nest and eggs, and she sailed away on her undulating woodpecker flight into the mists that always hang over Mississippi Valley mornings.

/w.

WE ALL KNOW that in 1776 Thomas Jefferson did *not* write: "We hold these truths to be self-evident, that all *species* are created equal." Notwithstanding Carolus Linnaeus classifying humans as animals, or Native American instincts about our kinship with other creatures, the idea that all species

were on a level playing field was manifestly *not* self-evident at the founding of the United States. Nonetheless, in the past half century—and to
some extent led by the United States—people around the world began to
dedicate themselves to an idea at least as radical as universal suffrage. We
accorded the right, by law and policy, for existing species to remain alive
alongside us.

This principle may be off the radar for most people, in part because
its thrust is largely about others. But make no mistake: protecting other
species from going extinct, and devising imaginative ways to help them
recover when they're threatened, is a titanic development in American
cultural life. Unease about passenger pigeons and bison first focused
national attention. Yes, we'd once lost mammoths, but that was so long
ago it was hard to wrap the mind around. Pigeon extinction happened
under our watch. Whooping cranes, trumpeter swans, heath hens, condors, eagles all faced serious trouble. Was this going to be the legacy of
the world's greatest nation, that in becoming a success it obliterated the
continental species that had once made it a biological marvel?

For decades compassion for wild animals had swirled visibly in American culture because Thoreau, Hornaday, Seton, and Florence Merriam
had readers and followers. But with the dawn of the 1960s, the world
cracked. Americans were reading Aldo Leopold, who encouraged us to
think in *bio*centric terms. The country also grappled with deadly smog
events in the Northeast and California in the 1950s, oil spills that wrecked
beaches on both the Pacific and Gulf Coasts, and pollution that was so
toxic in the Midwest that rivers actually caught fire. Economic success
gained without paying even slight attention to consequences led to disasters we hadn't anticipated, not just for pigeons or buffalo but for us. That
got our attention.

You didn't have to be there, on the scene in the 1960s, or attending the
rallies for the first Earth Day in 1970, for the Age of Ecology to capture
you. Like the other grand cultural upheavals of the time—Civil Rights
struggles in particular—the Age of Ecology was an awakening that has
reverberated across every decade since. Once culture rang that bell it
couldn't be un-rung. It is fitting that the science that had undermined
the predator war lent us a symbolic name for environmentalism's emergence. The Age of Ecology's simplest idea was that every species deserves
the right to exist. For too many species it was already too late. But for

the birds and animals still with us, morning sunlight was about to come streaming in.

Two ways of thinking compassionately and sympathetically about other animals came to the fore in the sixties. One derived from the nineteenth-century British philosopher Jeremy Bentham and the ethics of morality. Its focus was on the rights of individual animals and particularly of domesticated ones raised to feed humans, plus animals used experimentally in labs. The Australian writer Peter Singer became the most visible figure of expanding moral treatment to individual animals with a movement that found more fertile ground internationally, although PETA (People for the Ethical Treatment of Animals) and the Animal Liberation Front got plenty of headlines in the United States.

With America's peculiar history of destruction, it was the second idea—the defense of *species*—that captured us in the Age of Ecology. True, we had identified with Tito, Ernest Thompson Seton's coyote, with Martha the last passenger pigeon, and with Adolph Murie's individualistic wolves. But wildlife management had always been about elk or deer as species, not individual elk or some particular whitetail deer. Leopold's "Thinking Like a Mountain" was a plea for protecting holistic nature, very different from Peter Singer's focus on the individual animal. Many Americans had experienced a surge of pride when we'd created refuges to save trumpeter swans and whooping cranes, but again, those were for aggregates of birds. Expanding ethical treatment in America had always accorded rights to groups—white men in the Jefferson-Jackson tradition, African Americans with the Thirteenth, Fourteenth, and Fifteenth Amendments, women with the Nineteenth Amendment—so extending the circle of ethics to species made sense. Of course, like groups of people, species are made up of individuals, creatures (as today's biologists are reluctantly conceding) who are distinctive, who obviously experience personal joy in being alive.

For humans, *personhood* is the locus of individual rights. For nonhumans, a focus on species was always going to be problematic. My dog, lying at my feet as I'm writing this, is a member of the genus *Canis*, sure enough. He's also an individual dog named Kodi, an Alaskan malamute whose life has been unique and who is not like any other dog.

Nonetheless, protecting the rights of species was the track America was on. A breakthrough toward wider application of species protections came in 1964 with creation of the significant but barely remembered Land and

Water Conservation Fund. It collects money from offshore oil-drilling royalties and spends it (sometimes) to enhance outdoor possibilities around the country. But one part, inserted at the suggestion of a new "Office of Endangered Species" within the Fish and Wildlife Service, tapped Conservation Fund money for acquiring or expanding refuges for species that were in trouble. William Hornaday had set this ball rolling with the National Bison Ranges. As time went on, elk, pronghorns, trumpeter swans, and whooping cranes had all gotten refuges. Now employees of this new Fish and Wildlife office began to lay the groundwork to roll a much bigger ball much faster.

Another act, signed into law on the same day as the Land and Water Conservation bill, became a linchpin of what we might call America's original Green New Deal. The Wilderness Act of 1964 not only designated millions of acres of wild public lands as continued habitat for animals like grizzly bears that required truly wild country, it paid tribute to the history that wildlands and wild animals had played in the American story. These two acts were part of a quarter century of landmark environmental legislation. They began with efforts in the 1950s to clean up the country's air, that eventually targeted water and toxic-waste cleanup, and extended to preserving nature recreation by protecting wild and scenic rivers, laying out national hiking trails, and culminating in 1980 with a stupendous Alaska lands bill doubling all the lands previously protected in the national parks and Wilderness Act. Eventually including landmark, comprehensive laws to protect species from extinction, this grand environmental program was nonpartisan to its roots. In the sixties and at least for a while in the seventies almost all of us were environmentalists.

An environmental program this encompassing required an America strikingly different from the nineteenth-century nation, or even the twentieth-century America riven by world wars. By the 1960s there was a growing contingent of citizens who had the means and leisure to experience their country, and some were appreciating wild animals even when they weren't viewed over the sights of a rifle or shotgun. These were the people who'd bought Florence Merriam's books on birds and now avidly acquired and studied Roger Tory Peterson's field guides. While he was working on predators for the Biological Survey, Olaus Murie had taken the time to evaluate "the factions interested in [the] coyote question," as he phrased it. There were the stockmen, and the hunters, and the bureau's own field men, all of whom wanted predators dead. But Murie concluded there was an emerging group he called "the Nature Lovers." To the annoyance of his superiors, the scientist insisted that this new group actually represented the future, and an enlightened future at that. "I firmly believe that it is working against the best interests of humanity to . . . ridicule those who see beauty in a coyote's howl," he'd written.

Just who were these people who appreciated the beauty in a coyote's howl? Animals had a wide, natural constituency. Many in that constituency were biologists and ecologists. Barry Commoner (who eventually would run for president) was an academic biologist. Well-educated middle-class women segued seamlessly from conservation to environmentalism, especially in the wake of Rachel Carson's book *Silent Spring*. Government and wildlife professionals hovered around the edges, slightly alarmed at the emergence of new, activist groups like Defenders of Wildlife. Hunters, many of them rural, were even more suspicious. Once the backbone of conservation and saving wildlife, by the 1960s hunters were worried that urban environmentalists might object to shooting things. Thrilling to the beauty of a coyote's howl didn't come naturally, and fears about gun rights were losing them to environmentalism.

Finally there was that generation of us Baby Boomers who flooded the teach-ins on Earth Day. Awesome as they were, sex, drugs, and rock & roll weren't the whole story. We may have been privileged and coddled but we lived in a society of vast inequalities, visible everywhere you looked. War loomed over us like a Damocles Sword and that sharpened the instincts. Assuming that nature was the one respite, the one place of peace and redemption, it took only a little experience to realize that, like Thoreau, you were yourself that citizen you pitied.

/

IF YOU FELL INTO one of these groups and were intrigued with America's wild creatures, you were a ripe target for two new media breakthroughs that have fixated us since. In 1924 an animation genius named Walt Disney moved with his brother Roy from Missouri to Hollywood, and by the 1950s their series of beautiful and successful animated films had put the Disneys at the pinnacle of moviemaking. Both brothers had long been fascinated with wild animals in America's story, and beginning in 1948 they began their invention of a whole new film genre, the nature documentary, which spawned the first generations of Americans to have their nature aesthetic and philosophy shaped by film. Their first try at feature-length nature films was 1953's *The Living Desert*, directed by James Algar. Filmed near Tucson, it won that year's Academy Award for feature documentary and the International Prize at the Cannes Film Festival the next year. Algar went on to do multiple nature films for Disney, including *The Vanishing Prairie* in 1954 and the Arctic documentary *White Wilderness* in 1958.

By then Disney Studios had established itself in television with a weekly programming slot on Sunday nights that went on for three decades and wore various titles before becoming *The Wonderful World of Disney* in 1969. Week after week its episodes featured sympathetic takes on wild horses, bison, beavers, all straight out of America's story. Of course, in his personal life, kindly "Uncle Walt" kept the Cartoonists Guild out of his studios and supported both Joseph McCarthy and Barry Goldwater. But being a political conservative in the 1960s and 1970s didn't necessarily translate into being a knee-jerk foe of environmentalism. Disney and his nature films showed us that a fascination with the country's wild birds and animals and their stories was in fact apolitical, and paved the way for even more nature programming. No one who lived through the late twentieth century can forget Marlin Perkins's bizarre and kitschy *Wild Kingdom*, brought to us by an insurance company. As forerunners of twenty-first-century public television shows like *Nature* or David Attenborough's Earth documentaries, Disney and his imitators renewed interest in not merely the history of American animals but their survival in the modern world. I know. I was one of the fascinated.

A new category of books was changing us, too. Our century-plus of scorched-earth poison use was leading us to a crisis, however clueless we were about it. In the decades after World War II we possessed a raft of new, synthetic poisons, many of them products of the chemical revolution jump-started during the war. Corporations sold poisons like Compound 1080 and DDT as game changers in quality of life, "better living through chemistry," with little study of their wider effects, and all manner of inexperienced users were unleashing them on the landscape.

But other scientists did study them, and ultimately their conclusions couldn't be ignored. Much like the leadership at the Biological Survey

Rachel Carson on Hawk Mountain. Photograph by Shirley Briggs. Courtesy the Rachel Carson Foundation.

during the 1920s, corporate CEOs weren't used to public interrogation. They damned sure weren't used to being embarrassed by a scientist who was a *woman*. That establishment reaction is one of the things that made Rachel Carson's 1962 *Silent Spring* the culture-flipper it became. Carson was yet another ecologist who became a public social critic and one of many in a considerable line of powerful twentieth-century woman conservationists. This cohort included Florence Merriam, of course, but also more confrontational figures like Rosalie Edge and Marjory Stoneman Douglas. Edge, a New York suffragist and one of the most implacable wildlife conservationists in the decades after Hornaday, famously charged the Audubon Society with having too cozy a relationship with the firearms/hunting community. In 1934 she established the first refuge for birds of prey in America, Hawk Mountain Sanctuary, on the Appalachian Crest Flyway in Pennsylvania, a ridge where for years shooters blasted migrating eagles and hawks. Edge had fought alongside Joseph Grinnell against PARC's poisoning campaign, once remarking sarcastically that the bureau's Ira Gabrielson made killing a predator sound like a mystical experience for hunter and victim alike.

Like Rosalie Edge, Marjory Stoneman Douglas began her public life as a women's activist. She'd moved from her native Minnesota to Florida at an early age, become a journalist in Miami, then acquired international fame in 1947 with her book *The Everglades: River of Grass*, which figured prominently in the creation of Everglades National Park. Douglas lived to 108. As a voice for the Everglades well into her 80s, she not only preceded Rachel Carson as the grande dame of American conservationists, she also followed her. These were formidable advocates for nature and animals. Carson fit right in.

As an attack on the use of poison to "control" American nature, Carson's book had shock value, and it had a long tail. For one, *Silent Spring's* title and central metaphor were genius. The lyrical trilling of mockingbirds, hermit thrushes, canyon wrens, and other songbirds in springtime had always been part of the enchantment of American life. Why would any entity want to strike that from our experience? The chemical companies argued that they were aiming their chlorinated hydrocarbon, DDT—Carson's primary target—not at birds but at insects, "undesirables." Just as government agencies had heralded 1080 as the final solution to the country's predator problem, the chemical companies touted DDT as the beginning of a "pest-free" Green Revolution in agriculture. Wholesale spraying of DDT through towns and cities was going to confer benefits even beyond that, as the spray trucks would defeat Dutch elm disease and—a direct sensory appeal—beat back summertime mosquitos plaguing backyard barbecues, baseball games, and sunbathing.

A trained ecologist and fine writer (an earlier book, *The Sea around Us*, had won the National Book Award), Carson was an employee of the U.S. Fish and Wildlife Service. She was close enough to agency divisions like Animal Damage Control that she had an insider's sense of how its poisons were blanketing the American countryside. She also knew there had been little advance study of them. Carson became alarmed about DDT for the same reason the ecologists had called out using strychnine against wolves. Both produced shocking collateral damage among species that weren't intended targets. But the list of *intentional* targets always seemed to be growing. "As the habit of killing grows," Carson would write in *Silent Spring*, we "resort to 'eradicating' any creature that may annoy or inconvenience us."

While spokespeople for the chemical companies were telling a trusting public that DDT (once applied directly on troops to control lice) was harmless if you weren't an insect, Carson knew that dissolved in a spray it was highly toxic. She also knew the science was showing that DDT bioaccumulated in the fatty tissues of vertebrates. She discovered early in her research that DDT sprayed over a wildlife refuge near her Maryland home resulted in birds dying in horrible fashion. In the years after initial DDT use, she wrote, the Fish and Wildlife Service began to see "blank spots weirdly empty of virtually *all* bird life." There was more. Spraying DDT onto coastal marshes to kill mosquitos was likely to "make it necessary for us to find a new national emblem," as accumulating DDT in the fish bald eagles ate was wrecking the national bird's ability to reproduce.

Losing much of the country's bird life, even losing our national symbol, might not be enough to arouse the public. So Carson wasted little time taking readers to her most shocking argument. Anyone who got twenty pages into her book understood that in poisoning nature we were not just fouling the environment around us. We were poisoning ourselves, bioaccumulating DDT in our livers, kidneys, intestines, thyroid and adrenal glands, and testes. Mothers were passing the poison on to nursing children in their milk. Human self-interest was the one argument that made *Silent Spring* a blockbuster.

Despite the chemical and agricultural industries' best efforts to portray *Silent Spring*'s author as "hysterical," within a year the book had gone through three printings and forty state legislatures were entertaining pesticide bills. The book did produce a ban on DDT use in the United States. But in a larger way, Carson aroused in the American public a growing revulsion at the general poisoning of the natural world. Misguided government policies were placing "poisonous and biologically potent chemicals . . . indiscriminately into the hands of persons largely or wholly ignorant of their potentials for harm," she wrote. For forty years such poisoning had been about eradicating wolves and coyotes, in the process killing bears, eagles, condors, foxes, songbirds of every kind, indeed one species after another. It was as if North American evolution had floundered along for eons just waiting for us to show up and weed the place.

⁓

"THIS IS THE ENVIRONMENTAL awakening. It marks a new sensitivity of the American spirit and a new maturity of American public life. It is working a revolution in values. . . . Wild things constitute a treasure to be protected and cherished for all time. . . . The wonder, beauty, and elemental force in which the least of them share suggest a higher right to exist—not granted to them by man, and not his to take away."

It was an American president who delivered those ringing words, but this was not Lincoln at Gettysburg, freshly dedicating the country to the proposition that all of us were equal. These words, acknowledging "a higher right to exist" for America's wild animals, were spoken by Richard Nixon. The year was 1972 and the occasion was the announcement of an end to poisoning predators and raptors on the country's public lands. That unprecedented step was possible, Nixon told us, because we were in an "environmental awakening." The "new maturity" of which he spoke, the "revolution in values" he referenced, were possible because the American wild-animal story had brought us to this point. This period when the country did an about-face on our national animals was a special moment in time. It had never come before, and within another decade it would not be possible.

By the mid-1960s ecological thinking was rolling into a crest. In 1964 Interior Secretary Stewart Udall's office, inspired by 1940's Bald Eagle Protection Act, compiled a list of sixty-three American species of birds and animals that scientists believed were "rare" or "endangered." By 1966 scientists counted eighty-three. Udall called the bill that the Lyndon Johnson administration drew up to address these extinction fears the Endangered Species Preservation Act. Introduced into Congress by Democratic representative John Dingell of Michigan, the act established the legal category of an "endangered species," a global list of which a group of international scientists was already compiling in the so-called Red Data Book. The 1966 law began the official process of "listing" those in the United States as endangered, a start. But the law had no teeth and made harming listed species a crime only on a very narrow swath of land, the country's national wildlife refuges. Congress passed the act with little fanfare. The same blasé approach characterized 1969's Endangered Species Conservation Act, which also came out of the Johnson administration and added fishes, crustaceans, and invertebrates to birds, mammals, and reptiles as groups eligible for endangered classification.

By 1969 Richard Nixon was president of the United States but remarkably that did not mean the moment had passed. When Nixon took office the first Earth Day was only a year away. The president himself, of course, had not the slightest interest in wild creatures. On a spectrum of nature-loving American presidents, with Teddy Roosevelt and Thomas Jefferson occupying one end, Nixon and Donald Trump uphold the other. But Nixon understood a political bellwether, and even if he privately thought interest in wild animals was pathetic sentimentalism, he believed that if his administration publicly endorsed the green wave sweeping the country he might be able to swing votes toward the Republicans. If, that is—as an adviser put it—Nixon could "identify the Republican Party with concern for environmental quality." How controversial could this ecology thing be, anyway?

Knowing what we know now, it could hardly be more counterintuitive to have the Nixon Republicans in 1970 create the Environmental Protection Agency, with its requirement that nature-directed projects using federal funds conduct an "environmental impact statement" to assess potential danger to the natural world. But establishing the EPA didn't accomplish Nixon's political goal. When twenty million Americans turned out for the first Earth Day in history in April of that year, TV anchor Walter Cronkite reported the crowds to be "predominantly anti-Nixon." A political opportunist needed another issue, and with the Sierra Club, Defenders of Wildlife, and the Humane Society all suing the administration for poisoning coyotes without conducting an environmental impact study, Nixon decided to strike a deal. If the enviros would drop their lawsuit he would end the poisoning war against predators. Maybe that would allow him to surf the environmental crest. That is how ending a practice that had been routine since the early century, viewed askance by only a handful of ecologists, came to serve a presidential agenda in the new green world.

Nixon's "This is the environmental awakening" speech reflected not his own core values but instead the sentiments of its principal author, Republican Russell Train. It was Train who drew the task of turning Nixon into an environmentalist. A Rhode Islander and graduate of Princeton and Columbia Law School, he'd become a wildlife disciple as a result of photo safaris to Africa, a sad commentary on the loss of our own charismatic species. He was chairman of the Council on Environmental Quality at the time he wrote Nixon's speech, went on to become the second director of the EPA, and later founded the World Wildlife Fund. Train was a bona fide Republican environmentalist when that label was not an oxymoron.

It is highly unlikely Nixon had ever read Aldo Leopold or encountered the word "biocentrism" without stumbling over its pronunciation. But he did not shy from the line about other species' "higher right to exist." He didn't flinch from Train's phrasing of the poisoning question, either: "The old notion that 'the only good predator is a dead one' is no longer acceptable as we understand that even the animals and birds which sometimes prey on domesticated animals have their own value in maintaining the balance of nature," he told Congress. The speech acknowledged this was an issue the administration was joining—"The widespread use of highly toxic poisons to kill coyotes and other predatory animals and birds is a practice which has been a source of increasing concern to the American public"—but now Nixon was ready to make it his own: "I am today issuing an Executive Order barring the use of poisons for predator control on all public lands." Henceforth, supposedly, only individual, offending animals would be targeted. A month later the EPA stopped all interstate shipments of predator poisons, and in rural, conservative communities the shit hit the fan. Western Woolgrowers went so far as to bring torn and bloody lamb remains to public hearings.

Conservative outrage over the poisoning ban did not derail the crown jewel of the Age of Ecology, though. Waiting in the wings was a law the U.S. Supreme Court would later call "the most comprehensive legislation for the preservation of endangered species ever enacted by any nation." This was the watershed Endangered Species Act of 1973. It was at once the high tide of the first Green New Deal, and also the law that ruined environmentalism as a nonpartisan, mom/apple pie movement. Yet it gave no hint of such a future when it glided to passage almost without a ripple in the body politic, passing in the Senate by 92–0 and in the House by a vote of 390–12. How threatening to America could it be to protect the higher right of species to exist?

Introduced by Democratic senator Pete Williams of New Jersey just as eighty nations assembled in Washington for a convention on the international trade in endangered species of animals and plants, the 1973 law pole-vaulted beyond its two predecessors. As an indication of how far the country's attitudes had evolved, only decades before, the United States had taken no action whatsoever as one wild species after another had disappeared or almost become extinct. Now we were enacting a measure unprecedented in world history. The ESA's opening line pulled no punches: "The Congress finds and declares that various species of fish, wildlife, and plants in the United States have been rendered extinct as a consequence of economic growth and development untempered by adequate concern and conservation." When would Americans have ever said such a thing before?

The act of 1973 derived its potency—and the source of many of the controversies to come—from three of its sections. Section 4 gave the secretaries of Interior and Commerce the mandate to list species, subspecies, and "distinct population segments" as either threatened or endangered *solely by relying on the best science available*. What eventually shocked conservatives and economic interests was discovering that the law outright prohibited a consideration of economics in determining listings. (In the law's initial court challenge, the Supreme Court observed that Congress clearly intended it to "halt and reverse the trend toward species extinction, whatever the cost.") Section 4 also required federal agencies to identify and protect an endangered species' *habitat*. Even more remarkably, it specified that the agencies should draw up and pursue *recovery plans* for endangered species. Section 7 compelled federal agencies to forego projects that might imperil a listed species. And section 9 prohibited the "taking" of any individual of an endangered species, which courts subsequently interpreted to mean degrading a species' habitat, including on private land. A historian of the act describes section 9 as "perhaps the most powerful regulatory provision in all of environmental law."

There are any number of events or laws that are candidates for the prize of America's very best idea. But for the priceless genetic legacy of wild species that evolved in North America's deep time and existed here in health long before we ever arrived—and for us, too, human animals who get to live amid the continent's evolutionary richness—surely the Endangered Species Act of 1973 is America at its best.

∿

THE GREATEST PROMISE of the Endangered Species Act was the renewed chance to experience an America we assumed we'd never have again. Most Americans thought we'd never see a big cat like a jaguar in the wild. We believed we'd hear wolves howl only in TV documentaries. We guessed we might never again see a bald eagle, or a California condor, and that grizzly bears wouldn't last far into the twenty-first century. Growing up in Louisiana, on the banks of a classic southern swampland called the Black Bayou, I never once saw an alligator, never heard one roar. By the 1950s market hunting for their hides had nearly destroyed William Bartram's fire-breathing beasts. The year I graduated high school, 1967, America's alligators went on the endangered species list.

Here's a take-home. Endangered species protection and recovery turned on its head a horrific four-hundred-year wild-animal story in America. And despite the controversies and political battles over snail darters, northern spotted owls, and (today) wolf recovery, the reality for most recovery programs under the ESA has been feel-good celebration. The remarkable rescues of bald eagles, peregrine falcons, and California condors may not please everybody, but they thrilled most of us. Ecologies benefited. So did America's vision of itself.

The eagle story alone symbolizes how the modern United States has struggled to protect wildlife, even its national symbol. By the sixties bald and golden eagles were poised to replicate the fate of passenger pigeons and Carolina parakeets. Realizing this shook up both liberals and conservatives. This, after all, was our national bird, made so by a proclamation by the nation's Founders in the 1780s. In fact, given the market slaughter of ducks and shorebirds critical to their diet, bald eagles across the East and South began their decline simultaneously with that national coronation. Then the old-growth trees eagles preferred for nesting started falling to loggers. Agriculturalists shot the giant raptors by the thousands simply because they regarded birds of prey as suspicious characters around domestic fowl, calves, and lambs. In the East, eagles died from ingesting poison from lead shot used in waterfowl hunting. In the West, poison baits aimed at wolves, then coyotes, absolutely devastated carrion-eating eagles. Widespread acknowledgment of dwindling or nonexistent eagles finally led to the federal prohibition against killing bald eagles in 1940.

By 1962 Congress expanded that law to include golden eagles, who were undergoing a shocking decline, especially in western locales like Colorado, Wyoming, and West Texas. However, the new federal law didn't deter the livestock industry's efforts to wipe out what ranchers called "the vicious and relentless Mexican eagle" (i.e., golden eagles) just

American golden eagle. Photograph by Dan Flores.

as they had done with wolves. The West's "eagle war" led to an infamous
episode near Casper, Wyoming, in 1970, when ranchers armed with sixty-
five pounds of thallium sulfate (already banned because of its cruelty)
laced three dozen sheep carcasses with it and killed at least twenty-four
bald and golden eagles. Boy Scouts on a weekend field trip found the protected birds heaped into piles. "Eagle hearings" before Congress that year
produced a pilot who said that over the few months he had flown shooters
in pursuit of eagle kills, they had shot five hundred of the birds out of the
sky, balds and goldens alike.

Long before that, as Carson intuited, DDT had become part of the
story. Sprayed across marshes and swamps to control mosquitos, the pesticide accumulated in bald eagles from the fish they ate and so thinned the
birds' eggshells that breeding eagles were unable to reproduce. No one
knows an actual figure, but at the start of the 1800s there may have been
100,000 nesting bald eagles across America. By 1963 a census indicated
only 487 nesting pairs. That shocker led Interior Secretary Udall to list
bald eagle populations below the fortieth parallel as endangered under
the Endangered Species Act of 1966. Five years later, in 1972, Congress
outlawed the use of DDT, and the year after, bald eagles became one of

the first species listed following the ESA of 1973 (bald eagles weren't listed in Alaska and received a "threatened" designation in the upper Great Lakes and Pacific Northwest).

With official listing in place, the Fish and Wildlife Service commenced bald eagle recovery with no clear sense how it would go. Banning DDT had been a critical step, allowing the birds to breed successfully again. Captive-breeding programs produced young birds ready to release into the wild. And the Service (along with state game wardens) engaged in committed patrols and protection efforts. By 2006 the Service estimated the number of nesting pairs in the Lower 48 had grown to 9,789. The following year it deemed bald eagles recovered sufficiently to delist them, making our picturesque national symbols one of the poster species for endangered species recovery programs.

The ESA requires continued monitoring of recovered species, and the Fish and Wildlife Service does so with eagles. The 2009 estimate assumed 143,000 bald eagles lived or nested in the United States, including Alaska. Golden eagles are still protected under the 1962 amendment to the Eagle Protection Act as well as the Migratory Bird Act and are listed as endangered. In 2014 the Service estimated golden eagle numbers across the whole of the country, including Alaska, at 39,000 birds. No longer blasted by aerial gunners, golden eagles are nonetheless declining. Since 2016 an Eagle Rule Revision has enforced half-mile buffers around nesting sites. Some 500 golden eagles a year die of electrocution on powerlines. Others collide with the spinning props of wind generators, and poisoning by lead shot continues to be a threat to all eagles. But because they go after roadkill, the biggest threat to goldens is vehicle collisions. Having witnessed this myself on a miles-long straightaway in Utah's Great Basin, I've taken to dragging roadkill off highways when I've seen eagles on it.

Condors also magnify whatever problems exist in the ecologies they inhabit. Condors had ranged across the continent among all the predators of the Pleistocene, narrowing their focus to washed-up carrion on the Pacific Coast only with the great extinction crash twelve thousand years ago. When the ever-pessimistic William Hornaday was assembling the material for *Our Vanishing Wild Life*, he asked Joseph Grinnell for what he assumed would be the eulogy for California condors. But Grinnell was optimistic. Condors in 1912 were "still fairly common" in Ventura, Santa Barbara, San Luis Obispo, and Kern Counties, he told Hornaday. With

Adult California condors. Courtesy Shutterstock.

their soaring habits, nearly ten-foot wingspans, and naked heads, condors had a presence no other bird could match. But Grinnell's take came before Merriam's Biological Survey began to broadcast poison baits to wipe out wolves and coyotes. PARC operatives told themselves condors would vomit up poison baits, but in fact the big birds weren't just taken out by ingesting lead shot, they also died from predacides. Researchers using beetles to strip the flesh from dead condors found the carcasses so poison saturated they couldn't keep the beetles alive. Nature did not build condors to recover from human threats.

Wild condors often don't mate until six or seven years old, and commonly produce a single offspring a year. The giant vultures were listed early, in 1967, but it was barely quick enough. By 1982 there were only 22 of them left on the planet. In 1987 the recovery program decided that with a mere 27 condors alive it would capture every wild condor for a captive-breeding program. It was a Hail Mary pass in the final seconds of condor life, but it worked. With broadcast coyote poisoning largely at an end and many states moving to replace lead shot with steel pellets, threats to condors began to ebb. So in the early 1990s the Fish and Wildlife Service began releasing condors into the wild again. Their recovery has necessitated some remarkably imaginative efforts to get wild, self-sustaining populations nesting, including enticing chicks to imprint on biologists dressed in condor suits. But with more than 400 condors now released in Arizona, Southern California, Baja, and—with the assistance of the Yurok tribe—in Redwoods National Park in the Pacific Northwest, California condors ply American skies once again. The goal now is both a West Coast and an inland population, each with 150 birds and 15 pairs of breeders, with a reserve population remaining in captivity in case things go south.

〰️

BUT SUCCESS WITH EAGLES and condors (and peregrine falcons) didn't translate into bringing back every dramatic, declining American bird. The story of the ivory-bill has been a tragic comedown.

Growing up in Louisiana in the 1960s, I felt the mirage-like presence of ivory-billed woodpeckers every daybreak when I heard the sounds of the forest coming to life. I knew barbershop locals who snorted that ivory-bills were as common as robins. And I could feel my heart pounding in my chest every time a strapping black-and-white woodpecker flapped through the trees and lit upright on a bald cypress, the sound of its nails scrabbling for purchase as audible as its wingbeats on still air. But I knew better: by the sixties I was among pileated woodpeckers, not ivory-bills.

In 2005 I was living in Montana, but even from that distance I felt a personal, schoolboy excitement when a Cornell group made the spectacular announcement that it had found ivory-bills alive in the White River and Cache River National Wildlife Refuges in Arkansas, only 150 miles north of where James Tanner had studied the giant woodpeckers sixty-five years earlier. When National Public Radio and *Science* magazine broke the story, it went worldwide in a day. In a press conference Secretary of Interior Gail Norton told the country, "I cannot think of a single time we have ever found a species once thought extinct and now found in existence." The evidence for that remarkable claim was a four-second video. Google it sometime, then have a look at Arthur Allen's ivory-bill moving pictures from 1935. The truth was that the 2005 video was far too blurry and brief to convince ornithologists (David Sibley, for one) who weren't on the Cornell team. Within a year NPR ran a retraction. Cornell responded by sending an even larger team back to Arkansas in 2006, and again in 2007. It was a state-of-the-art search. They aimed remote time-lapse video cameras at woodpecker roosting holes. They broadcast and amplified recorded ivory-bill calls from 1935 into the swamps. They did aerial surveys with ultralights, and marched nearly fifty people abreast to create a "human mist net." Search efforts "were high."

They found no ivory-bills. The 2006 search yielded "no additional confirmation" and the 2007 one "no definitive evidence." The search broadened, ultimately, to eight states and 523,000 acres and lasted until 2010. It targeted the likeliest locales—the Pearl River in Louisiana, the Choctawhatchee River in Florida, and the new Congaree National Park in South Carolina, which preserves the largest remnant old growth left in the South. There was no evidence of a surviving population. Of course people all over the South continue to report seeing ivory-bills, but it's also likely that twelve millennia ago Clovis people imagined they had seen mammoths, only to find nothing there. Cornell now says, "The Ivory-billed Woodpecker is probably extinct." On September 29, 2021, the U.S. Fish and Wildlife Service officially declared the ivory-billed woodpecker lost to America and the planet.

≈

OF THE SPECIES presently vulnerable to extinction in the United States, the list is active and growing, but because of the success of some recovery plans it is also a revolving door. In 2021 the U.S. Fish and Wildlife Service's threatened and endangered lists (including Hawai'i and U.S. territories) were at 78 mammals, 106 birds, 38 amphibians, and 48 reptiles. In the list of amphibians and reptiles are numerous species of sea turtles, desert tortoises, skinks, and snakes, among the latter a rattlesnake, the New Mexican ridge-nosed. The American crocodile, with fewer than a thousand remaining in Florida, is there. Even with a current population of 5 million, the alligator remains listed as threatened because of its "similarity of appearance" to crocodiles. Among the mammals are longtime market targets like sea otters and the Canada lynx. There is that predator of prairie dog towns Audubon introduced to science in the 1840s, the black-footed ferret, within a whisker of extinction because of our herculean efforts to silence dog towns with poison.

A striking number of the listed are some of the most famous and iconic animals from the past five centuries of American history: two wolves (the red wolf and the Mexican gray wolf), grizzly bears and polar bears, the jaguar, and the only cougar left in the East, the Florida panther. Several, like the Mexican wolf and the Florida panther, are distinctive subspecies. Among those are the Sonoran pronghorn, the tiny whitetail deer of the Florida Keys, and a bighorn sheep subspecies in California.

The story is the same with threatened/endangered birds, with whooping cranes and California condors topping the list, joined by a Florida population of caracaras ("Mexican eagles"), Eskimo curlews, Florida scrub-jays, a subspecies of sandhill crane, a falcon species called the northern Aplomado, both the northern and Mexican spotted owls, all manner of warblers and flycatchers, even (in the West) the yellow-billed cuckoo. A great many specific populations of both mammals and birds appear now as "experimental, nonessential," which effectively provides "management options" in bureaucratic parlance (that *can* mean killing troublesome individuals) that full endangered species designation doesn't. California condors in the Southwest and Northwest, and many localized populations of whooping cranes, are listed this way. For better or worse, gray wolf recovery has relied very much on that designation. That kind of endangered species recovery dates to 1978 amendments to the ESA to allow certain exemptions to the act, almost always to the advantage of economies rather than ecologies. Those changes gave us the new category of "nonessential" populations of endangered animals.

Like every aspect of American life in the twenty-first century, the protection of animals that evolved in America is now hostage to the political war between Right and Left. By the 1980s, Republicans who had once voted for the Endangered Species Act had decided that American freedom, by which they meant the market's freedom to exploit nature, was threatened by "the specter of environmentalism." Unless you've been living under a rock you've likely either heard about, witnessed, or lived one or more battles of the most visible and controversial endangered species recovery of them all. I mean, of course, our war over restoring endangered wolves to a nation where we were so stunningly successful in wiping them out.

STEPPING ASHORE in the early 1600s, Old Worlders had been bitter to discover wolves on the continent. That's truly curious since as hunters humans watched and learned from wolves for hundreds of thousands of years. Domesticated wolves became our first companion animals, our oldest friends and helpmates, our templates for domestication. But wild wolves ceased to be our blood brothers when, seven to eight thousand years ago, we started taming and herding ungulates. Native people in America never took that step with bison or bighorn sheep or pronghorns and consequently they lived among wolves without animus, admiring and learning from them, seeking them as totems. But Europeans landed with all their long centuries of wolf hatred at full boil. That accompanied a recent Old World memory of success at eradicating them.

Then there was that shepherd's religion we brought along, which called something else into play. Supposedly wolves were heavenly revenge for Adam's Fall. But anyone who looked deeply into a wolf's eyes saw there a reminder of an ancient human past that Judeo-Christians shied from acknowledging. In his 1920s novel *Steppenwolf*, the German writer Hermann Hesse put this succinctly: "He calls himself part wolf, part man. . . . With the 'man' he packs in everything spiritual and sublimated or even cultivated to be found in himself, and with the wolf all that is instinctive, savage and chaotic." Wolfiness was the animal nature humans denied or at least believed we'd escaped.

So Americans yielded ourselves up to an unexamined wolf hatred. Yet by the 1960s our best natural scientists were telling us we'd been wrong. The continent's ancient top predators had shaped every niche, every population, from the way other predators like cougars or coyotes or foxes functioned to what kinds of prey animals there were and how many, to the numbers of beavers in the streams, the songbirds in the valleys, the kinds of trees and grass that grew. If we wanted modern, healthy versions of old-time America's ecologies, we had to bring wolves back. Aldo Leopold's Land Ethic almost came down to a single, delimited insight. Imagining a biologically healthy America meant we had to start thinking like mountains that had wolves.

Except in Alaska, though, we'd endangered all our wolves, everywhere. Both the eastern timber wolf and the red wolf were listed in 1967 under the ESA of 1966. The western gray wolf's official listing came in 1973, and Fish and Wildlife relisted the other two within a week of the

1973 act taking effect. The southwestern subspecies of gray wolf, the perilously scarce Mexican wolf, acquired its listing in 1976.

There was one wolf exception in the Lower 48. Back in 1959, inspired by Adolph Murie's *The Wolves of Mount McKinley*, a young graduate student from working-class roots named David Mech had conducted the first of three winters of field trips to a 210-square-mile island in Lake Superior. Isle Royale had become a national park in 1940, and Mech's project was to study an ongoing natural history experiment there. In the early 1900s moose swam from lake shorelines out to the island. Lacking predators, their numbers predictably underwent two massive irruptions and die-offs. Then, in 1949, gray wolves got to Isle Royale, probably by crossing the winter-frozen lake. Like Adolph Murie, Mech did the kind of wolf study others dreamed of, getting to see firsthand this evolving ecological interaction. At the time a few hundred wolves remained in the upper Midwest, most of them in Minnesota. This midwestern wolf population the Fish and Wildlife Service declared threatened in 1978.

When Mech's *The Wolves of Isle Royale* appeared as a National Parks publication in 1966, his insights spread through the scientific community. Murie had been right. Wolves mostly killed the young, injured, and very old. But there was more. Mech was the first naturalist to report how difficult the life of a wolf really was. Although Isle Royale's wolves killed and ate beavers, almost their entire focus was on big ungulates. Wolves did not survive on rabbits. Yet of the moose they pursued, even big wolf packs succeeded in making kills only 8 percent of the time! For many readers Mech's insights induced a newfound compassion for the life of a wolf.

By the 1980s, as the Fish and Wildlife Service was finalizing wolf recovery plans, a majority of Americans was internalizing a changing attitude toward wolves. Wealthier, better-educated, younger Americans (westerners and Alaskans among them) were becoming wolf advocates. In the wake of Farley Mowat's *Never Cry Wolf* and the charming movie Disney made of it, many were downright enchanted with them, even when biologists snickered that no wolves actually lived the way Mowat's animals did. Older Americans, the less educated, people who lived rurally or raised stock—and many hunters—thought the idea of recovering wolves bat-shit crazy and the animal's new human allies wolf-loving hippies. It was a divide that predicted yet another theater in a politicized future.

America's wolf-recovery moment faced complications both practical and bizarre everywhere. For starters, there were those who didn't believe either red wolves or Mexican gray wolves actually existed. The red wolf, for its part, looked suspiciously like a long-legged coyote. Was it merely a gray wolf–coyote hybrid? As for the Mexican wolf, it seemed barely distinguishable among the bewildering two dozen wolf subspecies twentieth-century canid zoology offered up. But in the twenty-first century biologists have drastically pruned the canid tree. The genetic revolution shifted the ground even more, and in 2012 Fish and Wildlife issued a peer-reviewed monograph offering an entire rethinking of speciation among American wolves. *An Account of the Taxonomy of North American Wolves* claimed that the red wolves and timber wolves of eastern America were not gray wolves but ancient continental wolves out of an indigenous line that had also produced coyotes: "Coyotes, *C. rufus*, and *C. lycaon* are modern representatives of a major and diverse clade that evolved within North America." It concluded that modern gray wolves had indeed returned here from Eurasia, arriving in America in three different waves some twenty thousand years ago, the last wave of which had brought Mexican wolves to the Southwest.

This taxonomy finally explained a phenomenon biologists had first confronted in the 1960s. In the West gray wolf–coyote hybridization was rare. The animals seemed to detest each other. But as the West's predator war drove coyotes to colonize across the country, canid standoffishness turned into hook-up willingness in the East. On the Louisiana-Texas boundary and along the Canadian border, coyotes interbreeding with remnant wolves produced new, quite successful eastern canids many would call "coywolves." Maybe the arrival of coyotes in Manhattan's Central Park early in the twenty-first century really was, as one New Yorker put it, "the end of civilization." Those same coyotes, though, were making a version of wolf recovery possible in an East that had lacked wolves since the 1930s.

Red wolf–coyote hybridization, however, threatened to become an ESA crisis. What if the red wolf wasn't actually a species but just a coyote with wolf genes? In my early teens I was lucky enough to witness what were unmistakably coyotes arriving in the Red River Valley of Louisiana. But on two occasions I came face-to-face with leggy, golden-eyed animals that looked awfully wolfy to a fifteen-year-old. If coyotes and red wolf–coyote hybrids *were* swamping red wolf genetics, much as we'd once done with Neanderthals, the mixture sure made for beautiful creatures. In 1970, Ron Nowak—another Louisianian poised to become Fish and Wildlife's endangered species guru—figured there were at most only three hundred true red wolves left.

After blessedly rejecting a "dog fence" reminiscent of the dingo fence across Australia, along with a firing-squad "buffer zone" to keep migrating hybrids and coyotes away from red wolves, in 1973 Fish and Wildlife tried another approach. It made biologist Curtis Carley the field coordinator for the Red Wolf Recovery Team. Carley came up with a template using morphology and recorded howls to try to identify actual red wolves for placement in a captive-breeding program. His method for identifying a true red wolf wasn't exactly rocket science, but in 1974 the recovery team set up facilities at Point Defiance Zoo in Tacoma, Washington, to breed the wolves he'd selected for returning to southern preserves presumably far from coyotes. Of the fifty zoo-kept "red wolves" Carley's team examined, they identified *one* they thought a pure red wolf. They shocked zoo personnel by insisting they destroy every animal they considered a hybrid!

Carley's trappers caught another 400 canids, but of those only 10 percent made the cut. Ultimately they identified just 14 animals, selected out of 450, to begin red wolf recovery. All those others, betrayed by the sin of miscegenation and coyote blood? Destroyed, every one. Still, red wolf recovery—if the red wolf was really a thing—was underway, with eventual releases in 1987 in Great Smoky Mountains National Park and on the Alligator River National Wildlife Refuge in North Carolina. Given that red wolves aren't celebrated as a success the way bald eagles are, you might guess their twenty-first-century arc has taken some strange turns.

Mexican wolf recovery started in the late 1970s with issues that essentially seemed insurmountable. The problem was, there weren't any Mexican wolves anymore, at least nearly none, certainly no packs that might be captured and transplanted. The best guess, as a result of the feds' war of attrition, was that maybe only five wild Mexican wolves remained alive on either side of the border in the 1970s. But between 1978 and 1980, Roy McBride, an old-hand federal predator hunter, managed to capture three males and, in a stroke of luck, a ten-year-old pregnant female ("Nina"), in Mexico. Eventually the breeding program acquired two more wolf lines from captive animals, the Ghost Ranch and Aragón lineages. But Mexican wolf recovery faced the hurdle of a genetic bottleneck of only three lines of animals. So trappers scoured nearby Mexico for a few more animals to diversify the gene pool.

Struggling to find wolves was one thing. But the 1982 recovery plan, with no criteria for delisting since the animal was so close to extinction, faced other hurdles. No one in the Southwest seemed to want the animals. During the Reagan administration Fish and Wildlife actually told target states they could say no to Mexican wolves. Big Bend National Park was a prime release site, but Texas leaped at the chance to decline wolves. The best New Mexico could do was to offer up the military bombing range north of White Sands National Park, or at least it did until stock interests persuaded the state to withdraw even that possibility. But after first equivocating, Arizona eventually proffered the Blue Range, a large National Forest Primitive Area east of Phoenix that merged into the famous Gila Wilderness in New Mexico.

A key move in releasing Mexican wolves into the wild was labeling the population "experimental and nonessential," meaning the wolves would not have full endangered species protections. Another came when Hank Fischer of Defenders of Wildlife told southwestern stockmen that his group would pay market compensation for any livestock wolves killed. The nearly unbelievable moment came in March of 1998, when release-pen doors high in the Blue Mountains opened and eleven Mexican wolves loped off to be wild wolves once more. Amid all the sound and fury, at least wolves were finally back in the evolutionary homeland of Earth's canids.

Far to the north a different gray wolf was also on its way back. The 1978 recovery plan for the eastern timber wolf began with fewer than a thousand wolves in Minnesota and almost none in Wisconsin and Michigan. Twenty years later, with 2,300 animals, some dispersing as far south as Iowa and as far west as the Dakotas, wolves had exceeded the plan's goal of 1,250 animals in Minnesota and 100 in the other two states. By 2003 there were 3,800 wolves in the western Great Lakes country and in that year the feds reclassified all wolves in the eastern half of the United States as threatened rather than endangered. That was not destined to last.

Meanwhile, the wolf recovery that thrilled people across the world was unspooling to international headlines in the Rocky Mountains. Because

I was living in Montana at the time, with a girlfriend who worked for Hank Fischer and Defenders, I had a sideline seat for what seemed the game of the century. Prior to this moment, the last time anyone had seen a wolf even in our one wildlife park, Yellowstone, was 1926. But soon after endangered species legislation passed in the 1960s, biologists in various agencies and universities began conferring about how to get wolves back. The 1975 northern Rocky Mountain wolf recovery plan identified Yellowstone and Glacier parks, along with the biggest wilderness area in the Lower 48, Montana's Bitterroot Mountains west into Idaho, as logical recovery zones. The plan: release fifteen wolves a year for five years until there were at least ten breeding pairs in each zone.

Before wolf recovery in the blue-green world of the Rocky Mountains could proceed, though, on their own, gray wolves began drifting across the border from Canada. In 1979 a silver female wolf researchers named Kishenena established a territory on the northwest edge of Glacier Park. The next year the mere presence of a big male called the Bearpaw wolf created consternation on the Rocky Mountain Front. In a reprise of our long wolf story, a Montanan shot the Bearpaw wolf dead. In the spring of 1986, Glacier wolves the American press labeled "naturalized" and named the Magic pack whelped the first litter of wolf pups born in a western U.S. national park since the 1920s. These colonizing wolves had full protection under the ESA and that complicated recovery plans as much as stockman outrage and political grandstanding. (Wyoming's Dick Cheney quipped that if America was going to return wolves to the West, we might as well draw up a recovery plan to get sharks into the Great Salt Lake.)

Bald eagles never faced anything like the uproar that ensued over wolf reintroduction. Neither had peregrine falcons, or eastern brown pelicans, even alligators. Western ranchers acted as if bringing back wolves would repudiate their entire history. The 1993 Northern Rockies wolf recovery plan had opted to classify *released* wolves as "experimental, nonessential" animals. That meant allowing ranchers to kill wolves threatening their stock. The big picture was sharpening now. Regional and national hearings over the plan produced the most public comments logged on any American environmental proposal, ever. As a calculus of how America had changed, 80 percent of it was in favor of wolf reintroduction.

It was an exciting history to live. The second week of January 1995, the first eight gray wolves captured in the Alberta Rockies came through

Missoula bound for acclimatization pens in Yellowstone's Lamar Valley. Three days later four of the first wolves headed for Idaho were blessed by the Nez Perces on the local airport tarmac, then trucked to Salmon for hard release in the Frank Church–River of No Return Wilderness. That happened on January 14. Two packs bred in their pens in Yellowstone that February, so pups were on the way. On the spring equinox, March 21, 1995, Yellowstone's new wolves sniffed at the opened doors of their enclosure, then bounded one after the other into the nearby trees. That spring and the following one, 1996, Yellowstone released forty-one wolves into the park. Officials released another thirty-five wolves into Idaho. The Rockies' wild ungulates, whose every sinew and tendon and muscle, whose speed and eyesight and startle reflexes were refined by millennia of wolf pursuit, must have sensed what was coming. Yellowstone's three thousand bison paid little attention, but the twenty-five thousand elk that had survived a recent irruption and die-off were reported to be nervous as hell.

Five years after the first wild pack since the 1920s—appropriately named the Leopold pack—formed in Yellowstone, a denning pair of wolves bore pups in the Bitterroot Valley, where I lived. Not long after that the Three-Mile pack emerged in the Sapphire Mountains just above my place. Wolf howls in evening twilight and wolf prints along the rivers had finally returned to places like Yellowstone and the Bitterroot.

By this point in the wolf-recovery story, watching gray wolves in Yellowstone Park was already a $5-million-a-year business, drawing tourists from around the globe. The famous cascade of ecological effects in the wake of returning top predators was now underway, altering and restoring Rocky Mountain animal life. Skirmish lines over delisting wolves and turning their management over to the states were forming. After Europeans had killed wolves for eight thousand years in the Old World and their descendants had erased wolves from coast to coast in America, we had finally set our national will against the wrongheadedness of trying to turn our Wild New World into a carbon copy of the Old.

⁂

PART OF THE MYTHOLOGY of American exceptionalism is that by getting to start over we avoided old mistakes. But that particular myth was severely tested by the continent's wild animals, especially the large, charismatic ones. Bears, big cats, wild horses, and bison exist in a nation whose human species still can't quite figure out whether it wants them in our present or our memories.

Grizzly and polar bears, we have to admit, represent America at just about its wildest. Magnificent examples of continental evolution and history, our big bears deserve a future. If we really are serious that some part of our national character was forged by wild nature, imposing and potentially dangerous creatures like these remain important and nearly essential. But they're in peril. In 2008 America's polar bears became the first species to be listed as threatened due to climate change, specifically a loss of sea ice that is shrinking polar bear habitat. Since their listing, 75 percent of the Arctic's volume of ice has melted and what's left is deteriorating in quality. When ESA biologists listed polar bears the Beaufort Sea population had dropped from 1,800 bears to 1,500 just since the 1990s. Polar bears can't sustain a continuing decline like that.

As critical as big bears or cats are to fully functioning ecologies—the so-called Greater Yellowstone Ecosystem specifically maps out a grizzly bear's world—our lives in places where cougars or giant bears reside requires a level of alertness to nature not all humans enjoy. In Alaska between 2000 and 2017, humans experienced sixty-six bear attacks, hospitalizing sixty-eight people and killing ten. Approaching the end of our Arctic Wildlife Refuge float at the Beaufort Sea in 2019, our guides' polar bear talk was dead-on direct: "If Nanook sees you he's likely to come knocking on the door." I lived and hiked in Montana for twenty-two years without any bad outcomes with bears or lions, but almost every year someone got in trouble with a grizzly, or a young cougar had to be chased away from a camp or a scout troop. Some people are fine with this. Others aren't. When Peter Matthiessen wrote *Wildlife in America* in 1959, his take on the grizzly bear then was that "clearly there is no place for it in settled country." In that same vein, Mike Davis's *The Ecology of Fear*, a treatise about living with nature in Southern California, titles its chapter about mountain lions in the hills around LA "Maneaters of the Sierra Madre."

Having reduced 50,000-plus grizzly bears in the Lower 48 to about 600 by the 1970s, with only 136 left in Yellowstone, we granted their populations threatened status under the ESA in 1975 and initiated the Grizzly Bear Recovery Plan in 1993. The largest population of remaining grizzlies in the West now, about 1,000, lives in the Northern Continental Divide Ecosystem, from Glacier Park southward down the mountains to Missoula. These are the grizzlies who make news by attempting to migrate onto their historic range on the Great Plains, or westward toward the Pacific. In 2021 a female grizzly the press named Ethyl trekked 2,800 miles across the region. The Yellowstone Ecosystem population has now grown beyond 700 bears, some of them also probing eastward now, and friends of mine who are runners in Cody, Wyoming, say they're *really* alert these days. Because the two primary bear populations are separated by 250 miles and can't exchange genes, the grizzly-recovery committee has wanted since the 1990s to return grizzlies to the Bitterroot Mountains as a linking population. A release of "nonessential" bears there was four months away when the Bush administration took office and shut it down in 2001. Grizzly recovery in the North Cascades Ecosystem suffered a similar fate, with bear releases almost underway, then shut down when Donald Trump took office.

As grizzlies have recovered and begun trying to reoccupy their former range, delisting has dominated the discussion. The Fish and Wildlife Service initially declared Yellowstone's population recovered in 2007, but a federal judge overturned that ruling. In 2017 the Service again announced delisting for the Greater Yellowstone population. Wyoming and Idaho at once planned grizzly bear hunts, but a federal court again reversed the decision. Grizzlies remain protected now, but as with wolves, state hunting seasons for them—the "most thrilling moments of an American hunter's life," Teddy Roosevelt claimed—are likely in their future. Likely, that is, if states are able to disentangle wildlife management from raw-meat politics, which threatens the vision of the Endangered Species Act.

America's big cats, where they exist at all, live in the shadows. Four hundred years ago mountain lions, which can weigh up to 170 pounds, had the most extensive range of any mammal in the Wild New World, living in every kind of ecosystem from Tierra del Fuego to the Yukon. We bountied cougars, killed them on sight, and both private and government hunters pursued them with hounds until we'd exterminated them almost everywhere east of the Mississippi. Meanwhile the wildlands they need steadily shrank. The result was predictable. In 2011 the Fish and Wildlife Service officially declared the eastern mountain lion subspecies extinct. Although everywhere reduced in numbers, mountain lions in the West are not endangered. Occasionally a long-distance wanderer even turns up in the East, most recently in Connecticut in 2021, and most western states still allow a hunting season. Where I live in New Mexico state wildlife officials tell us there are 3,500 cougars in our midst and allow an annual take of 600. Both figures strike many of us as too high.

On America's eastern side, keeping lions a part of nature has come down to saving a southern subspecies known as the Florida panther. Originally a mountain lion of the Gulf Coast swamplands from Louisiana to Florida, this cougar wells up to me personally out of the deep memories of childhood, through stories from my Louisiana ancestors who regaled a wide-eyed five-year-old about "panthers" screaming in the nights. My grandfather knew real panthers in the greenwood, but by the time I was in high school our regional mountain lions were already an endangered species. That happened in 1967, when only 20 of them remained, all in the Florida Everglades. It turned out our granddaddies had been too efficient at taking them out. With fewer than two dozen left, a death like that of the last mammoths, through loss of genetic diversity, seemed almost certain. Then, in 1995, biologists released 8 female lions in Florida they'd trapped from a related subspecies in East Texas. At least 20 kittens from these females added genes to Everglades lions. By 2015 their population had grown to 230 animals.

The Florida Panther Recovery Plan is asking southerners to forget their granddaddy's stories. It envisions the Gulf Coast cougars recovered when there are three distinct populations of 240 lions each, with sufficient habitat for them to hunt and breed successfully. The Florida Panther National Wildlife Refuge in the Everglades near Naples will serve that population, but that is only one recovery zone. The state's 2021 creation of a Florida Wildlife Corridor, a nearly eighteen-million-acre swath of greenbelt extending across the state, is intended in part to help Florida panthers colonize north and east. But what really may be necessary to recover the Deep South's mountain lion are wildlands corridors that enable the cats in the Everglades to start pushing westward once again. Toward my home state, where panthers once screamed in the nights.

A recovered Florida panther population may not get America any native black cats—cougars don't appear to have a melanistic phase—but one other big-cat recovery just might. As was evident in the early twentieth-century Southwest, jaguars (which can shade to a spotted

Female jaguar. Courtesy Shutterstock.

black) are one of our high-drama big mammals, at the northern extent of their vast range here. They're also another carnivore stockmen and government hunters casually destroyed. Not officially listed as a U.S. endangered species until 1997, jaguars didn't acquire a recovery plan until 2018. As with gray wolves, jaguar recovery depends on the nations bordering the United States having preserved big animals far better than we have. The 2018 plan really offers pretty faint hope for the return of El Tigre to the United States. Encouraged that seven male jaguars have ventured into the United States since 1996, the architects of the recovery plan don't envision a hard release of captured jaguars. They hope instead the cats will reoccupy their former range via two different corridors that could connect populations in Mexico to the Southwest.

The problem is that those jaguar-migration corridors reach the United States exactly where the country has been erecting its border wall against human migrants. The Fish and Wildlife Service's recovery is a forty-year proposition, with delisting happening only in the event a breeding population of female cats arrives, bringing genetic diversity. If we continue building a wall to repel humans but still want jaguars, we'll need to revisit that release choice, and two prominent environmental groups have already done so. The Center for Biological Diversity and Defenders of Wildlife have counterproposed a 329,000-square-mile recovery area in New Mexico and Arizona that significantly overlaps the *wolf* recovery zone. Rather than relying on migration, they propose a hard release of a dozen jaguars to begin recovery in an area that could eventually support 100 to 150 of the big felines.

It will be beyond ironic if the only way we can have Mexican cats come home is to haul them through a border wall that holds Mexican humans at bay.

≈

TWO OF AMERICA'S most charismatic big mammals don't need the Endangered Species Act to rescue them. In this century wild horses and buffalo are doing fine, maybe too well. Wild animals don't necessarily require sharp, white teeth for some to hate them, though. No wild horse or buffalo ever ate a cow. But sometimes they eat grass or drink water a cow might want. That seems about all that's necessary to despise both.

Wild horses have a unique ability to arouse ire. It doesn't much matter whether you regard them as a native animal that evolved here, with an asterisk for their ten-thousand-year absence, or as an invasive, domesticated beast gone feral, with an asterisk for their fifty-six million years of evolution here. Wild horses' well-publicized population increases, and government's inability to figure out what to do with them, has everyone uneasy. With the passage of the Wild Free-Roaming Horses and Burros Protection Act (1971—arch-enviro Richard Nixon again), we acknowledged that because of their role in history, wild horses are "an integral part of the natural system of the public lands." They have several national refuges devoted to them. A personal favorite is the Pryor Mountains Wild Horse Range, where Spanish mustangs that preserve the zebra-circled legs and backline stripes of horse antiquity graze the Martian-red canyons of southern Montana. But except in Theodore Roosevelt National Park in North Dakota, horses no longer live in their evolutionary grasslands/tundra home. Instead they're in the deserts west of the Rockies, wastelands where it wasn't so easy to hunt them down. That's been one problem.

The other wild-horse problem is that in deserts without predators, horses are in a constant state of demographic explosion. When the 1971 law passed there were about twenty thousand of them, not far off biologists' projected carrying capacity of twenty-six thousand. But without helicopter cowboys filling pet-food contracts, with wolves absent and mountain lions kept at a low ebb, horse numbers began to boom. Their normal rate of increase is about 15 to 20 percent a year, a growth rate that anciently enabled them to survive weather anomalies and wolves, cheetahs, and hyenas taking out colts and the old. In sagebrush deserts our wild-horse herds aren't replicating the equine Pleistocene. They're replicating irrupting elk in twentieth-century Yellowstone Park.

Since wild-horse hunting seasons aren't an option, the best idea the Bureau of Land Management has been able to devise for its scores of "Herd Management Areas" is rounding up as many as eleven thousand horses a year and corralling them in vast, dusty holding pens. By 2021 those pens held a phenomenal fifty-five thousand wild horses, with more than eighty thousand still out in the deserts. Having failed for four decades to find enough human adopters to take horses off their hands, in 2019 BLM sweetened the pot by *paying* adopters $1,000 per horse to haul them away. The outcome has been entirely predictable. Adopters with horse trailers are pocketing payment for the maximum number (four) and taking any animal they can get. Most give the horses minimal care for a year to get title. Then they sell them at auction, inevitably to "kill buyers" with pet-food contracts. In America this has been the wild horse's purpose and value for a century.

Short of full cougar and wolf recovery in horse country—the best idea—there's at least a better idea. In 2013 a National Academy of Sciences team studied the modern wild-horse disaster and made an entirely logical proposal. The BLM should stop holding horses in corrals and should move at once to dart captured horses with contraceptives (porcine zona pellucida for mares and chemical vasectomies for stallions), then release them. The $100 million a year we spend to corral horses could go toward contraceptives for the vast numbers still out there. Horses in country too rough for capture could maintain the population. That would preserve bands of wild herd stallions and their harems of mares, the wild-horse natural history that so intrigued Thomas Jefferson. And it would keep one of North America's grandest evolutionary contributions from living out their lives in holding pens or dying for our pets.

ONE EVENING IN THE 1990s I had dinner with Fred DuBray of the Cheyenne River Sioux Reservation. A rancher and Vietnam vet, Fred was one of the start-up founders of an organization now called the Intertribal Buffalo Council that had successfully released bison onto tribal lands all over the country. Fred was a storyteller, and over our meal he told me a story that we both knew laid out what was at stake in recovering bison.

When he and other visionaries had the idea of trying to recover buffalo herds in Indian country, he told me, a Lakota woman elder had taken them by the arm and said, in effect, "Best you ask the buffalo if they *want* to come back." That got my attention. I found myself unconsciously holding my breath while Fred took a couple of bites, sipped his drink, dabbed the corner of his mouth with his napkin. I felt like I was balancing on the lip of a canyon.

"Well?" I finally got out, "what happened?"

"Well," he replied, "we did a ceremony and asked them."

"So what did they *say*?"

Another interminable pause. Fred looked up from his meal.

"They said they did want to come back. But they said they didn't want to come back and be cows. They said they wanted to be buffalo. They said they wanted to be wild again."

TWO DECADES LATER I'm looking out on the coiling Missouri River in central Montana as a gusty wind in turn bends and releases the sea of grasses carpeting a place called Sun Prairie. Watching those yellow grasslands undulate in the breeze, I'm thinking no country in America is so minimalist or has such a power to turn one inward. With bison now roaming dozens of tribal pastures and grazing on hundreds of ranches in the United States—more buffalo are in America now than at any time since the 1880s—where I'm standing has drawn international admiration for another unfolding bison story. Nearby are two gangs of burly bison bulls, and in the opalescent distance is a far larger cow herd with frolicking red calves. They represent Fred DuBray's ceremonial vision for the animals the United States, all too belatedly, has made our National Mammal. If bison are going to be truly wild anywhere, this is the place.

It is May of 2019 and Sara and I are staying at a place in the middle of this prairie called the Enrico Education and Science Center. It's clearly a whirlwind spot. Researchers from the Smithsonian are on hand to study prairie ecology, and they are joined by a big group of volunteers from the Pacific Northwest here to dismantle fences. SUVs with license plates from sixteen states are parked outside. Two weeks earlier Sara and I had been in Manhattan, at the American Museum of Natural History, helping the organization that owns this part of the Great Plains give its annual Ken Burns Conservation Prize to Native writer N. Scott Momaday. The organization is called American Prairie. Among some, its two decades have made it more famous for the controversies enveloping it than its vision. But the vision is really compelling.

American Prairie has an idea that is on the scale of our national parks or wilderness system, a twenty-first-century opportunity to correct the error Hornaday and the Bison Society made with buffalo a century ago. Withholding bison from the wild and enabling the privatization of almost every inch of the Great Plains meant that neither America nor the animals ever got the kind of great game parks that thrill visitors to Kenya, Tanzania, or South Africa. AP is doing its best to reverse that historical mistake, and it's doing so with the very free-enterprise market that for centuries has been hip-deep in driving wildlife extinctions. AP has so far bought ranches totaling 420,000 acres that it hopes eventually to meld with the 377,000-acre Upper Missouri River Breaks National Monument upstream and the 916,000-acre Charles M. Russell National Wildlife Refuge downstream.

The epic idea is to re-create the American Serengeti with a preserve of some 3.5 million acres, half again as large as Yellowstone. By taking down the internal fencing on its properties, AP managers want to see twelve thousand or so bison roam free across private and public lands. Even more exciting, they want the full suite of Great Plains wildlife back. Not just bison and elk, pronghorns and bighorns, prairie dogs and ferrets, but the gray wolves and grizzlies that are already venturing out from the Rockies 150 miles to the west. By 2021 grizzlies had gotten as close as 50 miles to AP boundaries.

Two months later I was at Barley & Vine restaurant in Bozeman, this time dining with Sean Gerrity, who with biologist Curtis Freese imagined American Prairie into reality back in 2001. Sean is a man of accomplishment but it weighs lightly on him. Handsome, optimistic, self-deprecating, in all respects great fun to spend an evening with, he was (and is) to AP what John Muir was to the Sierra Club. A native Montanan who grew up in Great Falls, he was educated at Montana State, then discovered in California that somehow in the world of business consulting his psychology degree translated into a fortune for himself and his family. Returning to Montana in the late 1990s, he and his friend Freese began driving around the plains country and imagining how past might be prelude.

I don't know if it was the second gin and tonic or twenty years of living in Montana that prompted my educated guess. "It's the Lewis and Clark world you want back, isn't it? They left us one of the best descriptions of ancient Native America we have from anybody."

There was no hesitation in Sean. "Absofuckinglutely. All those animals! Those guys saw a Garden of Eden. I grew up hearing about it. Montana, the U.S., we should have it again!" With the support of the National Geographic Society, the Smithsonian, the American Museum of Natural History, and most of the regional Native tribes, American Prairie is now a world leader in carrying out historic rewilding. That reputation has sent Sean across the globe speaking about AP's vision for a wilder Earth even Europe now wants to embrace.

Not everyone here in Montana is thrilled with that future. Plenty of local ranchers, with the backing of a sharply more conservative state legislature in recent years, have fought bison and AP on every front. They hate seeing buffalo replace cattle, despise seeing fences come down. In short, they hate seeing buffalo go wild again. The Montana livestock industry has long insisted on killing bison when they leave Yellowstone for fear bison will spread brucellosis to cattle—a disease cattle *gave* to bison, a disease elk also carry, for which there is no field example of infection from bison. While Hornaday's old Bronx Zoo, now called the Wildlife Conservation Society, has assisted the Blackfeet in releasing eight hundred buffalo on tribal lands, Montana's state government has denied tribal requests to return bison herds to Glacier Park and to the C. M. Russell Wildlife Refuge adjoining American Prairie. When AP applied to the Bureau of Land Management to graze bison with the cattle permits

attached to ranches they'd bought, the legislature passed a joint resolu-
tion in opposition, then tried to confer to counties a local option to ban
bison. BLM approved the permits.

Sean did admit to me: "Those 'Save the Cowboy—Stop American Prairie' signs ranchers have tacked on their fences? Damn. That's pretty good
stuff!"

Yes, it is. It's an imagery that's worked for a very long time. But bringing back the wild diversity that once made this country the envy of the
globe by this point in our history is the model America ought to be offering up, and at long last it is.

HOW ARE YOU
ENJOYING THE
ANTHROPOCENE?

A few years back I made my pilgrimage to Thoreau's Walden Pond. To be honest, I'm not sure I could have warmed to the puritanical New Englander, who (as Emerson said of him) controverted every suggestion, for whom sensuality was an arrow that never left his quiver. But Thoreau does speak to me. Like him, I explored and floated the river I grew up on from its headwaters to its mouth. Inspired by him, I built a house and ranch in the wilds of Montana with my own hands and with found money. Thoreau's enchantment with the natural world resonated enough with me that I've spent a good bit of my life seeking out my own Mount Katahdins. In all this I'm one among the legions who have found Thoreau a lodestar, and one of all those to come who will discover themselves in his voice.

Walking counterclockwise around Walden Pond that perfect October day, listening to my feet crunch its shoreline gravel and watching as leaves of red and yellow, maple and birch, settled soundlessly onto its surface, I understood more about what's always been at stake in this country. Thoreau's spreadsheets of phenology, the timing of bird migrations, the births of young creatures he so meticulously recorded, explain part of his horror that in 1850s America he wasn't able to experience an entire heaven and entire earth. His regret, his near anguish about all the natural sights and signals his human predecessors in America had stolen from him, is the exact void we now stare into. Except in the twenty-first century the void has become far more vast.

As I picked my way through Walden's woods from Thoreau's cabin site to the rail lines he despised, I concluded I was testing my disappointment against his, probing at my feelings to see how my biophilia matches Thoreau's sense of tragedy. Even with our recent successes and possibilities, the big picture for many American creatures and for the continent's evolutionary legacy is frightening. With endangered species recovery and the modern genetic revolution, though, we may be oddly more optimistic than Thoreau despite the profound problems we and other living species now face. Still, his crisis is ours, both collectively and personally. I have one life and occupy one brief span. Like Thoreau, in my time I want my world to be as rich as I know it can be. I don't like thinking that self-absorbed demigods have cheated me of many of its delights.

It occurred to me circumnavigating Walden Pond that there might be an argument for blissful ignorance. Our minds almost seem programmed to forget. Possibly it's psychologically adaptive that the world we find ourselves in, which in truth is drastically diminished by our hand and further threatened, seems normal and complete. I once lived in Great Plains country that only a century before was defined by buffalo, pronghorns, wild horses, and wolves in unbelievable numbers. Few people who live there today imagine that version of the space they occupy. If the generations that witnessed passenger pigeons blotting out the sun and saw Carolina parakeets flashing tropical color through somber forests have passed and their stories lost, then a world without them is free to strike us as entirely normal. Unless we know better. Unless we haven't forgotten.

ALL OF US now should have this narrative uploaded. We are Earth's spawn, part of the river of life natural selection has shaped just for this planet. That premise itself is a profound one in the context of astrobiology, which is increasingly cognizant that the "sweet spot" for hosting an Earthlike biosphere rarely appears in our inventory of exoplanets. In the wake of one of our planet's grand evolutionary resets, the Chicxulub impact, North America spun up more than sixty million years of indigenous life-forms and became a destination for migrants from around the world. One of those was us, a species out of Africa that spread across most of the rest of Earth before finally arriving here. Given America's isolation, plus our more than two million years perfecting a lifestyle as cooperative predators, our arrival was an ecological shock wave. Widespread extinctions followed. In the aftermath of that chaos, we settled in among the animals that survived and for the next hundred centuries fashioned a sustainable continent. Then others of us from the Old World arrived, people whose access to extensive cultural exchange gave them domesticated animals and religious and economic ideas wherein animals were not kin but resources. Five hundred years later we find ourselves in an America drastically simplified of its life-forms and genetic diversity. We have lately awakened to save as many creatures as we can. But the future does not look good.

Where has this Big History gotten us? According to a National Academy of Sciences study in 2018, as our predatory genus *Homo* spread around Earth, the planet lost some three hundred mammal species and with them two and a half billion years of unique evolutionary genetics. One way modern conservation measures a nation's wildlife heritage is by assessing the "phylogenetic diversity" within its wildlife lineages. A significant number of the species North America lost during the Pleistocene extinctions—ground sloth, ungulate, and elephant lineages in particular—were not just the largest creatures, they were old and distinctive. Altogether, Pleistocene mammal losses erased a shocking two billion years of Earth's genetic diversity. Human-caused extinctions over the past five centuries cost the planet another half-million years of distinctive genetics. As the authors put it, "This means that prehistoric and historic extinctions were close to worst-case scenarios." Re-creating that much genetic information will take evolution millions of years. Which is to say that this Anthropocene epoch, this Sixth Extinction we set in motion fifteen thousand years ago, is unfixable. It is humanity's original sin.

But anguish and mourning won't get us very far in this present and future. We Americans should be proud of our efforts in recovering as many imperiled species as we have. We've innovated all manner of strategies and technologies to assist wild animals—banding birds, radio-collaring mammals, using motion-activated cameras to capture their movements, establishing connectivity and migration corridors for animals like pronghorns, building highway wildlife crossings, sequencing animal genomes.

We're also trying to understand and possibly redirect the downstream consequences of losing species. Scientists trace the twentieth-century explosion of Lyme disease in New England, for example, to the loss of the passenger pigeon. Without pigeons, hardwood mast fed a burgeoning white-footed mouse population, the reservoir for the ticks that carry the Lyme bacterium. Similarly, our extirpation of wolves didn't just launch a century of ungulate irruptions and die-offs, it provided an empty canid niche for coyotes to lope into and contributed to their spread. Pushing sea otters to the brink nearly destroyed kelp forests, a critical ecosystem that absorbs and sequesters atmospheric carbon.

Ecologists are frantically trying to figure out connections like these, because our good fortune at the expense of the rest of the planet's life still threatens vast numbers of Earth's wild creatures. In another pair of National Academy of Sciences studies in 2017 and 2020, ecologist Gerardo Ceballos and Stanford's Paul Ehrlich laid out the crisis nearly eight billion humans are creating for wildlife. Using the database assembled by the International Union for Conservation of Nature (IUCN) Red List of Threatened Species and Birdlife, they argue that a third of the planet's twenty-seven thousand vertebrate species are in decline and a quarter may face extinction. Habitat partition from human activity, plus an assault Americans long specialized in—the economic market in wildlife—again are the culprits, with looming climate change in the wings. The spillover disease COVID-19, which has unleashed misery on humans and is now spreading to animals like deer and the cat and weasel families, is simultaneously a kind of revenge and a cautionary tale from this continuing market in wildlife. Our disruption of ecologies around the world isn't just threatening wildlife extinctions. It's posing an existential threat to our own species.

America's set-aside of a large public lands system and recent decades of more enlightened policies have the United States in better shape than much of the world, with comparatively few of our animals making global alarm lists. There are concerns beyond our lengthy lists of threatened and endangered species, though. Half the species in North America have lost more than 40 percent of their original ranges, and almost a quarter have now lost 80 percent. Having once-widespread animals like mountain lions surviving only in distinct population pockets, as with the Florida panther, is worrisome. Discrete local populations go extinct more frequently than entire species do. Another North American trend has millions of us anxious. Our bird life is in crisis. In the past half century avifauna numbers have plummeted by three billion birds, a stunning 29 percent decline from the decade of *Silent Spring*. Much of that collapse is in just twelve bird families, including sparrows, warblers, blackbirds, and finches. As Rachel Carson well understood, just going about our ordinary lives birds and birdsongs are things we notice when they vanish.

In this moment of our history we've saved California condors and gray wolves, but what guarantees do they have in the future? To evaluate recovery projects, in 2021 the IUCN developed a new "Green Status" for 181 global species, including some of America's most charismatic ones. Recovering a bird like the condor to its baseline range and ecological niche isn't easy in the modern world, but its continued survival looks promising. International conservation still regards the condor as critically endangered, with a Green Status rating of "largely depleted." But the condor recovery plan is robust and the giant birds may eventually reach a 75 percent full recovery. The IUCN also rates the return of the gray wolf in America as "promising." Our keystone predators are unlikely to reacquire their full historical range, but from bases in the Northern Rockies, Southwest, and upper Midwest gray wolves are colonizing many parts of their former range. Wolves are now in Washington and Oregon, and California has a resident population in 2022 of three packs, among them a male who in 2021 traveled downstate 935 miles, to within 70 miles of Los Angeles, before being hit by a car. That almost equaled the journey of the female wolf from Yellowstone who made it a thousand miles down to the Grand Canyon in 2014.

That wolf fell to a hunter in Utah, which vows no tolerance for wolves. But the wild canids now nearly surround Utah. In 2020 Coloradans voted to welcome wolves back to their state, the place where Bill Caywood snuffed out Rags and Greenhorn just a century ago.

※

WHEN AMERICANS THINK about our history with wild animals, we reflexively blame losses on big abstractions, things beyond our control. Extinctions were "inevitable." Animals disappeared because of the "spread of civilization" or "habitat loss" or because they were "too wild to be suited to modern conditions." Losing species that had been in America for millions of years just happened. No one, no institution, was to blame. At least that's what we tell ourselves.

The truth is very different. The staggering losses in the first five thousand years of the Anthropocene pitted a superpredator species—us—against prey animals that either had no experience with us or whose evolutionary defenses were keyed to older dangers and were unable to protect them against our spread and efficiency. But animals confronting us across the past five centuries have faced something different. Since then we've thought of living creatures as mere resources in an economy designed to enrich us, and that has produced one ugly, depraved story after another, a history of inhumanity perpetrated by ordinary Americans in the name of freedom and the market, its cruelty and barbarism as often as not endorsed by government and sometimes even carried out by its agents. This is how we de-buffaloed, de-pigeoned, de-wolfed America.

And privileging economics over every other value remains our present. It defines the political battles we're waging over animals right now. Conservative forces running the federal government between 2017 and 2021 employed every possible stratagem to undo protections for wild animals in the name of freeing the American economy. The "businessman's" administration of Donald Trump thus modified the Migratory Bird Treaty Act of 1918 by striking "accidental avian losses" from the law. Under that revision British Petroleum would have suffered no consequences for the tens of thousands of birds killed by its Gulf Oil Spill, an accident that resulted in a fine of $100 million. For decades conservatives have wanted to drill for oil in the country's crown jewel for wildlife, the Arctic National Wildlife Refuge. Trump tried to make those dreams come true by opening ANWR to leasing. Chafing that the Endangered Species Act lists species based solely on best science, his administration issued a statutory rule inserting economic impacts into listings. And it instructed Fish and Wildlife to remove climate change as a factor in endangered species listings and recovery.

Today more than 500 species are under consideration for listing—two examples are the greater sage grouse and the wolverine—but compared to Obama, who protected 340 species, Trump's administration listed a grand total of 20. The grouse, whose population has suffered a 90 percent decline from climate change and oil/gas leases, and the wolverine, once poisoned nearly to oblivion and now down to only 300 individuals, were not among them.

✵

THE WOLF STORY that has loomed over America's relationship to nature for the past five centuries continues, if anything even more politicized now than it was in the 1920s and a century later even more complex. A 2018 genetic tour de force by the National Academy of Sciences gave both Mexican and red wolves legitimacy as true subspecies worthy of protection, but both are perilously endangered. On the pair of North Carolina national wildlife refuges where they're found today, the wild population of red wolves has dropped to a mere twenty-five animals. Researchers have discovered since 2018 that there are still wild canids on the Louisiana-Texas border that carry as much as 10 percent red wolf ancestry and infrequently as much as 75 percent. Those long-legged, yellow-eyed hybrids I saw as a teenager are still out there. But whether their faint wolf genetics will earn them protection is a question we haven't yet answered.

Mexican wolf recovery is also struggling. Only 241 Mexican wolves roam the Southwest in 2023. In 2015 a new ESA rule enlarged their recovery zone to extend to the boundaries of Texas and California, then strangely chose Interstate 40 as the northern perimeter. Those boundaries plus a maximum wolf population of 325 animals remained in the new Mexican Wolf Recovery Plan of 2017. I asked Dave Parsons, Mexican Wolf Recovery Coordinator from 1990 to 1999, for his thoughts on the future of Mexican wolves. An advocate for establishing additional populations in the Southern Rockies and in the Grand Canyon, with a total population of at least 750 wolves across the Southwest, Dave thinks the new plan is both "unscientific and antithetical to full recovery of the lobos."

"So what do you think the potential is for recovery and eventual delisting?" I asked.

To which the one-time director of Mexican wolf recovery responded: "I don't expect to see it in my lifetime."

Gray wolves in the upper Midwest and the Northern Rockies are especially noteworthy nowadays as the pawns and victims of America's political/cultural wars. Conservatives have a lot to say about wolf recovery, but the shorthand is, they're against it. Once despised as a biblical

Endangered red wolves. Courtesy Shutterstock.

curse, wolves have somehow become surrogates for a hated government. Isle Royale Park, where disease and inbreeding had collapsed the wolf population to only two animals by 2016, was a cautionary tale about how wolf recovery can go south. But in 2020 the Trump administration included Midwestern gray wolves among thirteen delisted species whose management it wanted to turn over to the states. A hurriedly planned and disastrous wolf hunt in Wisconsin was an immediate result, confirming scientists' fears that the states weren't ready to manage wolves. A court ruling in late 2021 returned management of all wolves in the Lower 48, except those in the Northern Rockies, to the federal government.

Yet anything Wisconsin could do, the Rocky Mountain states stood ready to exceed by an order of magnitude. Under federal direction the gray wolf story in the Northern Rockies had been a worldwide poster example for wildlife recovery. Despite disease—the Montana law that once called for veterinarians to infect coyotes with sarcoptic mange spread that disease throughout the Rockies—wolf recovery there was a grand success, their numbers reaching twenty-six hundred between 1995 and 2021. Ecologists and environmentalists were thrilled, but even with state compensation programs for their losses, ranchers never stopped complaining.

Watching wolves in Yellowstone National Park.   Photograph by Dan Flores.

And despite growing elk numbers, hunters resented the supposed threat to "their game," marching in protests in the streets of Missoula, Cody, and Boise carrying signs proclaiming the way to Make America Great Again was to return to a West with no wolves in it.

Management of Rocky Mountain wolves had passed to the states in 2010 by an act of the Republican House (the first time in history *Congress* removed a species from ESA protection), and with every election installing ever-more-conservative legislatures in the mountain states, the stage was set for reopening the wolf war. Taking wolf management out of the hands of their scientists and wildlife agencies, Montana and Idaho legislatures have reprised the goals of the 1880s with an overlay of twenty-first-century technology. State wolf management now means hiring professional hunters, paying bounties, luring wolves with bait, hunting them after dark with night-vision equipment, shooting them from planes and helicopters, running them down with ATVs and snowmobiles, and using wire snares that tighten at every move a panicked wolf makes to slowly strangle their victims. Ignoring a century of ecological science, the states' goal is to reduce wolf populations to just above the obligatory minimum number of animals that keeps the Endangered Species Act at bay.

Because Yellowstone is owned by the public and managed by a federal agency, wolf recovery in the park has unfolded at least partially shielded from the viciousness in the surrounding states. Yellowstone is now the premier public wolf-viewing locale on the planet, today producing $35 million annually for surrounding communities. With multiple prey animals and several competing predators, the park's wolves have enormously magnified our knowledge about wolf behavior and culture in conditions that mirror Native America. The way wolves produce ecosystem-wide effects that flow in all directions has been one of the park's more compelling lessons. When they were released, wolves pushed coyotes out of a primary canid niche and coyote numbers dropped by half, then stabilized. But magpies, ravens, eagles, grizzly bears, and even cougars have all seen their numbers grow in the park. At one wolf kill a biologist counted 347 ravens and 49 bald eagles partaking of the largesse.

The most surprising park makeover the wolf presence produced has been the recovery of aspens, cottonwoods, and willows, accompanied by beavers, songbirds, and other species of riparian ecologies. This wolf-induced "trophic cascade" is a phenomenon on YouTube, although its exact causes remain unresolved. Fewer browsing elk to eat down woody vegetation clearly figures. But elk numbers had been low before without producing this cascade, which initially led biologists to argue for a "landscape of fear," or elk anxiety from the mere presence of wolves. Subsequent researchers haven't found much evidence for that. Nonetheless, the same kind of ecological re-creation has happened in Canadian parks like Banff and Jasper when wolves returned. With numbers once as high as twenty-five thousand, an elk decline that began before wolves were in the park reached a low of five thousand in 2012. But since then elk herds have risen toward ten thousand, a level biologists believe may be a wolf-elk equilibrium.

With fewer elk now, in open country like the Lamar Valley bison numbers have increased four times over. Bison can overgraze but they don't browse woody vegetation, so the riparian cascade continues. At this point in Yellowstone's fascinating wolf-sponsored recovery, bison—some of them dragged down by wolf packs but most scavenged—make up 25 percent of the wolf diet in the Lamar Valley. The remarkable transformation goes on.

THE COMPASSIONATE IMPULSE that lies deep within us and is so profoundly evident in the animal art of a Chauvet Cave from thirty-five thousand years ago sometimes seems difficult to locate in the modern human
animal. The critical quality of humanity's ancient impulse for biophilia,
though, must have been compassionate empathy, an ability to reach outside the self and identify with others. Today that doesn't seem anywhere
in sight in the garroting of wolves, or the running down of panicked
animals with snowmobiles, or in the "recreational shooting" of eagles
that's now growing in parts of the country. Empathy is definitely missing in the prurient strangeness of Texas's yearly "Rattlesnake Roundups,"
where like frontier crowds watching a hanging, audiences fire cell-phone
cameras as a quarter-million rattlesnakes die beneath the vaporous musk
the terrified rattlers emit. Any element of biophilia is hard to locate in
so-called modern "hunting contests," where contestants win prizes (and
bets) for killing the most coyotes, the biggest coyote, often the *smallest*
pup, with special prizes for the youngest killer. Then, in the only nation
where such contests happen, discard the bodies and go for beer.

As British ecologist Melanie Challenger has written, "The world is
now dominated by an animal that doesn't think it's an animal." Among
residents of Western nations, fewer Americans, around 60 percent, accept
evolution than is the case for almost any other First World country (a third
of us even in that group endorse the idea that evolution is "directed" by
a deity). In 2018, fifteen-year-olds in the United States ranked eighteenth
in the world, behind almost every major European and Asian country,
in their grasp of basic science, including evolution. Vast numbers of us
apparently still believe in a human exceptionalism based on the idea
that of all the planet's creatures, only humans had the luck to be earthly
clones of some sky god. Our species has existed for three hundred thousand years yet we're still struggling to think clearly. No deity chose us
for world domination. Nor did evolution. We were omnivorous primates
with hands adapted for tree life who took to open country because of the
carnivorous economy it offered, whose perfection of hunting big, dangerous prey and avoiding predators refined our cooperative social skills.
Protein grew our brains and the hunting lifestyle transformed our verbal-
izations into rich language. Then we learned from one another for thousands of generations. So here we are.

In fairness, liberal humanists have also struggled to see humans as animals. An anthropologist with whom I used to debate this issue always insisted that culture, culture, culture—a so-called cultural intelligence hypothesis—meant we were vastly different from every other species. Vastly! she said. We *are* remarkable, the paragon of animals, in Hamlet's words, and perhaps our apprehension of nature *is* becoming godlike. We've figured out evolution, split the atom, grasped plate tectonics, discovered genes, are able to predict climate change, and can shoot a dune buggy fifty million miles into space and hit Mars every time.

But the latest science suggests we're remarkable because we ramified precisely the characteristic traits of higher life—consciousness, communication, and cultural transmission—that are widespread among our fellow species. Humanity appears to have emerged gradually, glacially, without some grand leap out of animal existence. Today we tinker with genes and ponder astronomical physics with exactly the brains and intelligence and language two million years of cooperative hunting of animals gave us. But as our numbers have grown, from something like 4 million of us ten thousand years ago to 7.9 billion as of early 2022, as we invented writing, printing, computers, the internet, our ability to transmit culture and knowledge across distances of time and geography has without question made us a global paragon. Nonetheless, the paragon is still an animal. We still catch spillover diseases from bats and birds. We still have sex, still relieve ourselves. We give birth just like other mammals. Our genes, our limbic systems, and our brains even now reflect millions of years spent among wild animals. We still can't give up our predator's diet. And we continue to die.

The earthly template we're built on is the same one that life-forms all around us employ. Our self-denial about our animal nature has been problem enough for us but catastrophic for other animals because the denial allows us to emphasize all the qualities we believe they lack. This is deadly hubris. In 2012 a global team of neuroscientists issued the Cambridge Declaration on Consciousness, a scientific consensus about perceptions of awareness and identity among other species. As they wrote, "non-human animals have the neuroanatomical, neurochemical, and neurophysiological substrates of conscious states." All mammals, all birds, even invertebrates like octopuses share the trait we've long thought belonged only to us. With animals like wolves, for example, neuroscientists had already demonstrated that they possess the same "triune brain" we do, the same neural networks as us. They communicate with a wolf language. They preserve memories, use their imaginations, employ reason. Animals like wolves have emotional lives, feel pain, joy, misery, delight, and take pleasure in being alive. They possess a theory of mind, ascribing actions to others and anticipating actions from them. Now, with the Cambridge Declaration, we believe they also have a sense of self, an inner life.

Americans a century ago observed that the last herds of wild bison had grown remarkably wary, that the final few wolves were astonishingly difficult to trap or poison. What we were observing was animal culture in action. We've long and mistakenly thought culture was our provenance alone. But in the mid-twentieth century it began to dawn on ornithologists studying birdsong that there were actually *regional* versions, handed down by local populations and distinct from the songs members of the same species sang elsewhere. We even understand now that an early signal of a threatened bird population is difficulty mastering the local song for want of mentors.

Since this epiphany, researchers have found evidence of animal culture everywhere. Animals teach one another to use tools, learn migration paths, develop preferences in forage or prey or nesting sites or mates that get handed down through generations. Just like us. Cultural teaching seems especially common among predators, whose young—like ours—must undergo a long process of learning about the world. As a 2021 overview in the journal *Science* put the matter: "Culture permeates the lives of a great diversity of animals, with far-reaching implications for evolutionary biology, anthropology, and conservation." That this truth is but a recent breakthrough reveals the depths of our myopia.

※

THEN THERE'S INDIVIDUALITY. It's one of our evolutionary biases that compassion comes more easily with individuals than with groups. Individually is how we best relate to one another. It's obviously how we see our companion animals. Even biologists, who for a century focused exclusively on species, increasingly are studying individuality. Differentiating personalities among wild canids has become an intriguing new angle that encourages Wildlife Services (the current name for what we once called

PARC and later Animal Damage Control) to target offending animals rather than wolves or coyotes generally in stock depredations. Personality differences also explain why some coyotes and even some predatory birds are so successful in cities and others aren't.

Like the ranchers who knew and named wolves and grizzlies a century ago, most of us have no problem bestowing names on the individual animals we come to know. When a red-tailed hawk became a celebrity bird in and around Manhattan's Central Park in the 1990s, he quickly became "Pale Male." All his several mates (First Love and Blue among them, Octavia most recently) became specific birds. New Yorkers also named at least some of the coyotes that have become residents of the boroughs, like 1999's Otis, or Hal in 2006. In 2020–21, New Yorkers were charmed by a charismatic female barred owl—"Barry," naturally—who frequented Central Park. When she was hit by a maintenance vehicle 250 people gathered in the park for a memorial service. A cougar roaming Griffith Park became another Los Angeles celebrity, "P-22." Even biologists sometimes can't resist. In the wake of the Oscar-winning film *Gladiator*, Montana researchers conferred the name Maximus on the grizzly they believed the largest in the state. A model bear who steered clear of humans and trouble, Maximus tragically ended up shot for no apparent reason, his body left to rot.

In the years since wolf recovery, some of the wolves of Yellowstone National Park have become among the most famous celebrity animals in the world. Even when they bear numbers (rather than names) linked to the telemetry of their radio collars, some Yellowstone wolves are now the subjects of book-length biographies. No one has made the intimate lives of individual wolves available to so many people as has a red-haired, cherubic naturalist named Rick McIntyre. Rick is this century's Ernest Thompson Seton, a comparison I know he relishes. In 2013 Sara and I were lucky enough to spend three perfect autumn days with Rick, watching wolves and coyotes and falling easily into his obsession. A New England native, Rick worked at various parks in the West before moving to Yellowstone full-time in 1999, when he rented a cabin in Silver Gate, just east of the park's grand Lamar Valley. You don't encounter many people this committed. His daily goal was never to leave the park without seeing a wolf, which at one point he pushed to 891 days in a row. That fall of 2013 his focus was on the Junction Butte pack, led by an uncollared alpha male Rick called Puff because of his past bouts with mange. "Even though he doesn't give the appearance of being a really strong wolf, he is," Rick said of this tall, thin alpha. "He's become maybe the best hunter in Yellowstone."

I know that Doug Smith, Yellowstone's Wolf Recovery Team leader, is uneasy about the celebrity-wolf phenomenon, but even he knows it's created wolf aficionados like never before. The way to grasp our newfound intimacy with the park's wolves is to read Rick's books, a publishing project called The Alpha Wolves of Yellowstone series. The books track a kind of *Game of Thrones* story (although the wolves tend to be far more sympathetic than the show's characters) featuring the ongoing battle among wolf packs for control of the choice hunting grounds in Yellowstone. It's the riveting, day-to-day experiences of individual wolves, all with distinctive personalities—some confident, competent, empathetic, loving to their mates, others violent, sneaky, not so admirable—that make McIntyre's project a masterwork on a par with Jane Goodall's celebrated time with wild primates.

McIntyre's saga begins with the story of wolf 8, an undersized young male whose displays of courage and mercy became (in Rick's telling) a model for wolf 21, the first real celebrity wolf in Yellowstone. With his lifelong mate, 42, wolf 21 came to lead the famous Druid pack in Lamar, which at one point included thirty-seven adults and pups! The relationship of that pair and its pack became the subject of an Emmy-winning film documentary in 2003. Then there was the gray male, 302, who became a legendary Lothario, irresistible to females in several packs, and who sired a very large number of pups, the most of any male wolf in Yellowstone so far. Rick's books and Bob Landis's several documentary films about these animals have made 2011's theatrical film *The Grey*, starring Liam Neeson and a man-eating, monstrous "omega" wolf, all the more astonishingly tone-deaf and absurd.

The only gray wolf ever to land an obituary in the *New York Times* was a female Rick spent nearly seven years watching. This was 06, not a telemetry number but the year of her birth. A gray alpha whose intelligence and benevolent character took her to the leadership of the Lamar pack, 06 merited a biography with journalist Nate Blakeslee's *American Wolf* in 2017. Based heavily on McIntyre's notes, it's appropriately a dual biography, devoting as much space to McIntyre himself as to legendary 06. Employing guile and courage to maintain her pack's territory in the face of expanding wolves like the Mollies—a large pack of buffalo-hunting wolves from farther south in the park—06 captured the imagination of the world. She'd died the fall previous to my 2013 time with Rick, in the kind of outcome everybody dreaded. With pressure from the Mollies, who killed nine wolves from various packs, in 2012 06 began leading her pack east, which unfortunately took them out of the park just as wolf-hunting season began.

Of course a wolf who had never experienced gunfire could not have known this place and this time were different. A hunter cut their tracks on Chief Joseph Highway, walked into the woods and blew a dying rabbit call and saw two wolves step into the open. Walking to where his shot had dropped the gray female he watched wide-eyed as eleven wolves emerged from the woods, formed a semicircle around her and began to howl. Even to her killer this horrifying reaction sounded like "sheer overwhelming sadness."

Discovering that other animals are like us, individuals who possess consciousness and transmit culture, isn't just a revolution in evolutionary biology. Anthropologist Franz Boas concluded a century ago that human cultures aren't all aimed at one destination, and that same insight can help guide us with respect to other animals. Biologists Marc Bekoff and Frans de Waal are both modern advocates for the abilities of animals, but like Boas with human culture, they want us to see animals as complete as they are, not pitiable "dumb brutes" wishing they were us. *Planet of the Apes*, in other words, got it wrong from the start. Make chimps and orangutans smarter and they don't become like humans. Instead they become remarkably good at doing what chimps and orangutans do.

And it is fascinating, is it not, that understanding this can also carry us back in time, to when animals were magical and kin?

🐾

SINCE WE ARRIVED in America twenty-three thousand years ago the place has changed.

Mammoths and saber-toothed cats are gone and 330 million humans have taken up residence. Great auks, heath hens, passenger pigeons, native parrots, giant woodpeckers have vanished. It's not easy to see a wild bison today. Eagles and condors and alligators are back, but William

Bartram's alligators now share their swamps with a veritable zoo of "new immigrants" from the rest of the world. Five centuries after Old Worlders introduced hogs, the feral creatures have become one of the most intractable wildlife problems of the twenty-first century. There are now between six and nine million of them in forty-two states and they're still spreading. They're the mammalian equivalent of Burmese pythons in Florida—nothing seems to stop them. The Florida Fish and Wildlife Conservation Commission in 2021 finally banned breeding and owning six different species of exotic pythons, along with green anacondas, monitor lizards, green iguanas, and all species of tegu lizards. The Argentine black-and-white tegu, a lizard that grows to the size of a collie, nonetheless is spreading out of Florida north and west along the Gulf Coast. In the 1730s Mark Catesby told the world America was tropical, but this unprecedented mixture of species courtesy of the exotic-pet trade is pure ecological madness.

Some things haven't changed. Because we have been hunters our entire history, some of us still hunt animals, although the numbers of Americans who do so is now down to about 6 percent. Still, the hunt answers an evolutionary call, and it continues as a wildlife-management tool, although one (as Leopold always said) not so precise as predation. True tender carnivores in pursuit of sacred game to prepare and eat don't arouse much hostility among those who understand that we evolved to do this very thing. That's why most environmentalists, despite hunter fantasies to the contrary, are not usually anti-hunting.

The mammoth in the room, looming over all us animals, is the superheating planet. Privileging economics over all else has given humans the richest world (measured by global GDP) but the poorest (measured by biodiversity) since we emerged as a species. But along with laying waste to wild animals, our economic success has pumped so much carbon dioxide, methane, and nitrous oxide into the biosphere that we're close to replicating the skies of Pliocene Earth of four million years ago, when conifer forests blanketed the Arctic plain to the very edge of the Beaufort Sea and sheets of rain regularly drenched much of North America. No one knows if we'll replicate that, but at Thoreau's cabin site at Walden the temperatures are now 4.3 degrees Fahrenheit warmer than when he longed to experience an entire heaven and earth.

Despite all the technology we can bring to bear on wildlife and ecology these days—automated data collection and recognition algorithms are revolutionizing ecological studies, making $4,000 radio collars look like metal bird bands—we've set in motion centuries of upcoming change with virtually no knowledge about how wild species will fare. Or whether they'll survive at all. Scientists think Earth harbors 8 million distinct life-forms. In 2019, when the IUCN issued a report on just 100,000 of those, it calculated 27,159 were in trouble or already gone. Most are plants but there were 1,223 mammals and 1,492 birds in its lists. A study in *Nature* the next year stated the obvious: "Climate change is projected to become a leading driver of biodiversity loss." But how that might play out remains a mystery. *Nature*'s authors predicted that by 2050 climate change "at the very least . . . will drive a sudden shift across many ecological assemblages" and that such rapid disruptions would pose risks to ecosystems formed on a more temperate Earth.

As a result of our visionary efforts to set aside vast public wildlands plus a half-century effort on behalf of endangered species, our wild animals in the United States are in better shape than those elsewhere in the world. A country that produces 25 percent of climate-altering emissions with a mere 4 percent of the global population obviously has a moral obligation to nations and species across the planet. But part of the problem with climate change is our evolutionary past. For hunter-gatherer minds, addressing a problem with weather is a job for shamans or deities to take on. The task seems beyond technological fixes or some adjustment to daily life. Plus, we Americans have never been good at accepting blame for screwing up the world. Surely the gods, or the government, or the Chinese, or the sun! must be doing this. It can't be us.

It's already apparent that a warming climate is producing weird and troubling effects on species and habitats. The white-nose syndrome that has killed seven million American bats and the fungal disease now loose on twenty-five species of our snakes are both imports that are probably exacerbated by climate change. *Mycoplasma bovis*, a pathogen we identified in 2013 that cattle transmit to bison, has killed up to 25 percent of some buffalo herds. A warmer climate amplifies its effects. Winters that once were cold enough to limit insect pests are now demonstrably so mild they allow pine bark beetles and spruce budworms to destroy forests across the country. Prolonged droughts are unleashing uncontrollable fires, and warmer seas translate into Gulf and Atlantic Coast hurricanes unlike anything Mark Catesby ever witnessed in the 1730s. One thing the delegates at the 2021 global climate conference in Glasgow, Scotland, could agree on was attempting to slow or end deforestation. Not only are forests habitat for so many of the planet's threatened species, they're the planet's primary carbon sinks that scrub from the skies a good percentage of the warming gases our carbon economy is spewing.

But as ecologists are realizing, natural selection is a hell of a tool. We humans may have a difficult time hanging on to our situation, certainly the one we built for a more temperate world, but scientists are already studying evidence that wild animals are adjusting. Our Anthropocene may be punctuating the equilibrium, but for a certain Darwinian mind, Earth's response is going to be the show of the millennium. Employing natural selection to adapt and mobility to migrate, nature will always find a way. Some creatures are going to cope in dramatic and surprising ways. Some are going to hunker down in refuges. Some—with our awareness and help, hopefully not so many as we fear—will become extinct. The climate change we're producing will tragically cost Earth existing genetic diversity. But it may also generate new.

We find evidence of all these trends every day. Many American birds—white and brown pelicans on the coasts; in the Southwest, white-winged doves and mockingbirds—are visibly moving north. The Audubon Society, which has calculated climate-induced range shifts for 604 species of American birds, says bird winter ranges are now shifting at about half a mile a decade. Animals and trees are also moving up, to higher elevations, spoiling all those careful life-zone calculations C. Hart Merriam plotted out a century ago. Some biologists believe between 25 percent and 85 percent of all Earth's species are now on the move. Where they end up is going to scramble every ecosystem we know, but at least the federal agencies managing animals and habitats are no longer operating under mandates to ignore climate change. This fast-moving future is going to favor rapid adaptations, leave species like mountain pikas, red squirrels, and many snakes, turtles, and tortoises in shrinking refugia, and no doubt encourage speciation. It will also favor generalists over specialists, a reason the coyote seems a happy camper in this phase of the Anthropocene. The Dude abides.

IN 2005 we paragons of animals effected a scientific breakthrough that might actually promote us to the status of something like gods. In that year geneticists discovered Cas9, an enzyme that bacteria evolved to protect themselves against viruses. In our hands Cas9 made it possible to "edit genomes" by slicing away genes and replacing them. This discovery turned visionaries to an idea that had its origins—unfortunate for just about any idea—in Germany in the 1930s. If we could edit genes, was it possible to use related animals to re-create species we once drove to extinction? With the genomes of woolly mammoths and passenger pigeons reconstructed from excavated and collected specimens, the editing technology of CRISPR ("clustered regularly interspaced short palindromic repeats") seemed to offer breathtaking possibilities for bioengineering. Using as surrogates Asian elephants and band-tailed pigeons, the surviving species most closely related to the ones we lost, we ought to be able to insert into their genomes the genes that made woolly mammoths and passenger pigeons what they were.

So far we don't have either, or heath hens or Carolina parakeets or ivory-bills. But the possibilities are real enough to raise modern-world questions. Could vast flocks of bioengineered passenger pigeons work in an America crossed by thousands of airplanes a day? And will "de-extinction" (that's the term of art) become a fundamental part of species recovery?

With good reason, *Jurassic Park* comes to mind. Bioengineering does have a sci-fi quality about it, yet has already produced a remarkable success for one of the rarest mammals in America, with only 604 living individuals. Black-footed ferrets are hanging on by a thread and their breeding programs suffer from a severely pinched genetic diversity. But in 2021 a San Francisco lab that is the brainchild of futurist Stewart Brand, Revive & Restore, successfully cloned a black-footed ferret they named Elizabeth Ann from tissues researchers had preserved in the 1980s. The cloned female not only increased the ferret-recovery founder population from seven to eight, she fortuitously turned out to possess a large number of alleles the others lacked.

I turned to Ben Novak, Revive & Restore's lead scientist, to get a sense of what kinds of de-extinction miracles await us. A Montanan who studied with paleogeneticist Beth Shapiro (author of *How to Clone a Mammoth*) at UC–Santa Cruz, Ben did his graduate work on passenger pigeons. But

John Woodhouse Audubon, Black-Footed Ferret.   Courtesy the National Gallery of Art.

when I asked him how close we were to having passenger pigeons again, he had to admit that re-creating extinct birds was turning out to be more difficult than it at first seemed. Initially the "Great Passenger Pigeon Comeback" hoped to hatch designer passenger pigeons as early as 2025. Now the timeline has gotten blurry. The problem hasn't been editing the genomes of band-tailed pigeons with passenger pigeon genes. The barrier has been the reproductive step to produce the bioengineered results. Bandtails and passenger pigeons were 97 percent identical, Ben tells me, and what he wants is "a hybrid that will look and behave like and be a passenger pigeon." But "the only cell culture we can produce from a bandtail is a fibroblast cell, and we can't turn fibroblasts into a living bird." The avian genetic-rescue program—bringing back extinct birds like heath hens, or helping save endangered ones—depends on finding a recipe to grow wild-bird germ (or stem) cells.

Harvard's George Church, who has been working on cloning mammoths long enough to have become "the mammoth guy" in the eyes of the public, now has funding from a new company called Colossal. The "Woolly Mammoth Revival" envisions repopulating Alaska, Canada, and Siberia with herds of proxy mammoths, created by inserting cold-adapted mammoth genes into the genomes of closely related Asian elephants. "Church has already isolated sixty genes that could make a proxy Asian elephant/mammoth survive cold weather," Ben Novak tells me. The reproductive step is again the problem. Culturing stem cells seems easier with mammals, but finding surrogate mothers in an endangered Asian elephant population is problematic enough that Church is experimenting with an artificial uterus. Still, Ben says, "we could have a mammoth within five years, although it's more likely fifteen or twenty years out."

Suspend disbelief and it dazzles. But, why do this? When it comes to the ecological world, "because we can" probably isn't good enough. There are remains of mammoths and mastodons all over the Lower 48. One of the warm-adapted mammoth species would be easier to design. Instead, de-extinction scientists imagine herds of proxy woolly mammoths engineering the northern tundra into grasslands, sequestering carbon for us. But as even Ben admits to me, climate change may be too far advanced for that now. Passenger pigeons, Carolina parakeets, ivory-bills are one thing. But given our difficulty rewilding with living megafauna—with wolves, bison, wild horses, grizzlies, cougars—the romance of a mammoth revival wanes some. If wild horses have become a modern problem because they lack their original predators, project that logic to a woolly mammoth ecosystem. Saber-toothed cats and short-faced bears, anyone?

҈

HOW ONE RELATES to the big historical story here is probably a personal thing. Christopher Cokinos's *Hope is the Thing with Feathers* and Charles Bergman's *Wild Echoes* are memoirs in pursuit of loss. I can appreciate that. I've paid attention to know what's gone from my own world. I missed the last passenger pigeons in my part of America by exactly forty-six years. Late in their life on Earth the birds appeared several times near where I would grow up. In 1881 they flourished in enormous numbers on the Colorado River near Austin, Texas. In East Texas, near Tyler, a hundred miles from my ancestors, fourteen pairs nested in 1887. In 1896 a single pigeon in a flock of mourning doves was shot in Cameron Parish.

Scene from ancient America now continuing in Yellowstone Park: gray wolf and grizzly on an elk kill, 2018. Photograph courtesy Doug Smith, National Park Service.

That was the last passenger pigeon killed in my state. Observers saw the last wild pigeons in Louisiana on Prairie Mer Rouge, about 125 miles east of home. That was in the winter of 1902–3. All four of my grandparents were alive then. I got *that* close.

Rather than mourn, I've tried to experience what's left. That's how I've attempted to cope with Thoreau's anguish, and mine, over the America we never got to see.

So I've journeyed to the Arctic Wildlife Refuge to witness the grand summer migration of the Porcupine caribou herd. I've spent afternoons in Nebraska watching rivers of gangly sandhill cranes pass overhead, enveloped in the aural waves of their primeval fluting, once catching the rare treat of a white whooping crane flying with them. I've hiked through grizzly country and met wild bears, once watching a big sow throw haymakers at circling gray wolves over ownership of an elk as eagles and magpies hopped beyond reach and the bear's three cubs sat shocked and bug-eyed. I've crept close enough to a bugling bull elk to hear the gurgling in his throat at the end of his whistles, and stood among grazing buffalo in all of Hornaday's first bison refuges from a century ago. On my afternoon runs in Montana a bald eagle used to cock a reptilian eye as I passed beneath her cottonwood perch, and I returned the favor when she and her mate waddled about the pastures in early spring, eating new calf afterbirth.

I've long been drawn to pronghorns, those gorgeous Pleistocene survivors, and one time saw a young buck turn himself sideways running fifty miles per hour (clocked by car) so he could dart between strands of wire fence, the impact a *thwanging* explosion of white hairs drifting in the wind as he trotted off, daydreaming of cheetahs. Sitting on a yellow raft pontoon in the bottom of the otherworldly Grand Canyon I've watched a California condor soaring the-blue-that-will-always-be-there. In the Nevada desert I've nestled in the sagebrush with binoculars and laughed to myself as a wild stallion, so black he had to turn at an angle before I could see his sleek musculature, ran himself ragged chasing equine Don Juans away from his harem of flirtatious mares.

Trumpeter swans sailing Montana's seven-thousand-foot Centennial Valley have awed me. So have sea otters along California's Central Coast. So have red-tailed hawks a biologist friend and I banded, one of them taking flight while my hands still grasped its feet, the closest I've ever come to flying. So have winter dancers at Jemez Pueblo in New Mexico, who perfectly, uncannily, mimic the movements of mule deer, eagles, buffalo. Sipping coffee in my kitchen I've watched the miracle of coyotes gliding past the windows the way coyotes go through the world, every hair on their pointed muzzles and upright ears crisply backlit by raking chromium-yellow light. I've never seen a cougar. Anywhere. I'm heading to Florida to look, hoping for one glimpse of my natal Southland's secretive big cats.

I've been wolf lucky, though. Except for two decades I spent in Texas, I've had wolves everywhere I've lived. I grew up with red wolves, at least rangy red wolf–coyote hybrids. I got to experience gray wolf recovery in Montana. Where I live today, near Santa Fe, I know that Mexican wolves recently released into New Mexico's Black Range are close. In my one state that lacked wolves, I introduced them. In the 1990s I raised a pair of wolf-malamute hybrids and their litter in Yellow House Canyon in West Texas. That turned out to be a sin against wolves I didn't intend to commit. I've since moved on to wolfy Alaskan malamutes.

But "my wolves" live in my mind. I can still conjure crouching in the dirt of their enclosure on a summer day as they repeatedly "charged" me, hurtling with such velocity and lack of wasted motion that while I should remember knife-edge detail, all I can pull up is a slur of Cubist images, snatches of red, open mouths, tongues streaming saliva, yellow eyes gyring in on me, the rush carried by a primal, thumping rhythm punctuated by the *clack* of jaws as the female snaps at my ponytail when she surges past like a wave of black water.

The prescription I've come to seems to be this. Know the heaven and earth that was, but experience the world that is.

BIBLIOGRAPHY

Alagona, Peter. *After the Grizzly: Endangered Species and the Politics of Place in California.* Berkeley: University of California Press, 2013.

Allender, Matthew, et al. "Ophidiomycosis, An Emerging Fungal Disease of Snakes: Targeted Surveillance on Military Lands and Detection in the Western US and Puerto Rico." *PLOS One* (October 8, 2020): https://doi.org/10.1371/journal.pone.0240415.

Alroy, John. "A Multispecies Overkill Simulation of the End-Pleistocene Megafaunal Mass Extinction." *Science* 292 (June 2001): 1893–96.

Andrei, Mary Anne. *Nature's Mirror: How Taxidermists Shaped America's Natural History Museums and Saved Endangered Species.* Chicago: University of Chicago Press, 2020.

Angulo, Elena, et al. "Review: Allee Effects in Social Species." *Journal of Animal Ecology* 87 (January 2018): 47–58.

Anonymous. "An Ojibwe Recounts the Arrival of 'Heavenly Visitors' to His People." In *Major Problems in American Indian History: Documents and Essays,* edited by Albert Hurtado and Peter Iverson. Lexington, MA: D. C. Heath, 1994.

Ardrey, Robert. *African Genesis: A Personal Investigation into the Animal Origins and Nature of Man.* New York: Dell, 1961.

Audubon, John James. *Audubon and His Journals.* Vol. 2. Edited by Maria Audubon. New York: Charles Scribner's Sons, 1897.

Barlowe, Arthur. *The First Voyage Made to the Coasts of America . . . Written by One of the Said Captaines, and Sent to Sir Walter Raleigh.* Electronic edition. University of North Carolina at Chapel Hill, 2002.

Barr, James. "Man and Nature: The Ecological Controversy and the Old Testament." In *Ecology and Religion in History,* edited by David Spring and Eileen Spring. New York: Harper and Row, 1974.

Barrow, Mark. *Nature's Ghosts: Confronting Extinction from the Age of Jefferson to the Age of Ecology*. Chicago: University of Chicago Press, 2009.

Bartram, William. *The Travels of William Bartram*. Edited by Francis Harper. New Haven, CT: Yale University Press, 1958.

Bechtel, Stefan. *Mr. Hornaday's War: How a Peculiar Victorian Zookeeper Waged a Lonely Crusade for Wildlife That Changed the World*. Boston: Beacon Press, 2012.

Bednarz, James. *The Mexican Gray Wolf: Biology, History, and Prospects for Reestablishment in New Mexico*. Endangered Species Report 18. Albuquerque, NM: U.S. Fish and Wildlife, 1988.

Beeland, T. Delene. *The Secret World of Red Wolves: The Fight to Save North America's Other Wolf*. Chapel Hill: University of North Carolina Press, 2013.

Bekoff, Marc. *The Emotional Lives of Animals*. New York: New World Library, 2008.

Belue, Ted. *The Long Hunt: Death of the Buffalo East of the Mississippi*. Mechanicsburg, PA: Stackpole Books, 1996.

Bennett, Matthew, David Bustos, Daniel Odess, et al. "Walking in Mud: Remarkable Pleistocene Human Trackways from White Sands National Park (New Mexico)." *Quaternary Science Reviews* 249, no. 1 (December 2020): article 106610.

Bennett, Matthew, David Bustos, Jeffrey Pigati, et al. "Evidence of Humans in North America during the Last Glacial Maximum." *Science*, September 24, 2021.

Benson, Maxine, ed. *From Pittsburgh to the Rocky Mountains: Major Stephen Long's Expedition, 1819–1820*. Golden, CO: Fulcrum Press, 1988.

Berger, Joel, and Carol Cunningham. *Bison: Mating and Conservation in Small Populations*. New York: Columbia University Press, 1994.

Bergman, Charles. *Wild Echoes: Encounters with the Most Endangered Animals in America*. New York: McGraw-Hill, 1990.

Berlandier, Jean Louis. *Journey to Mexico during the Years 1826 to 1834*. Translated by Sheila Ohlendorf, Josette Bigelow, and Mary Standifer. 2 vols. Austin: Texas State Historical Association, 1980.

Blakeslee, Nate. *American Wolf: A True Story of Survival and Obsession in the West*. New York: Crown, 2018.

Blockstein, David, and Stanley Temple. "Fauna in Decline: Extinct Pigeon's Tale." *Science* 345 (September 2014): 1129.

Bodmer, Karl. *Karl Bodmer's America*. Introduction by William Goetzmann, annotations by David C. Hunt and Marsha V. Gallagher. Omaha, NE: Joslyn Art Museum, 1984.

Boehme, Sarah, ed. *John James Audubon in the West: The Last Expedition, Mammals of North America*. New York: Harry Abrams, 2000.

Botkin, Daniel. *Discordant Harmonies: A New Ecology for the Twenty-First Century*. New York: Oxford University Press, 1990.

———. *Our Natural History: The Lessons of Lewis and Clark*. New York: G. P. Putnam's Sons, 1995.

Brannen, Peter. "The Terrifying Warning Lurking in the Earth's Ancient Rock Record." *The Atlantic*, March 2021. https://www.theatlantic.com/magazine/archive/2021/03/extreme-climate-change-history/617793/.

Bray, Kingsley. "Lone Horn's Peace: A New View of Sioux-Crow Relations, 1851–1858." *Nebraska History* 66 (Spring 1985): 29–47.

Brinkley, Douglas. *The Wilderness Warrior: Theodore Roosevelt and the Crusade for America.* New York: HarperCollins, 2009.

Brown, David. *The Grizzly in the Southwest.* Norman: University of Oklahoma Press, 1985.

———, ed. *The Wolf in the Southwest: The Making of an Endangered Species.* Tucson: University of Arizona Press, 1983.

Brown, Joseph Epes. *Animals of the Soul: Sacred Animals of the Oglala Sioux.* Rockport, MA: Element, 1992.

Browning, Meshach. *Forty-Four Years, The Life of a Hunter: Being Reminiscences of Meshach Browning, a Maryland Hunter and Trapper.* 1859. Reprint, n.p.: Pantianos Classics.

Bryan, Liz. *The Buffalo People: Prehistoric Archaeology on the Canadian Plains.* Edmonton: University of Alberta Press, 1991.

Budiansky, Stephen. *The Covenant of the Wild: Why Animals Chose Domestication.* New Haven, CT: Yale University Press, 1992.

Burgio, K., et al. "Lazarus Ecology: Recovering the Distribution and Migratory Patterns of the Extinct Carolina Parakeet." *Ecology and Evolution* 7, no. 14 (July 2017): 5467–75.

Burlingame, Merrill. "The Buffalo in Trade and Commerce." *North Dakota Historical Quarterly* 3 (July 1929): 262–91.

Byers, John. *American Pronghorn: Social Adaptations and the Ghosts of Predators Past.* Chicago: University of Chicago Press, 1997.

Cabeza de Vaca, Álvar Núñez. "The Narrative of Alvar Nunez Cabeca de Vaca." In *Spanish Explorers in the Southern United States, 1528–1543,* edited by Frederick Hodge, 12–126. New York: Charles Scribner's Sons, 1907.

Callicott, J. Baird. "Animal Liberation: A Triangular Affair." *Environmental Ethics* 2 (1980): 311–38.

Calloway, Colin. *One Vast Winter Count: The Native American West before Lewis and Clark.* Lincoln: University of Nebraska Press, 2003.

Carhart, Arthur. "Poisons: The Creeping Killer." *Sports Afield* (November 1959): 56–57, 127–29.

Carson, Rachel. *Silent Spring.* 40th anniversary ed. New York: Houghton-Mifflin, 2002.

Cartmill, Matt. *A View to a Death in the Morning: Hunting and Nature through History.* Cambridge, MA: Harvard University Press, 1993.

Catlin, George. *Letters and Notes on the Manners, Customs, and Conditions of the North American Indians.* 2 vols. 1841. Reprint, New York: Dover, 1973.

Ceballos, Gerardo, Paul Ehrlich, and Rodolfo Dirzo. "Biological Annihilation via the Ongoing Sixth Mass Extinction Signaled by Vertebrate Population Losses and Declines." *Proceedings of the National Academy of Sciences* 114 (July 2017).

Ceballos, Gerardo, Paul Ehrlich, and Peter H. Raven. "Vertebrates on the Brink as Indicators of Biological Annihilation and the Sixth Mass Extinction." *Proceedings of the National Academy of Sciences* 117 (June 2020).

Cervantes, Fernando. *The Devil in the New World: The Impact of Diabolism in New Spain.* New Haven, CT: Yale University Press, 1994.

Challenger, Melanie. *How to Be Animal: A New History of What It Means to Be Human.* New York: Penguin, 2021.

Chambers, Steven, et al. *An Account of the Taxonomy of North American Wolves from Morphological and Genetic Analysis.* North American Fauna no. 77. Washington, DC: United States Department of Interior Fish and Wildlife Service, 2012.

Chapman, Joseph, and George Feldhamer, eds. *Wild Mammals of North America: Biology, Management, and Economics.* Baltimore: Johns Hopkins University Press, 1982.

Chauvet, Jean-Marie, et al. *Dawn of Art: The Chauvet Cave, the Oldest Known Paintings in the World.* New York: Harry Abrams, 1996.

Christian, David. *Maps of Time: An Introduction to Big History.* 2nd ed. Berkeley: University of California Press, 2011.

Clark, Tim, and Denise Casey. *Tales of the Grizzly: Thirty-Nine Stories of Grizzly Bear Encounters in the Wilderness.* Moose, WY: Homestead Publishing, 1991.

Cokinos, Christopher. *Hope Is the Thing with Feathers: A Personal Chronicle of Vanished Birds.* New York: Tarcher/Penguin, 2009.

Coleman, Jon. *Vicious: Wolves and Men in America.* New Haven, CT: Yale University Press, 2004.

Collins, G. S., et al. "A Steeply-Inclined Trajectory for the Chicxulub Impact." *Nature Communications* 11 (May 2020): article 1480. https://doi.org/10.1038/s41467-020-15269-x.

Cong. Globe, 42d Cong., 2d Sess., February 14, 1872.

Cong. Rec., 43d Cong., 2d Sess., February 2, 1874; March 10, 1874; 44th Cong., 2d Sess., February 23 and 25, 1876.

Connolly, Guy. "The Effects of Control on Coyote Populations: Another Look." In *Symposium Proceedings—Coyotes in the Southwest: A Compendium of Our Knowledge,* 23–29. Lincoln, NE: Wildlife Damage Management, 1995.

Cook, John. *The Border and the Buffalo.* Topeka, KS: Crane, 1907.

Cooper, James Fenimore. *The Pioneers, or, The Sources of the Susquehanna.* New York: Grosset & Dunlap, 1930.

Corbey, Raymond, et al. "The Acheulean Handaxe: More Like a Bird's Song Than a Beatles' Tune?" *Evolutionary Anthropology* 25, no. 1 (2016): 6–19.

Cornell Laboratory of Ornithology. *Summary and Conclusions of the 2005–6 Ivory-Billed Woodpecker Search in Arkansas.* Ithaca, NY: Submitted to the U.S. Fish & Wildlife Service, 2007.

Coupland, Robert. "The Effects of Fluctuations in Weather upon the Grasslands of the Great Plains." *Botanical Review* 24 (May 1958).

Covone, Giovanni, Riccardo Ienco, Luca Cacciapuoti, and Laura Inno. "Efficiency of the Oxygenic Photosynthesis on Earth-Like Planets in the Habitable Zone." *Monthly Notices of the Royal Astronomical Society* 505, no. 3 (August 2021): 3329–35.

Cronon, William. *Changes in the Land: Indians, Colonists, and the Ecology of New England.* New York: Hill and Wang, 1983.

Crosby, Alfred. *Ecological Imperialism: The Biological Expansion of Europe, 900–1900.* New York: Cambridge University Press, 1986.

———. "Virgin Soil Epidemics as a Factor in the Aboriginal Depopulation in America." *William and Mary Quarterly* 33 (April 1976): 289–99.

Cutright, Paul Russell. *Lewis and Clark: Pioneering Naturalists*. Urbana: University of Illinois Press, 1969.

———. *Theodore Roosevelt: The Making of a Conservationist*. Urbana: University of Illinois Press, 1985.

Cuvier, Georges, et al., eds. *The Animal Kingdom: Arranged in Conformity with Its Organization*. 16 vols. London: Printed for G. B. Whitaker, 1827–35.

Dant, Sara. *Losing Eden: An Environmental History of the American West*. 2nd ed. Lincoln: University of Nebraska Press, 2022.

Darwin, Charles. *From So Simple a Beginning: The Four Great Books of Charles Darwin*. Edited by Edward O. Wilson. New York: W. W. Norton, 2006.

Dasmann, Raymond. "Environmental Changes before and after the Gold Rush." *California History* (Winter 1998/99): 105–22.

Daubenmire, Rexford. "The Western Limits of the Range of the American Bison." *Ecology* 66 (April 1985): 622–24.

Davis, J. W., et al. *Infectious Diseases of Wild Mammals*. Ames: Iowa State University Press, 1981.

Davis, Matt, Soren Faurby, and Jens-Christian Svenning. "Mammal Diversity Will Take Millions of Years to Recover from the Current Biodiversity Crisis." *Proceedings of the National Academy of Sciences* 115, no. 44 (October 30, 2018): 11262–67.

Davis, Mike. *Ecology of Fear: Los Angeles and the Imagination of Disaster*. New York: Metropolitan Books, 1998.

Dawkins, Richard. *The Selfish Gene*. New York: Oxford University Press, 1989.

Dax, Michael. *Grizzly West, New West: The Failed Attempt to Reintroduce Grizzly Bears in the Bitterroot Mountains*. Lincoln: University of Nebraska Press, 2015.

DeBuys, William, and Craig Allen. "A Historical Chronology of Events and Observations for the Pecos Wilderness in the Territorial Period." *New Mexico Historical Quarterly* 90 (Fall 2015): 415–87.

Dehler, Gregory. *The Most Defiant Devil: William Temple Hornaday and His Controversial Crusade to Save American Wildlife*. Charlottesville: University of Virginia Press, 2013.

Demuth, Bathsheba. *Floating Coast: An Environmental History of the Bering Strait*. New York: W. W. Norton, 2019.

Denevan, William. "The Pristine Myth: The Landscapes of the Americas in 1492." *Geographical Review* 86 (July 1996): 385–97.

Denig, Edwin. *Five Indian Tribes of the Upper Missouri: Sioux, Arickaras, Assiniboines, Crees, Crows*. Edited by John Ewers. Norman: University of Oklahoma Press, 1961.

Derrida, Jacques. *The Animal That Therefore I Am*. Edited by Marie-Louise Mallet. Translated by David Wills. New York: Fordham University Press, 2008.

de Steiguer, Edward. *Wild Horses of the West: History and Politics of America's Mustangs*. Tucson: University of Arizona Press, 2011.

DeVoto, Bernard. *Across the Wide Missouri*. Boston: Houghton-Mifflin, 1947.

———, ed. *The Journals of Lewis and Clark*. Boston: Houghton-Mifflin, 1953.

de Waal, Frans. *Are We Smart Enough to Know How Smart Animals Are?* New York: W. W. Norton, 2016.

———. *Good-Natured: The Origins of Right and Wrong in Humans and Other Animals*. Cambridge: Harvard University Press, 1995.

Diamond, Jared. *Guns, Germs, and Steel: The Fates of Human Societies.* New York: W. W. Norton, 1997.

Dillehay, Tom. "Late Quaternary Bison Population Changes on the Southern Plains." *Plains Anthropologist* 19 (August 1974): 180–96.

Dobak, William. "The Army and the Buffalo: A Demur—A Response to David D. Smits's 'The Frontier Army and the Destruction of the Buffalo: 1865–1883.'" *Western Historical Quarterly* 26 (Summer 1995): 197–202.

Dobyns, Henry. *Native American Historical Demography.* Bloomington: Indiana University Press, 1976.

Dodge, Richard Irving. *The Plains of the Great West and Their Inhabitants, Being a Description of the Plains, etc., of the Great North American Desert.* 1877. Reprint, New York: Archer House, 1959.

Dolin, Eric Jay. *Fur, Fortune, and Empire: The Epic History of the Fur Trade in America.* New York: W. W. Norton, 2010.

Doroff, Angela, et al. "Sea Otter Population Declines in the Aleutian Archipelago." *Journal of Mammology* 84 (February 2003): 55–64.

Dunlap, Thomas. "'The Coyote Itself': Ecologists and the Value of Predators, 1900–1972." *Environmental History Review* 7 (Spring, 1983): 54–70.

———. *Saving America's Wildlife: Ecology and the American Mind, 1850–1990.* Princeton, NJ: Princeton University Press, 1988.

Egerton, Frank, ed. *History of American Ecology.* New York: Arno Press, 1977.

Emory, William. *Report on the United States and Mexican Boundary Survey.* 2 vols. Washington, DC: Government Printing Office, 1855.

Erdoes, Richard, and Alfonso Ortiz. *American Indian Trickster Tales.* New York: Penguin, 1998.

Fagan, Brian. *The First North Americans: An Archaeological Journey.* London: Thames and Hudson, 2011.

Farr, William. "Going to Buffalo: Indian Hunting Migrations across the Rocky Mountains: Part 2, Civilian Permits, Army Escorts." *Montana, the Magazine of Western History* 54 (Spring 2004): 26–43.

Fiddes, Nick. *Meat: A Natural Symbol.* London and New York: Routledge, 1991.

Fischer, Hank. *Wolf Wars: The Remarkable Inside Story of the Restoration of Wolves to Yellowstone.* Helena, MT: Falcon Press, 1995.

Flannery, Tim. *The Eternal Frontier: An Ecological History of America and Its Peoples* (New York: Atlantic Monthly Press, 2001).

Fleharty, Eugene. *Wild Animals and Settlers on the Great Plains.* Norman: University of Oklahoma Press, 1995.

Flippen, J. Brooks. *Nixon and the Environment.* Albuquerque: University of New Mexico Press, 2000.

Flores, Dan. *American Serengeti: The Last Big Animals of the Great Plains.* Lawrence: University Press of Kansas, 2016.

———. "Bison Ecology and Bison Diplomacy: The Southern Plains from 1800 to 1850." *Journal of American History* 78 (September 1991): 465–85.

———. *Coyote America: A Natural and Supernatural History.* New York: Basic Books, 2016.

————, ed. *Jefferson & Southwestern Exploration: The Freeman & Custis Accounts of the Red River Expedition of 1806.* Norman: University of Oklahoma Press, 1984.

————. "Essay: The Great Plains 'Wilderness' as a Human-Shaped Environment." *Great Plains Research* 9 (Fall 1999): 343–55.

Frison, George. "Paleoindian Large Mammal Hunters on the Plains of North America." *Proceedings of the National Academy of Sciences* 95 (November 1998): 14576–83.

Fulton, Tara, et al. "Nuclear DNA from the Extinct Passenger Pigeon (*Ectopistes migratorius*) Confirms a Single Origin of New World Pigeons." *Annals of Anatomy* 194, no. 1 (March 16, 2011): 52–57.

Furtwangler, Albert. *Acts of Discovery: Visions of America in the Lewis and Clark Journals.* Urbana: University of Illinois Press, 1993.

Gabler, Neal. *Walt Disney: The Triumph of the American Imagination.* New York: Vintage Books, 2006.

Gehrt, Stanley, Seth Riley, and Brian Cypher, eds. *Urban Carnivores: Ecology, Conflict, and Conservation.* Baltimore: Johns Hopkins University Press, 2010.

Gelabert, Pere, et al. "Evolutionary History, Genomic Adaptation to Toxic Diet, and Extinction of the Carolina Parakeet." *Current Biology* 30 (January 2020): 108–14.

Gillespie, Angus, and Jay Mechling, eds. *American Wildlife in Symbol and Story.* Knoxville: University of Tennessee Press, 1987.

Glutton-Brock, Juliet. "Aristotle, the Scale of Nature, and Modern Attitudes to Animals." *Social Research* 62 (Fall 1995): 421–40.

Goetzmann, William. *Army Exploration in the American West, 1803–1863.* New Haven, CT: Yale University Press, 1959.

————. *Exploration and Empire: The Explorer and the Scientist in the Winning of the American West.* New York: W. W. Norton, 1966.

Goetzmann, William, and Kay Sloan. *Looking Far North: The Harriman Expedition to Alaska, 1899.* Princeton, NJ: Princeton University Press, 1982.

Goldfarb, Ben. *Eager: The Surprising, Secret Life of Beavers and Why They Matter.* White River Junction, VT: Chelsea Publishing, 2018.

Goldman, E. A. "The Predatory Mammal Problem and the Balance of Nature." *Journal of Mammalogy* 6 (February 1925): 28–33.

Goode, George Brown, ed. *The Published Writings of Spencer Fullerton Baird, 1843–1882.* Washington, DC: Government Printing Office, 1883.

Gould, Stephen Jay. *Wonderful Life.* New York: Pantheon, 1991.

Grace, Molly, et al. "Testing a Global Standard for Quantifying Species Recovery and Assessing Conservation Impact." *Conservation Biology* 35 (December 2021): 1833–49. https://doi.org/10.1111/cobi.13756.

Graeber, David, and David Wengrow. *The Dawn of Everything: A New History of Humanity.* New York: Farrar, Straus and Giroux, 2021.

Grayson, Donald. *Giant Sloths and Sabertooth Cats: Extinct Mammals and the Archaeology of the Great Basin.* Salt Lake City: University of Utah Press, 2016.

Greenberg, Joel. *A Feathered River across the Sky: The Passenger Pigeon's Flight to Extinction.* New York: Bloomsbury, 2014.

Gregg, Josiah. *Commerce of the Prairies*. Edited by Max Moorhead. Norman: University of Oklahoma Press, 1954.

Greiser, Sally Thompson. "Predictive Models of Hunter-Gatherer Subsistence and Settlement Strategies on the Central High Plains." *Plains Anthropologist* 20 (1985): 1–134.

Grimm, David. "In Dogs' Play, Researchers See Honesty and Deceit, Perhaps Something Like Morality." *Washington Post*, May 19, 2014.

Grinnell, George Bird. *The Passing of the Great West: Selected Papers of George Bird Grinnell*. Edited by John Reiger. Norman: University of Oklahoma Press, 1972.

Grinnell, Joseph, and Tracy Storer. "Animal Life as an Asset of National Parks." *Science* 44 (1916): 375–80.

Hall, Raymond, and Keith Kelson. *The Mammals of North America*. 2 vols. New York: The Ronald Press, 1959.

Hamalainen, Pekka. *The Comanche Empire*. New Haven, CT: Yale University Press, 2008.

———. *Lakota America: A New History of Indigenous Power*. New Haven, CT: Yale University Press, 2019.

Hames, Raymond. "Game Conservation or Efficient Hunting." In *The Question of the Commons*, edited by Bonnie McKay and James Atcheson, 92–120. Tucson: University of Arizona Press, 1987.

Hampton, H. Duane. *How the U.S. Cavalry Saved Our National Parks*. Bloomington: Indiana University Press, 1971.

———, ed. *Life and Death at the Mouth of the Musselshell: Montana Territory, 1868–1872*. Stevensville, MT: Stoneydale Press, 2011.

Hanson, Thor. *Hurricane Lizards and Plastic Squid: The Fraught and Fascinating Biology of Climate Change*. New York: Basic Books, 2021.

Hariot, Thomas. *A Briefe and True Report of the New Found Land of Virginia: of the Commodities and of the Nature and Manners of the Naturall Inhabitants: Discouered by the English Colony*. 1588. Electronic edition, University of North Carolina at Chapel Hill, 2003.

Harris, Marvin, and Eric Ross. *Death, Sex, and Fertility: Population Regulation in Preindustrial and Developing Societies*. New York: Columbia University Press, 1987.

Harrod, Howard. *The Animals Came Dancing: Native American Sacred Ecology and Animal Kinship*. Tucson: University of Arizona Press, 2000.

Hauser, Marc. *Wild Minds: What Animals Really Think*. New York: Henry Holt, 2000.

Haynes, Gary. *The Early Settlement of North America: The Clovis Era*. Cambridge: University of Cambridge Press, 2002.

Hays, Samuel. *Beauty, Health, and Permanence: Environmental Politics in the U.S., 1955–1985*. New York: Cambridge University Press, 1987.

Heise, Ursula. *Imagining Extinction: The Cultural Meanings of Endangered Species*. Chicago: University of Chicago Press, 2016.

Heppenheimer, Elizabeth, et al. "Rediscovery of Red Wolf Ghost Alleles in a Canid Population along the American Gulf Coast." *Genes* 9, no. 618 (December 2018). www.mpdi.com/journal/genes.

Herman, Daniel. "From Farmers to Hunters: Cultural Evolution in the Nineteenth-Century United States." In *A Cultural History of Animals in the Age of Empire*, edited by Kathleen Kete, 47–71. New York: Oxford/Berg, 2007.

————. *Hunting and the American Imagination*. Washington, DC: Smithsonian Institution Press, 2001.

Hodgson, Derek. "The Symmetry of Acheulean Handaxes and Cognitive Evolution." *Journal of Archaeological Science: Reports* 2 (2015): 204–8.

Holder, Preston. *The Hoe and the Horse on the Plains: A Study of Cultural Development among North American Indians*. Lincoln: University of Nebraska Press, 1970.

Hornaday, William. "The Extermination of the American Bison, with a Sketch of Its Discovery and Life History." In *Report of the United States National Museum for the year ending June 30, 1887 (Pt. 2 of the Annual Report of the Board of Regents of the Smithsonian Institution for the year ending June 30, 1887)*, 367–548.

————. *Our Vanishing Wild Life: Its Extermination and Preservation*. New York: Charles Scribner's Sons, 1913.

Horvath, Gabor, et al. "Cavemen Were Better at Depicting Quadruped Walking Than Modern Artists: Erroneous Walking Illustrations in the Fine Arts from Prehistory to Today." *PLOS One* (December 5, 2012). https://doi.org/10.1371/journal.pone.0049786.

Howard, James. "Yanktonai Ethnohistory and the John K. Bear Winter Count." *Plains Anthropologist* 21 (August 1976): 28–52.

Hunt, Alex. "Hunting Charles Goodnight's Buffalo: Texas Fiction, Panhandle Folklore, and Kiowa History." *Panhandle-Plains Historical Review* 77 (2004): 1–13.

Hyde, George. *Indians of the High Plains: From the Prehistoric Period to the Coming of Europeans*. Norman: University of Oklahoma Press, 1975.

Hyde, Lewis. *Trickster Makes This World: How Disruptive Imagination Creates Culture*. Edinburgh, Scotland: Canongate, 2008.

Irving, Washington. *The Adventures of Captain Bonneville, U.S.A., in the Rocky Mountains and Far West*. Edited by Edgeley Todd. Norman: University of Oklahoma Press, 1986.

Isbell, Lynne. *The Fruit, the Tree, and the Serpent: Why We See So Well*. Cambridge, MA: Harvard University Press, 2011.

Isenberg, Andrew. *The Destruction of the Bison: An Environmental History, 1750–1920*. New York: Cambridge University Press, 2000.

Jacobs, Wilbur. "The Tip of an Iceberg: Pre-Columbian Indian Demography and Some Implications for Revisionism." In *The Fatal Confrontation: Historical Studies of American Indians, Environment, and Historians*. Albuquerque: University of New Mexico Press, 1996.

James, Edwin. "Account of an Expedition from Pittsburgh to the Rocky Mountains Performed in the Years 1819, 1820." In vol. 17 of *Early Western Travels*, edited by Reuben Gold Thwaites. Cleveland: Arthur H. Clark, 1906.

Jennings, Jesse. *Prehistory of North America*. New York: McGraw-Hill, 1968.

Johnson, Eileen, ed. *Lubbock Lake: Late Quaternary Studies on the Southern High Plains*. College Station: Texas A&M University Press, 1987.

Jones, Karen. *Wolf Mountains: A History of Wolves along the Great Divide*. Calgary, AB: University of Calgary Press, 2002.

Jones, T. L., et al. "The Protracted Holocene Extinction of California's Flightless Sea Duck (*Chendytes lawi*) and Its Implications for the Pleistocene Overkill Hypothesis." *Proceedings of the National Academy of Sciences* 105, no. 11 (March 18, 2008): 4105–8.

Jordan, Douglas. "The Deadliest Flu: The Complete Story of the Discovery and Recon-
    struction of the 1918 Pandemic Virus." Centers for Disease Control and Prevention.
    Last reviewed December 17, 2019. https://www.cdc.gov/flu/pandemic-resources/
    reconstruction-1918-virus.html.

Jordan, Terry. *North American Cattle-Ranching Frontiers*. Lincoln: University of Nebraska
    Press, 1993.

Kastner, Joseph. *A Species of Eternity*. New York: E. P. Dutton, 1977.

Kay, Charles. "Aboriginal Overkill and the Biogeography of Moose in Western North
    America." *Alces* 33 (1997): 141–64.

Keiter, Robert, et al., eds. *The Greater Yellowstone Ecosystem: Redefining America's Wilderness
    Heritage*. New Haven, CT: Yale University Press, 1991.

Keith, Lloyd. *Wildlife's Ten-Year Cycle*. Madison: University of Wisconsin Press, 1963.

Kellert, Stephen, and Edward O. Wilson, eds. *The Biophilia Hypothesis*. Washington, DC:
    Island Press, 1993.

Kennett, D., et al. "Archaeogenomic Evidence Reveals Prehistoric Matrilineal Dynasty."
    Nature Communications 8 (2017): article 14115. https://doi.org/10.1038/ncomms14115.

Kete, Kathleen. "Animals and Human Empire." In *A Cultural History of Animals in the Age of
    Empire*, edited by Kathleen Kete, 1–24. New York: Oxford/Berg, 2007.

King, Philip. *Evolution of North America*. Princeton, NJ: Princeton University Press, 1959.

Klimova, Anastasia, et al. "Genetic Diversity and Demography of Two Endangered Captive
    Pronghorn Subspecies from the Sonoran Desert." *Journal of Mammalogy* 95 (December
    2014): 1263–77.

Koch, Alexander, et al. "Earth System Impacts of European Arrival and Great Dying in
    the Americas after 1492." *Quaternary Science Reviews* 207, no. 1 (March 2019): 13–36.
    https://doi.org/10.1016/j.quascirev.2018.12.004.

Kofalk, Harriet. *No Woman Tenderfoot: Florence Merriam Bailey, Pioneer Naturalist*. College
    Station: Texas A&M University Press, 1989.

Kolbert, Elizabeth. *The Sixth Extinction: An Unnatural History*. New York: Henry Holt,
    2012.

Koth, Barbara, et al. "Effects of Restoring Wolves on Yellowstone Area Big Game and Griz-
    zly Bears: Opinions of Fifteen North American Experts." In *Wolves for Yellowstone?*
    Mammoth, WY: Yellowstone National Park, 1990.

Krech, Shepard. *The Ecological Indian: Myth and History*. New York: W. W. Norton, 1999.

Kruuk, Hans. *Hunter and Hunted: Relationships between Carnivores and People*. New York:
    Cambridge University Press, 2002.

Lack, David. *The Natural Regulation of Animal Numbers*. London: Oxford University Press,
    1954.

LaPier, Rosalyn. *Invisible Reality: Storytellers, Storytakers, and the Supernatural World of the
    Blackfeet*. Lincoln: University of Nebraska Press, 2017.

Lawson, Merlin. *The Climate of the Great American Desert: Reconstructing the Climate of West-
    ern Interior United States, 1800–1850*. Lincoln: University of Nebraska Press, 1974.

Leakey, Richard, and Roger Lewin. *Origins Reconsidered: In Search of What Makes Us Human*.
    New York: Doubleday, 1992.

Lee, Richard, and Irven Devore, eds. *Man the Hunter*. New York: Aldine de Gruyter, 1968.

Leeds, Anthony, and Andrew Vayda, eds. *Man, Culture, and Animals*. Washington, DC: American Association for the Advancement of Science, 1965.

Leopold, Aldo. *The River of the Mother of God and Other Essays by Aldo Leopold*. Edited by Susan Flader and Baird Callicott. Madison: University of Wisconsin Press, 1991.

———. *A Sand County Almanac and Sketches Here and There*. New York: Oxford University Press, 1949.

Levy, Sharon. *Once & Future Giants: What Ice Age Extinctions Tell Us about the Fate of Earth's Largest Animals*. New York: Oxford University Press, 2011.

Librado, Pablo, et al. "The Origins and Spread of Domestic Horses from the Western Eurasian Steppes." *Nature* (October 2021). https://doi.org/10.1038/s41586-021-04018-9.

Linden, Eugene. *The Parrot's Lament*. New York: E. P. Dutton, 1999.

Linnæi, Caroli [Carolus Linnaeus]. *Systema naturae per regna tria naturae*. Edited by J. E. Gmelin. 4 vols. London, 1789.

Liu, Yichen, et al. "Insights into Human History from the First Decade of Ancient Human Genomics." *Science*, September 24, 2021, 1479–84.

Lopez, Barry. *Giving Birth to Thunder, Sleeping with His Daughter: Coyote Builds North America*. New York: Harper Perennial, 2001.

———. *Of Wolves and Men*. New York: Charles Scribner's Sons, 1978.

Lovejoy, Arthur. *The Great Chain of Being: A Study of the History of an Idea*. Cambridge, MA: Harvard University Press, 1964.

Lowery, George. *Louisiana Birds*. 3rd ed. Baton Rouge: Louisiana State University, 1974.

Lund, Thomas. *American Wildlife Law*. Berkeley: University of California Press, 1980.

MacFadden, Bruce. *Fossil Horses: Systematics, Paleobiology, and Evolution of the Family Equidae*. New York: Cambridge University Press, 1992.

MacPhee, Ross. *End of the Megafauna: The Fate of the World's Hugest, Fiercest, and Strangest Animals*. New York: W. W. Norton, 2019.

Mann, Charles. *1491: New Revelations of the Americas before Columbus*. New York: Vintage Books, 2005.

Martin, Calvin. *Keepers of the Game: Indian-Animal Relations and the Fur Trade*. Berkeley: University of California Press, 1979.

Martin, Paul. "40,000 Years of Extinctions on the 'Planet of Doom.'" *Palaeography, Palaeoclimatology, Palaeoecology* 82 (1990): 187–201.

———. *Twilight of the Mammoths: Ice Age Extinctions and the Rewilding of America*. Berkeley: University of California Press, 2005.

Martin, Paul, and Richard Klein, eds. *Quaternary Extinctions: A Prehistoric Revolution*. Tucson: University of Arizona Press, 1985.

Martin, Paul, and Christine Szuter. "War Zones and Game Sinks in Lewis and Clark's West." *Conservation Biology* 13 (Winter 1999): 36–45.

Matheson, Neill. "Ardent Creatures: William Bartram and the Limits of Compassion." In *American Beasts: Perspectives on Animals, Animality and U.S. Culture, 1776–1920*, edited by Dominik Ohrem, 225–44. Berlin: Neofelis Verlag, 2017.

Matthiessen, Peter. *Wildlife in America*. 2nd ed. New York: Viking, 1987.

Mattson, David, and Troy Merrill. "Extirpations of Grizzly Bears in the Contiguous United States, 1850–2000." *Conservation Biology* 16 (August 2002): 1123–35.

Mayer, Frank, and Charles Roth. *The Buffalo Harvest.* St. Paul, MN: Pioneer Press, 1995.

Mayr, Ernst. *The Growth of Biological Thought: Diversity, Evolution, and Inheritance.* Cambridge, MA: The Belknap Press, 1982.

McCabe, Richard, Bart O'Gara, and Henry Reeves. *Prairie Ghost: Pronghorn and Human Interaction in Early America.* Boulder: University Press of Colorado and the Wildlife Management Institute, 2004.

McCain, Emil, and Jack Childs. "Evidence of Resident Jaguars (*Panthera onca*) in the Southwestern United States and the Implications for Conservation." *Journal of Mammalogy* 89 (February 19, 2008): 1–10.

McCullough, David. *The Pioneers: The Heroic Story of the Settlers Who Brought the American Ideal West.* New York: Simon and Schuster, 2019.

McDaniel, Carl, and John Gowdy. "Markets and Biodiversity Loss: Some Case Studies and Policy Considerations." *International Journal of Social Economics* 25 (1998): 1454–65.

McDonald, David. "Food Taboos: A Primitive Environmental Protection Agency (South America)." *Anthropos* 72 (1977): 734–48.

McDonald, Jerry. *The North American Bison: Their Classification and Evolution.* Berkeley: University of California Press, 1981.

McIntyre, Rick. *The Reign of Wolf 21: The Saga of Yellowstone's Legendary Druid Pack.* Vancouver, BC: Greystone Books, 2020.

McKay, Bonnie, and James Acheson, eds. *The Question of the Commons: The Culture and Ecology of Communal Resources.* Tucson: University of Arizona Press, 1987.

McKay, John. *Discovering the Mammoth: A Tale of Giants, Unicorns, Ivory, and the Birth of a New Science.* New York: Pegasus Books, 2017.

Meachen, Julie, and Joshua Samuels. "Evolution in Coyotes (*Canis latrans*) in Response to the Megafaunal Extinctions." *Proceedings of the National Academy of Sciences* 109 (March 2012): 4191–96.

Mead, James. *Hunting and Trading on the Great Plains, 1859–1875.* Edited by Schuyler Jones. Norman: University of Oklahoma Press, 1985.

Mech, L. David. *The Way of the Wolf.* New York: Voyageur Press, 1992.

———. *The Wolf: Ecology and Behavior of an Endangered Species.* Garden City, NY: Natural History Press, 1970.

———. *Wolf Island: Discovering the Secrets of a Mythic Animal.* With Greg Breining. Minneapolis: University of Minnesota Press, 2020.

Mech, L. David, Douglas Smith, and Daniel MacNulty. *Wolves on the Hunt: The Behavior of Wolves Hunting Wild Prey.* Chicago: University of Chicago Press, 2015.

Meltzer, David. *The Great Paleolithic War: How Science Forged an Understanding of America's Ice Age Past.* Chicago: University of Chicago Press, 2015.

Merriam, C. Hart. *Results of a Biological Survey of the San Francisco Mountain Region and Desert of the Little Colorado, Arizona.* North American Fauna no. 3. Washington, DC: U.S. Department of Agriculture, 1890.

Mezrich, Ben. *Woolly: A True Story of the Quest to Revive One of History's Most Iconic Extinct Creatures.* New York: Atria, 2017.

Mighetto, Lisa. *Wild Animals and American Environmental Ethics.* Tucson: University of Arizona Press, 1991.

Mooney, James. *Calendar History of the Kiowa Indians*. Washington, DC: Smithsonian Institution Press, 1979.

Morcote-Rios, Gaspar, et al. "Colonisation and Early Peopling of the Colombian Amazon during the Late Pleistocene and the Early Holocene." *Quaternary International* 578 (March 2021): 5–19.

Morgan, Grace. *Beaver, Bison, Horse: The Traditional Knowledge and Ecology of the Northern Great Plains*. Regina, SK: University of Regina Press, 2020.

Mormann, F., Dubois, J., Kornblith, S., et al. "A Category-Specific Response to Animals in the Right Human Amygdala." *Nature Neuroscience* 14 (2011): 1247–49.

Moulton, Gary, ed. *The Journals of the Lewis and Clark Expedition*. Vol. 4, *April 7–July 27, 1805*. Lincoln: University of Nebraska Press, 1987.

Murie, Adolph. *Ecology of the Coyote in Yellowstone*. Fauna of the National Parks of the United States no. 4. Washington, DC: U.S. Department of Interior National Park Service, 1940.

———. *The Wolves of Mount McKinley*. Fauna of the National Parks of the United States no. 5. Washington, DC: U.S. Department of Interior National Park Service, 1944.

Murie, Olaus. *Food Habits of the Coyote In Jackson Hole, Wyo*. Bulletin 362. Washington, DC: U.S. Department of Agriculture, 1935.

Murphy, Sean, et al. "Substantial Red Wolf Genetic Ancestry Persists in Wild Canids of Southwestern Louisiana." *Conservation Letters* (December 4, 2018). https://doi.org/10.1111/conl.12621.

Nash, Roderick. *The Rights of Nature: A History of Environmental Ethics*. Madison: University of Wisconsin Press, 1989.

National Academies of Sciences, Engineering, and Medicine. *Evaluating the Taxonomic Status of the Mexican Gray Wolf and the Red Wolf*. Washington, DC: National Academies Press, 2019. https://doi.org/10.17226/25351.

National Research Council. *Using Science to Improve the BLM Wild Horse and Burro Program: A Way Forward*. Washington, DC: National Academies Press, 2013.

Nelson, Richard. *Make Prayers to the Raven: A Koyukon View of the Northern Forest*. Chicago: University of Chicago Press, 1983.

Nijhuis, Michelle. *Beloved Beasts: Fighting for Life in an Age of Extinction*. New York: W. W. Norton, 2021.

Nixon, Richard. "Special Message to the Congress Outlining the 1972 Environmental Program," February 8, 1972. Online by Gerhard Peters and John T. Woolley, American Presidency Project. https://www.presidency.ucsb.edu/node/255047.

Novak, Ben. "Deciphering the Ecological Impact of the Passenger Pigeon: A Synthesis of Paleogenetics, Paleoecology, Morphology, and Physiology." Master's thesis, University of California–Santa Cruz, 2016.

———. "De-Extinction." *Genes* 9 (2018): 1–33.

Nowak, Ronald. *Walker's Mammals of the World*. 6th ed. 2 vols. Baltimore: Johns Hopkins University Press, 1999.

Odum, Eugene. *Fundamentals of Ecology*. Philadelphia: W. B. Saunders, 1971.

Oelschlaeger, Max. *The Idea of Wilderness: From Prehistory to the Age of Ecology*. New Haven, CT: Yale University Press, 1991.

Olsen, Sandra, ed. *Horses through Time*. Boulder: Roberts Rinehart Publishers for the Carnegie Museum of Natural History, 1997.

Osburn, Alan. "Ecological Aspects of Equestrian Adaptations in Aboriginal North America." *American Anthropologist* 85 (September 1983): 563–91.

Ostler, Jeffrey. "'They Regard Their Passing as Wakan': Interpreting Western Sioux Explanations for the Buffalo's Decline." *Western Historical Quarterly* (Winter 1999): 475–97.

Ott, Jennifer. "'Ruining' the Rivers in the Snake Country: The Hudson's Bay Company's Fur Desert Policy." *Oregon Historical Quarterly*, 104 (Spring 2003): 166–95.

Pagnamenta, Peter. *Prairie Fever: British Aristocrats in the American West, 1830–1890*. New York: W. W. Norton, 2012.

Pavlik, Steve. "Will Big Trotter Reclaim His Place? The Role of the Wolf in Navajo Tradition." *American Indian Culture and Research Journal* 24 (Fall 2000): 1–19.

Perri, A. R., et al. "Dire Wolves Were the Last of an Ancient New World Canid Lineage." *Nature* 591 (2021): 87–91. https://doi.org/10.1038/s41586-020-03082-x.

Peterson, Rolf. *The Wolves of Isle Royale: A Broken Balance*. Minocqua, WI: Willow Creek Press, 1995.

Petersen, Shannon. "Bison to Blue Whales: Protecting Endangered Species before the Endangered Species Act of 1973." *Environs* 22, no. 2 (May 1999): 1–26.

Pielou, E. C. *A Naturalist's Guide to the Arctic*. Chicago: University of Chicago Press, 1994.

Ponting, Clive. *A Green History of the World*. New York: St. Martin's Press, 1991.

Ray, Arthur. *Indians in the Fur Trade*. Toronto: University of Toronto Press, 1974.

Reiger, John. *American Sportsmen and the Origins of Conservation*. New York: Winchester Press, 1975.

Riley, Glenda. *Women and Nature: Saving the "Wild" West*. Lincoln: University of Nebraska Press, 1999.

Rinella, Steven. *American Buffalo: In Search of a Lost Icon*. New York: Random House, 2009.

Ripple, William, and Robert Beschta. "Linking Wolves and Plants: Aldo Leopold on Trophic Cascades." *BioScience* 55, no. 7 (July 2005): 613–21. https://doi.org/10.1641/0006-3568(2005)055[0613:LWAPAL]2.0.CO;2.

Roberts, D. L., et al. "On the Functional Extinction of the Passenger Pigeon." *Conservation Biology* 31 (2017): 1192–95.

Robinson, Michael. *Predatory Bureaucracy: The Extermination of Wolves and the Transformation of the West*. Boulder: University Press of Colorado, 2005.

Rockwell, David. *Giving Voice to Bear: North American Indian Rituals, Myths, and Images of the Bear*. Niwot, CO: Robert Rinehart Publishers, 1991.

Rome, Adam. *The Genius of Earth Day*. New York: Hill & Wang, 2013.

Rosenberg, Kenneth, et al. "Decline of the North American Avifauna." *Science* 366, no. 6461 (October 2019): 120–24.

Rostlund, Erhard. "The Geographic Range of the Historic Bison in the Southeast." *Annals of the Association of American Geographers* 50 (December 1970): 395–407.

Rowlands, Mark. "Philosophy and Animals in the Age of Empire." In *A Cultural History of Animals in the Age of Empire*, edited by Kathleen Kete, 135–52. New York: Oxford/Berg, 2007.

Russell, Edmund. *Evolutionary History: Uniting History and Biology to Understand Life on Earth*. New York: Cambridge University Press, 2011.

Russell, Osborne. *Journal of a Trapper, 1834–1843*. Edited by Aubrey Haines. Lincoln: University of Nebraska Press, 1955.

Ruxton, George Frederick. *Adventures in Mexico and the Rocky Mountains*. London: John Murray, 1847.

———. *Life in the Far West*. Edited by Leroy Hafen. 1851. Reprint, Norman: University of Oklahoma Press, 1950.

Saikku, Mikko. "The Extinction of the Carolina Parakeet." *Environmental History Review* 14 (Fall 1990): 2–18.

Saunt, Claudio. *West of the Revolution: An Uncommon History of 1776*. New York: W. W. Norton, 2014.

Savage, Henry. *Discovering America, 1700–1875*. New York: Harper & Row, 1979.

Schmidly, David. *Texas Natural History: A Century of Change*. Lubbock: Texas Tech University Press, 2002.

———. *Vernon Bailey: Writings of a Field Naturalist on the Frontier*. College Station: Texas A&M University Press, 2018.

Schullery, Paul. *Lewis and Clark among the Grizzlies: Legend and Legacy in the American West*. Guilford, CT: Globe Pequot Press, 2002.

Schultz, P. H., and S. D'Hondt. "Cretaceous–Tertiary (Chicxulub) Impact Angle and Its Consequences." *Geology* 24 (1996): 963–67.

Senior, Matthew, ed. *A Cultural History of Animals in the Age of Enlightenment*. New York: Oxford University Press, 2007.

Seton, Ernest Thompson. *Lives of Game Animals*. 1929. Reprint, Boston: Charles T. Branford, 1953.

———. "Tito: The Story of the Coyote That Learned How." *Scribner's*, August 1900, 1–25.

———. *Wild Animals I Have Known*. 1898. Reprint, Columbia, SC: Feather Trail Press, 2010.

Shelford, Victor. *The Ecology of North America*. Urbana: University of Illinois Press, 1963.

———. "Preservation of Natural Biotic Communities." *Ecology* 14 (1933): 240–45.

Shepard, Paul. *The Others: How Animals Made Us Human*. Washington, DC: Island Press/Shearwater, 1996.

———. *The Tender Carnivore and the Sacred Game*. New York: Scribner's, 1973.

Sheridan, Philip, to E. D. Townsend. Telegram, Chicago, October 31, 1879. Letters Received, Adjutant General's Office, National Archives Record Group 94.

Shivik, John. *Mousy Cats and Sheepish Coyotes: The Science of Animal Personalities*. Boston: Beacon Press, 2017.

———. *The Predator Paradox: Ending the War with Wolves, Bears, Cougars, and Coyotes*. Boston: Beacon Press, 2014.

Slotkin, Richard. *Regeneration through Violence: The Mythology of the American Frontier*. Middleton, CT: Wesleyan University Press, 1973.

Smalley, Andrea. *Wild by Nature: North American Animals Confront Colonization*. Baltimore: Johns Hopkins University Press, 2017.

Smith, Brian, et al. "The Demography of Extinction in Eastern North American Birds." *Proceedings of the Royal Society B* (February 2021).

Smith, Douglas, and Gary Ferguson. *Decade of the Wolf: Returning the Wild to Yellowstone*. 2nd ed. Guilford, CT: Lyons Press, 2012.

Smith, Douglas, Daniel R. Stahler, and Daniel R. MacNulty, eds. *Yellowstone Wolves: Science and Discovery in the World's First National Park.* Chicago: University of Chicago Press, 2020.

Smith, Eric Alden. "Anthropological Applications of Optimal Foraging Strategy: A Critical Review." *Current Anthropology* 24 (December 1983): 628–32.

Spielmann, Katherine. "Late Prehistoric Exchange between the Southwest and Southern Plains." *Plains Anthropologist* 28 (November 1983): 257–79.

Spiro, Jonathan. *Defending the Master Race: Conservation, Eugenics, and the Legacy of Madison Grant.* Burlington, VT: University Press of New England, 2009.

Spring, David, and Eileen Spring, eds. *Ecology and Religion in History.* New York: Harper & Row, 1974.

Stahle, David, and Malcolm Cleaveland. "Texas Drought History Reconstructed and Analyzed from 1698 to 1980." *Journal of Climate* (January 1988): 59–74.

Stanford, Dennis, and Bruce Bradley. *Across Atlantic Ice: The Origin of America's Clovis Culture.* Berkeley: University of California Press, 2012.

Stanton, Jessica. "Present-Day Risk Assessment Would Have Predicted the Extinction of the Passenger Pigeon (*Ectopistes migratorius*)." *Biological Conservation* 180 (2014): 11–20.

Stoll, Mark. *Protestantism, Capitalism, and Nature in America.* Albuquerque: University of New Mexico Press, 1997.

Straus, Lawrence, et al. *Humans at the End of the Ice Age.* New York: Plenum Press, 1996.

Stroud, Patricia. *Thomas Say: New World Naturalist.* Philadelphia: University of Pennsylvania Press, 1992.

Stuart, David. *Anasazi America: Seventeen Centuries on the Road from Center Place.* 2nd ed. Albuquerque: University of New Mexico Press, 2014.

Swagerty, William. "Indian Trade in the Trans-Mississippi West to 1870." Vol. 6 of *Handbook of North American Indians: European-Indian Interaction,* edited by Wilcomb Washburn, 351–74. 20 vols. Washington, DC: Smithsonian Institution Press, 1978–89.

Sykes, Rebecca. *Kindred: Neanderthal Life, Love, Death and Art.* London: Bloomsbury Sigma, 2020.

Taliaferro, John. *Grinnell: America's Environmental Pioneer and His Restless Drive to Save the West.* New York: Liveright, 2019.

Tanner, Adrian. *Bringing Home Animals: Religious Ideology and Mode of Production of the Mistassini Cree Hunters.* New York: St. Martin's Press, 1979.

Tanner, James. *The Ivory-Billed Woodpecker.* New York: National Audubon Society, 1942.

Tattersall, Ian. *Masters of the Planet: The Search for Our Human Origins.* New York: Palgrave Macmillan, 2012.

Taylor, Alan. " 'Wasty Ways': Stories of American Settlement." *Environmental History* 3 (July 1998): 291–310.

Thomas, Jessica, et al. "Demographic Reconstruction from Ancient DNA Supports Rapid Extinction of the Great Auk." *eLife* 8 (2019): article e47509. https://doi.org/10.7554/eLife.47509.

Thomas, Keith. *Man and the Natural World: A History of the Modern Sensibility.* New York: Pantheon Books, 1983.

Thoreau, Henry David. *The Heart of Thoreau's Journals*. Edited by Odell Shepard. Boston and New York: Houghton-Mifflin, 1906.

Thornton, Russell. *American Indian Holocaust and Survival: A Population History since 1492*. Norman: University of Oklahoma Press, 1987.

Townsend, John Kirk. *Narrative of a Journey across the Rocky Mountains to the Columbia River*. Introduction by Donald Jackson. Lincoln: University of Nebraska Press, 1970.

Trefethen, James. *An American Crusade for Wildlife*. Alexandria, VA: Boone and Crockett Club, 1975.

Trigger, Bruce. "Early Native North American Responses to European Contact: Romantic versus Rationalistic Interpretations." *Journal of American History* 77 (March 1991): 1195–215.

Trisos, Christopher, Cory Merow, and Alex Pigot. "The Projected Timing of Abrupt Ecological Disruption from Climate Change." *Nature* 580 (April 23, 2020): 496–516.

Turvey, S. T., and C. L. Risley. "Modeling the Extinction of Steller's Sea Cow." *Biological Letters* 22 (March 2006): 94–97.

U.S. Fish and Wildlife Service. *Bald and Golden Eagles: Population Demographics and Estimation of Sustainable Take in the United States, 2016 Update*. Washington, DC: Division of Migratory Bird Management, 2016.

———. *Mexican Wolf Recovery Program. Final Environmental Impact Statement and Final Decision*. January 2015. www.fws.gov/southwest/es/mexicanwolf/nepa_713.cfm (removed from website as of March 2022).

———. *Monitoring Plan for the American Peregrine Falcon, a Species Recovered under the Endangered Species Act*. Portland, OR: U.S. Fish and Wildlife Service, Divisions of Endangered Species and Migratory Birds and State Programs, Pacific Region, 2003.

———. *Questions and Answers: Final Recovery Plan for the Jaguar*. U.S. Fish and Wildlife Service, April 24, 2019. https://www.fws.gov/southwest/docs/Jaguar_Recovery_PlanQ_A%20FINAL.pdf.

van der Valk, Tom, et al. "Million-Year-Old DNA Sheds Light on the Genomic History of Mammoths." *Nature*, 591 (February 2021): 265–69.

van Hoesel, Annelies, et al. "The Younger Dryas Climate Change: Was It Caused by an Extraterrestrial Impact?" *Quaternary Science Reviews* 83 (January 2014): 95–114.

Van Nuys, Frank. *Varmints and Victims: Predator Control in the American West*. Lawrence: University Press of Kansas, 2015.

Varner, Gary. "Environmental Ethics, Hunting, and the Place of Animals." In *The Oxford Handbook of Animal Ethics*, edited by Tom Beauchamp and R. G. Frey, 855–76. New York: Oxford University Press, 2011.

vonHoldt, Bridgett, et al. "A Genome-Wide Perspective on the Evolutionary History of Enigmatic Wolf-Like Canids." *Genome Research* 21, no. 8 (August 2011): 1294–1305.

Waller, Dennis. *Raven Tales: Stories of the Raven Based on the Folklore of the Tlingit, Haida, Tsimshian, Inuit, and Athapascan of Alaska*. Self-published, 2014.

Wallerstein, Immanuel. *The Modern World System II: Mercantilism and the Consolidation of the European World Economy, 1600–1750*. New York: Academic Press, 1980.

Ward, Peter. *The Call of Distant Mammoths: Why the Ice Age Mammals Disappeared*. New York: Springer-Verlag, 1997.

Warren, Louis. *The Hunter's Game*. New Haven, CT: Yale University Press, 1997.

Waters, Michael, Thomas W. Stafford Jr., and David L. Carlson. "The Age of Clovis—13,050 to 12,750 cal yr B.P." *Science Advances* 6, no. 43 (October 2020).

Waters, Michael, Thomas W. Stafford Jr., Brian Kooyman, and L. V. Hills. "Late Pleistocene Horse and Camel Hunting at the Southern Margin of the Ice-Free Corridor: Reassessing the Age of Wally's Beach, Canada." *Proceedings of the National Academy of Sciences* 112 (April 2015). https://doi.org/10.1073/pnas.1420650112.

Weatherford, Carole Boston. *The Carolina Parakeet: America's Lost Parrot in Art and Memory*. Minneapolis: Avian Publications, 2005.

Webb, Walter Prescott. *The Great Frontier*. Austin: University of Texas Press, 1964.

Weber, David. *The Taos Trappers: The Fur Trade in the Far Southwest, 1540–1846*. Norman: University of Oklahoma Press, 1971.

White, Richard. "Animals and Enterprise." In *The Oxford History of the American West*, edited by Clyde Milner, 247–84. New York: Oxford University Press, 1994.

———. *The Roots of Dependency: Subsistence, Environment, and Social Change among the Choctaws, Pawnees, and Navajos*. Lincoln: University of Nebraska Press, 1983.

White, Richard, and William Cronon. "Ecological Change and Indian-White Relations." In *Indian-White Relations*, vol. 4 of *Handbook of North American Indians*, edited by William Sturtevant, 417–29. 20 vols. Washington, DC: Smithsonian Institution Press, 1978–89.

Whiten, Andrew. "The Burgeoning Reach of Animal Culture." *Science* 372, no. 6537 (April 2021).

Whittlesey, Lee, Paul Schullery, and Sarah Bone. *The History of Mammals in the Greater Yellowstone Ecosystem, 1796–1881: A Cross-Disciplinary Analysis of Thousands of Historical Observations*. 2 vols. Manuscript report in possession of the author. Mammoth, WY: National Park Service, Yellowstone National Park, 2018.

Wildlife Management Institute. *The American Game Policy and Its Development, 1929–1930*. Washington, DC: Wildlife Management Institute, 1930.

Williams, Randall. "Drifting to the Right: Sport Hunting and the Politics of Nature, Guns, and Family." PhD diss., University of Montana, 2015.

Wilson, Alexander. *American Ornithology; or, The Natural History of the Birds of the United States*. 9 vols. Philadelphia, 1808–14.

Wilson, Edward O. *Biophilia: The Human Bond with Other Species*. Cambridge: Harvard University Press, 1984.

———. *Half-Earth: Our Planet's Fight for Life*. New York: Liveright, 2016.

———. *The Meaning of Human Existence*. New York: Liveright, 2014.

Wilson, Maurine, and Jack Jackson. *Philip Nolan and Texas: Expeditions to the Unknown Land, 1791–1801*. Waco, TX: Texian Press, 1987.

Wilson, Paul, Sonya Grewal, et al. "DNA Profiles of the Eastern Canadian Wolf and the Red Wolf Provide Evidence for a Common Evolutionary History Independent of the Gray Wolf." *Canadian Journal of Zoology* 78 (2000): 2156–66.

Wilson, Paul, and Linda Rutledge. "Considering Pleistocene North American Wolves and Coyotes in the Eastern *Canis* Origin Story." *Ecology and Evolution* 11 (July 2021): 9137–47. https://doi.org/10.1002/ece3.7757.

Wise, Michael. "Killing Montana's Wolves: Stockgrowers, Bounty Bills, and the Uncertain

Distinction between Predators and Producers." *Montana, the Magazine of Western History* 63 (Winter 2013): 51–67.

Wishart, David. *The Fur Trade of the American West, 1807–1840: A Geographical Synthesis*. Lincoln: University of Nebraska Press, 1979.

Wood, William. *New Englands Prospect*. 1639. Reprint, Amherst: University of Massachusetts Press, 1977.

Wood, W. Raymond, and Thomas Thiessen, eds. *Early Fur Trade of the Northern Plains: Canadian Traders among the Mandan and Hidatsa Indians, 1738–1818*. Norman: University of Oklahoma Press, 1985.

Wooler, Matthew, et al. "Lifetime Mobility of an Arctic Woolly Mammoth." *Science* 373, no. 6556 (August 13, 2021): 806–8.

Worster, Donald. *Nature's Economy: A History of Ecological Ideas*. 2nd ed. New York: Cambridge University Press, 1994.

———. "Other People, Other Lives." Chap. 3 in *An Unsettled Country: Changing Landscapes of the American West*, 55–90. Albuquerque: University of New Mexico Press, 1994.

———. *The Wealth of Nature: Environmental History and the Ecological Imagination*. New York: Oxford University Press, 1993.

Wulf, Andrea. *The Invention of Nature: Alexander von Humboldt's New World*. New York: Vintage Books, 2015.

Young, Stanley, and Edward Goldman. *The Wolves of North America*. 2 vols. 1944. Reprint, New York: Dover Press, 1964.

Zaraska, Marta. *Meathooked: The History and Science of Our 2.5-Million-Year Obsession with Meat*. New York: Basic Books, 2016.

Zontek, Ken. *Buffalo Nation: American Indian Efforts to Restore the Bison*. Lincoln: University of Nebraska Press, 2007.

Zubrow, Ezra. *Prehistoric Carrying Capacity: A Model*. Menlo Park, CA: Sunset Books, 1975.

Zweifel, Richard, ed. *Catalogue of American Amphibians and Reptiles*. New York: Society for the Study of Amphibians and Reptiles, 1971.

INDEX

Page references in *italics* refer to illustrations.